Springer Polar Sciences

Springer Polar Sciences is an interdisciplinary book series that is dedicated to research in the Arctic, sub-Arctic regions, and the Antarctic. In recent years, the polar regions have received increased scientific and public interest. Both the Arctic and Antarctic have been recognized as key regions in the regulation of the global climate, and polar ecosystems have been identified to be particularly susceptible to the ongoing environmental changes. These changes are having widespread implications for human communities, businesses, and governance systems and are interacting with demographic shifts, globalisation, resource development, cultural change, territorial disputes, and growing calls for self-determination in some regions. Consequently, the international efforts in polar research have been enhanced considerably, and a wealth of new findings is being produced at a growing rate by the international community of polar researchers and those who live in the region.

Springer Polar Sciences aims to present a broad platform that will include state-of-the-art research, bringing together both science, humanities, and perspectives rooted in indigenous and local knowledge to facilitate an exchange of knowledge between the various polar science communities. The series offers an outlet to publish contributions, monographs, edited works, conference proceedings, etc. Topics and perspectives will be broad and will include, but not be limited to, climate change impacts, climate change policy, environmental change, polar ecology, governance, health, economics, indigenous populations, tourism, resource extraction activities, and research design in polar regions. Books published in the series will appeal to scientists, students, polar researchers, community leaders, and policy makers.

Markku Lehtimäki • Arja Rosenholm
Elena Trubina • Nina Tynkkynen
Editors

Cold Waters

Tangible and Symbolic Seascapes of the North

Springer

Editors
Markku Lehtimäki
Faculty of Humanities
University of Turku
Turku, Finland

Arja Rosenholm
Faculty of Information Technology and
Communication Sciences
Tampere University
Tampere, Finland

Elena Trubina
UNC Global Research Institute
Chapel Hill, NC, USA

Nina Tynkkynen
Faculty of Social Sciences
Business and Economics
Åbo Akademi University
Turku, Finland

ISSN 2510-0475 ISSN 2510-0483 (electronic)
Springer Polar Sciences
ISBN 978-3-031-10148-9 ISBN 978-3-031-10149-6 (eBook)
https://doi.org/10.1007/978-3-031-10149-6

This Springer imprint is published by the registered company Springer Nature Switzerland AG
The registered company address is: Gewerbestrasse 11, 6330 Cham, Switzerland

Note on Transliteration

The Library of Congress transliteration system has been used in reference notes to Russian sources. Place and family names that are likely to be well known to non-specialist readers are given in the most commonly used form (Dostoevsky rather than Dostoevskii, Tchaikovsky rather than Chaikovskii, Gorky rather than Gor'kii, Yakutia rather than Iakutiia, Novaya Zemlya rather than Novaia Zemlia). Some Russian names are rendered in the form which the persons themselves prefer and use in the non-Russian contexts (Popogrebsky instead of Popogrebskii, Yuri instead of Iurii). This may lead to the same last names being transliterated differently (Yuryev instead of Iur'ev).

Northern Waters: From Terrestrial to Water-Bound Knowledge

The traditionally remote and inaccessible Arctic has a history of standing at the very center of the world. With the two superpowers standing face to face across the region, the Cold War made the Arctic Ocean "a modern-day equivalent of the Mediterranean of ancient times" (Zellen, 2009). Many countries found themselves in the orbit of the global struggle over the Arctic. When the Cold War ended, the confrontation was expected to turn into an era of peaceful cooperation with less international attention and tension (Palosaari, 2012). Yet, climate change and its impacts on the Northern regions' increased accessibility concerning natural resources became the driving force in maintaining the Arctic as a field where "geopolitical actors allegedly scramble to reterritorialize an opening Arctic space (and especially the Arctic Ocean) in pursuit of national security interests and resource competition" (Dittmer, 2011, p. 202; see also Keil, 2013).

Initially, the run on hydrocarbons resulted in international joint ventures on the economic side and the strengthening of Arctic governance in the political realm (Morgunova & Westphal, 2016). Oran Young (2012a) argued that the morphosis of sea ice into free water, that is, the recession of sea ice, first led to a reorientation from terrestrial issues towards marine issues, narrowing the scope of global attention, and "fueled worldwide interest in opening commercial shipping lanes in the Arctic and exploiting reserves of oil and gas that are becoming increasingly accessible. Enhanced prospects for ship-based tourism and industrial fishing have come into focus as well" (Young, 2012a, p. 167). This situation rapidly changed in 2014 in the aftermath of Crimea's annexation alongside increasing tensions between Russia and the West. Currently, the war in Ukraine threatens to bring the Arctic back to its Cold War role as a highly strategic and militarized zone. A loss of "cooperative spirit" can be observed. In response to Russia's invasion of Ukraine in February 2022, the seven members of the Nobel Peace Prize–nominated Arctic Council protested Russia's Chairmanship (2021–2023), announcing a temporary pause in their participation in all Council meetings.

Despite the changing geopolitical scene, climate change and receding sea ice are game changers that have come to stay at the center of debates concerning the Arctic. A temperature increase of 1.1 degree Celsius during the last 50 years has been recorded, and the Arctic summer sea ice is receding at a rate of 12% per decade (NSIDC, 2012; Lemke, 2011). Resources previously unattainable have become attainable: unusable routes quickly became usable for several months. With the abundance of natural resources, this accessibility has provided the impetus for many nations to increase their military and economic presence in the region. Apart from the Arctic Five – the nations with an Arctic coastal area and an exclusive economic zone (EEZ) extending into the region (Russia, Canada, Denmark [Greenland], Norway, and the United States [Alaska]) as defined by the UN Convention on the Law of the Sea – there are also unquestionably Arctic nations (Iceland, Sweden, and Finland) that lack an Arctic coastal area. These nations, and, to a growing extent, China, consider the Arctic endowed with strategic significance and, thereby, of national interest. Increased access to the Arctic increases its exploitation by resource extraction or introducing new transport lines. Thus, risks such as climate change, oil extraction, and shipping will also increase from the ongoing militarization of the region.

The renaissance of the interest in the Arctic and northern areas is also visible in the global media as spectacular events staged by various human and non-human actors, such as the hoisting of national flags on ice, planting them on the seabed, and the melting of polar sea ice (Lehtimäki et al., 2021, p. 1). In addition to state-led and corporate explorations of the Arctic, in the past 20 years, the Northern regions have seen a drastic increase in artistic representations and fictional explorations. As Jen Rose Smith (2021, p. 158) notes, "Melting ice has become a spectacle on a global scale […]. Within dominant conversations of the Anthropocene, ice is made a focal point as its melting is a danger to a temperate world." Indeed, the proliferation of images and catastrophic news puts the questions of the region's common aquatic and terrestrial space, state sovereignty, and the symbolic and economic value of the Arctic spaces on the global agenda. This mediation occurs in different forms, including national Arctic policies, texts and images in popular culture, museum exhibits, and news media. The news media plays a particular role with its potential to reach large audiences – directly or indirectly – through its narrowcasting and spillover to social media (Nilsson & Christensen, 2019, p. 6).

Whether the Arctic is in focus because of its potential for cooperation or conflict, the amount of knowledge about the Arctic is rapidly growing. Inevitably, the nexus of space, geopolitics, economic, military and security questions, and science becomes increasingly complex. Receding ice not only jeopardizes habitats but opens up the "lacunae" deemed well-hidden since the years right after World War II: when the military competition in the Arctic started unfolding, nobody could have predicted the speed with which ice would melt. "Cold waters" were and remain pivotal for the history of science during and after the Cold War: environmental sciences were practiced hand in hand with nuclear science and technology. More generally, scientific investigation of the polar areas was a significant part of defense; security strategies became a strategically important element in the US and Soviet defense policy and military subsidized research in meteorology, geology, seismology, and oceanography (Nielsen & Nielsen, 2021).

The Arctic has been extensively addressed by the international scientific community from natural scientific research to political science. The region(s) has been a focus of research, particularly regarding geopolitics, international governance, and global economic interests. Øyvind Østerud and Geir Hønneland (2014) note in their extensive review that the English language research regarding geopolitics and the international governance of the Arctic mainly focuses on how institutions can best be crafted to maintain peace and stability in the region – more or less refuting speculations about a "scramble for the Arctic." However, French research has been more concerned about the rivalry between states for strategic resources, downplaying the institutions of cooperation. With the vast opportunities and challenges posed by climate change and increasing globalization, the future of the Arctic concerning geopolitics and governance has become a particularly hot topic (see Arbo et al., 2013; Young, 2012b).

Yet, the Arctic change is more than just the physical change induced by climate change. This change comprises complex social and technological dimensions affecting the causes and effects of climate change within local contexts in economic, political, and cultural dimensions (Christensen & Nilsson, 2017). Geographical constructions such as "the state" are not fixed entities but constructions of human agency (Dodds et al., 2013; Burkart & Christensen, 2013). Thus, the approaches developed in humanist studies to understand these human dimensions are crucial. How the history of Northern waters is and has been deeply implicated in and shaped by human actors and ambitions, most tellingly in resource competition and the technological aspirations of modernization, is prominent. The elementary and constant transformation of water – liquid into ice and glaciers into floods – works against conventional but largely discredited divisions between "nature" and "culture" (Chen et al., 2013, p. 9). Contemporary scholars draw increasingly from the humanities and social sciences in reconsidering human intercourse with nature and water (Goodbody & Wanning, 2008). As Hartmut Böhme (1988, p. 11, 12, 16) points out in his classic *Kulturgeschichte des Wassers*, the human is simultaneously an initiator and object/recipient of water circulation and water technologies. People are inseparable from the ecosystem and water circulation; thus, we cannot separate the two sides of the physical and cultural presence of water, even if specific circumstances demand an emphasis or focus on one over the other.

Experiencing the maritime or riverine environment as a cultural phenomenon brings along a specific water-bound identity endowed with cultural and national consciousness emerging in the historical alteration of knowledge, ideology building, and local and global power structures. Moreover, the history of river narratives and various waterways and coastal borderlands as privileged contact zones (Cohen, 2006; Pratt, 1992) shows how any aquatic landscape can play a key role in cultivating cultural master narratives concerning life's circulation in a region where the Arctic seabed has become increasingly more significant in affective geopolitics (Dodds & Wilson Rowe, 2021).

Furthermore, the scholarly practices, including drilling ice cores, observing animals, analyzing trends and changes, discussing their significance for environmental policies and cultures and humanity's present and future, contribute to the renewed interest in the spatial and temporal dimensions of the links between Earth and its increasingly important part: the Arctic. These interdisciplinary results of investigating the Arctic suggest a delineation of the Arctic space/time as a tangible materiality and set of symbols. Thus, providing neat, discipline-specific characterizations of the region is difficult. We believe the strength of the present volume lies in that scholars working in humanities and social sciences joined forces in a multidisciplinary manner to tackle little-investigated aspects of the Arctic and the knowledge of it.

This volume aims to encourage interdisciplinary dialogue and emphasize the significance of humanist understanding among scholars exploring Arctic issues. We draw on previous research referring to the Northern Sea areas as a "resource frontier" (Van Alstine & Davies, 2016), "spatiality of hope" (Guerrieri, 2019), and "time machine" (Reutsch et al., 2018), where various physical but sociocultural changes occur rapidly and should urgently be addressed. The "resource frontier" metaphor captures the debates on the enlarged availability of Arctic riches, which will involve the growth of extractive industries in the region. The concept of "spatiality of hope" also apprehends the processes of frontier-making by examining the implication of maps and cartography in encouraging various hopes in map users. Connecting maps and hopes (to extract more oil or guard national sovereignty) allows demonstrating that maps mobilize and motivate while sometimes producing problematic promises. The "time machine" metaphor refers to the Baltic Sea, in particular, stating that it can serve as a mechanism to study and predict consequences and mitigate future perturbations of the social-ecological system due to its unique combinations of multistressor disturbances, strong scientific foundation, and early institutionalized cross-border environmental governance to address these disturbances (Reutsch et al., 2018). Consequently, the metaphors pinpoint that Northern Sea areas are particularly vulnerable – physically, socially, politically, and culturally. Simultaneously, these areas can offer valuable lessons on responding to these rapidly changing circumstances.

Cold Water in Focus

One emerging framework through which to address global concerns connected to the Arctic is to focus on the Northern waters and the drastic consequences arising from the melting and polluting processes affecting all life on Earth. Water is understood as a geographical and physical element and as part of the environment shaped by cultural methods, meaning that water is not alienated from its social significance. The current volume makes sense of the social significance of Arctic waters and interprets the Anthropocene in the Northern regions by exploring cold waters and Northern seascapes through that lens.

The volume focuses on knowledge, discourses, and representations regarding the Northern waters, including permafrost, sea areas, coastal areas, and estuaries (rivers and other minor water bodies) near the Northern coastlines. These areas are of globally emerging interest and significance, embodying a dialogue between water and

Earth. The volume's contextual background is provided by interdisciplinary approaches generated by the rapidly growing communities of scholars working towards the fundamental shift from terra-based towards aqua-based thinking (or *aquagraphy* in our coinage). The volume is informed by research conducted by several overlapping conceptual movements such as new materialism and post-humanism (e.g., Braidotti, 2013), oceanic studies (Steinberg, 2001; Blum, 2010), "blue humanities" (Gillis, 2014), and literature by human geographers (Peters & Anderson, 2014) and anthropologists (Driessen, 2004).

As the chapters draw a connection between culture and nature, attention is paid to the parallel realities of human and non-human lives in the shared regions of the Northern waters. The seascape is filled with human and non-human lives that are equally endangered and threatened by climate change and the toxic chemicals in the Northern waters. The threats are particularly felt by the Northern people and coast-dwelling communities living in a tense relationship with the natural world and non-human species facing the consequences of global warming and ecological pollution. The chapters pay attention to social and physical connectedness between humans and non-humans, articulating their encounters. The cases in the volume include the Baltic ringed seal, the gray seal, and whales as having a long history of interaction with human coast-dwelling communities where, as Virginia Richter (2016, p. 155) writes, "[whales] have become sources of food and fuel, mythological and literary figures and, finally, symbols of human ruthlessness and ecological endangerment."

The main framework is defined by the turn to Arctic and oceanic studies (Klein & Mackenthun, 2004). In the last few centuries, the signification of the sea has traveled from a temporal non-history to a location of reimagining, rewriting, and remembering historical processes. The sea has been restored as a space of antagonistic forces and a meeting place for different cultures. As Maija Ojala-Fulwood (2021, p. 24) shows in her article on the Arctic region in the early modern maps, the polar region has been powerfully visualized, mythologized, and imagined through its history of exploration and human–animal relations, which, as she suggests, has "long-standing implications on how the Arctic region has been understood."

Accordingly, this volume is motivated by the history of the Northern waters and the innovative ocean studies of the last decades (e.g., Rediker, 1987; Gilroy, 1993; Rice, 1996). The focal point is at the Northern waters and the seascape as part of modern history, which is already known by the historicization of the Mediterranean Sea, the Atlantic history, and the Pacific looming large in contemporary studies. We follow the idea implied by the blue humanities and ocean studies that the Northern waters of the Arctic Ocean and the Baltic Sea are not beyond history but are material and meaningful locations of historical meaning and cultural agency.

We use the environmentally oriented humanities as a mediator or "cultural interpreter" among different disciplines, facilitating interdisciplinary dialogue and research on cold waters. Here, the primary analytical interest lies in stories, theories, metaphors, and myths that are told, recycled, and reproduced. The chapters remind us that one of water's properties is to *circulate* – physically and symbolically. Water circulates not only in the physical world – from springs to rivers and rivers to oceans – but in the symbolic realms of philosophy, literature, and the aesthetic imagination embedded in yet evolving images and narratives (Strang, 2004, p. 119). The aquagraphical mapping of the Northern waters corresponds to the mental mapping and cultural production of specific hopes and anxieties, reflecting drastic changes in whole lifestyles of the Arctic's inhabitants because of climate change.

The focus is on how knowledge production is provided, generated, and circulated (Secord, 2004) by cultural technologies and products, such as print literature and digital and visual media. Yet, we also acknowledge that knowledge production in the humanities and most social sciences is a slow process of mediation and interpretation. This volume highlights cultural discourses and artistic representations concerning human experientiality and imagination in the context of cold waters.

Through examining the Northern water bodies, the volume's chapters contribute to the contemporary understanding of water's multiplicity and its refusal to remain passive or abstract. The idea in common is that water has a history: Water is not an abstraction or resource on which humans may draw at will but a dynamic process that, paradoxically enough, depends on humans using it. As Jamie Linton (2010, p. 3) writes, "[w]ater is what we make of it." Moreover, if human history is a history, among other things, of how we use water, this also covers the question of what we think of environmental water bodies, oceans, seas, lakes, and rivers.

In our coinage, water is a culture-generating category that comes close to the notions of how people "think with water" since "people think *with* water as well as *about* water, and these processes are often impossible to disentangle" (Linton, 2010, p. 38). The post-humanist and new materialist shift of dualities calls for ongoing dialogue between cultural and environmental studies, attention to relationality, and the implications of discursive and material power. Working against or through this dualism while recognizing water as a physical substance and imaginative power is what recent interdisciplinary ecocritical studies have done (DeLoughrey et al., 2015; Alaimo, 2012). Societies and water make each other in processes wherein water and society change. Cecilia Chen, Janine MacLeod, and Astrida Neimanis (2013, p. 5) have coined the term "hydrological turn" while emphasizing the ecopolitical dimensions of water, materiality, contextuality in time and place, and water's involvement in discursive practices and knowledge production.

This book has its own genealogy stemming from the research of Finnish scholars in research projects *Water as Social and Cultural Space: Changing Values and Representations*, *The Changing Environment of the North: Cultural Representations and Uses of Water*,[1] and *Living with the Baltic Sea in a Changing Climate: Environmental Heritage and the Circulation of Knowledge*,[2] and in publications on water in the social imagination (Costlow et al., 2017), social and symbolical meanings of water (Costlow & Rosenholm, 2017; Lehtimäki et al., 2018), visual representations of the Arctic (Lehtimäki et al., 2021), the Arctic and Baltic in the context

---

[1]Academy of Finland, Grant Number SA307840. Eeva Kuikka, Markku Lehtimäki, Arja Rosenholm, Anna Stammler-Gossmann, and Elena Trubina have written their chapters as part of the project.

[2]Academy of Finland, Grant Number SA315715. Jaana Kouri, Otto Latva, Tuomas Räsänen, Kirsi Sonck-Rautio, and Nina Tynkkynen have written their chapters as part of the project.

of international environmental regimes (e.g., Palosaari & Tynkkynen, 2015; Tynkkynen, 2016), and the linkages of the environmental changes with the meanings given to the Baltic Sea (Kouri et al., 2020).

Contributions in this volume are case studies covering environmental issues, Arctic-related international water governance, animal history and the history of science, and literary and visual representations of the Northern seascapes, including the region from the north of Scandinavia to Northern America and Russia. Five chapters of the book are devoted to the Russian Federation, the Arctic Ocean's largest seascape. The Russian case, among others, enables us to speak of the North as a transnational zone. Therefore, our approach advances the idea of multiple Norths. Each chapter advocates its own understanding of the Northern waters, including a multidimensional exploration of water in cultural, political, economic, and environmental life. Accordingly, the book is an interdisciplinary project at the intersection, combining methods from cultural geography and research on the human and non-human environment, as well as in literary, film, and cultural studies. All the chapters emphasize the dynamic and mobile nature of space on land and in water.

The book is structured in three parts: Part I focuses on how different human and non-human actors cope with and mediate the Arctic's changes, balancing the tragedies and opportunities. The focus is on the cold waters as an arena for competing perspectives and problem definitions, where not only human beings but non-human actors such as animals like seals or polar bears and material entities such as sea ice and oil have agency, connecting the abundancy of Arctic resources with the fragility of Northern waters. Part II addresses the hydrological space and politics by addressing the issues of different scales on which governance of the Arctic waters has been implemented. This localized and internationalized production of the Arctic hydrological space encompasses a spectrum of water-related imaginaries, ranging from anthropological interest in nostalgia for the formerly available lifestyle and fishing practices experienced by the inhabitants of the Finnish Turku archipelago to the complexities of the Arctic's changing geopolitics. The nexus of water, rivers, oceans, international politics, and local challenges is built to address the influence of the Arctic waters on the political and socioeconomic dynamics of their spaces. Part III emphasizes the symbolic-material entanglements of Northern waters as represented in literature and cinema. Here, cold water is analyzed in the context of local and indigenous cultures. The specific cultural relationship to the sea, lakes, and rivers depends on the geography and spatiality and each community's language, tradition, and memory. As the chapters suggest through their "aquagraphic" reading of the aesthetics and the materiality of the liquid element, water is a central part of modern history and a symbolic source of poetic imagination.

Part I: Mediating the Change

With the eventually drastic consequences emerging from the melting and polluting processes affecting all life on earth, the loss of Arctic ice and permafrost is tragic while opening windows of opportunity for many. Part I focuses on this ambivalence as well as the different human and non-human actors and how they mediate the change while coping with it – balancing the tragedies and opportunities posed by the change.

Human activities have become significant geological forces, particularly in the fragile North, such as land-use changes, deforestation, and fossil fuel burning. Therefore, the current geological epoch has been assigned the term "Anthropocene" (Crutzen, 2006). The Arctic region works as a prism through which the Anthropocene can be analyzed: because of human activities, the climate and the entire ecosystem in the Arctic are changing in rapid and complex ways, with effects encompassing the entire region. These changes have societal implications, and their management necessitates fundamental social changes alongside the more obvious technological ones. In the age of the Anthropocene, understanding humans' relationship with nature has become a necessary imperative of all scientific scrutiny.

Alternative, or rather, complementary, perspectives are among the popular efforts to reorient human–nature relations in a manner that is sensitive to the idea of the triple bottom line – environmental, social, and economic dimensions – associated with the sustainable development to live through the Anthropocene. In the context of the current volume, such implies giving a voice to non-human actors, including animals, and noting the agency of material entities, such as geomorphological constellations or sea ice. These perspectives are inspired by a new materialism, setting out the radical propositions that nature and culture are to be treated as parts of a continuum of materiality, not as distinct realms, and that the capacity for "agency" – the actions producing the social world – extends beyond human actors to the non-human and inanimate (e.g., Latour, 2005; Braidotti, 2013) – implicative of the privilege humans enjoy concerning the rest of the natural and social environment. Environmental social science, for instance, has long criticized society's anthropocentric notion and experienced theoretical turns towards materiality and the study of practice.

The Arctic has been treated as a backwater (see Economist 2012) or, as **Tuomas Räsänen** puts it in his contribution in this volume, a frontier environment with little or no function other than providing resources for human beings. Räsänen describes how the frontierization of Northern waters was symbolized by the Baltic seals, which were hunted because they preyed on fish and later became almost extinct because of toxic chemicals. Lately, the seals have suffered from the warming climate and iceless winters, becoming tragic examples of the changes in the Northern seas. Also, the attention devoted to these animals has increased as the Northern/Arctic regions have experienced a rise in human interest regarding business opportunities and geopolitics, accompanied by mushrooming artistic representations and fictional explorations.

Evolving developments potentially invite conflicting stakeholder values and encompass incomplete or contradictory processes of participation and empowerment. Yet, the recent developments are evidently driven by global economic imperatives rather than a quest for socioecological resilience within the Arctic itself (Young, 2012a, p. 168), as showcased in **Elena Trubina**'s chapter about BBC's nature documentary series on the Arctic and Antarctic environment and wildlife. This series concludes that "even the most remote Northern corners are artistically appropriated for corporate gain" and that "the alarmed statements about melting Arctic ice and the threat it poses for the planet coexist with the structures and processes which generate climate change." Even if many interpreted the recent economic developments as an empowering opportunity for the region's indigenous peoples (e.g., National Research Council, 2014), it is justified to ask to what extent they encompass on-the-ground experiences, such as local perspectives and the voices of those less powerful.

The human experience is needed for sense-making. However, the matter is a necessary part of the experience and has its own agency in the sense-making process. In this volume, the approach is demonstrated in **Maria Sakaeva**'s chapter, applying an anthropological approach to examine the perceptions and interpretations of the Pechora River basin following damages caused by oil in the Komi Republic, Russia. She demonstrates how oil's long-term presence has impacted how the inhabitants live with the river and how the river has become "victimized." In other words, the oil has agency and power over water, linking the abundance of Arctic resources and the fragility of Northern waters.

In **Otto Latva** and **Nina Tynkkynen's** chapter on the Arctic's plastic problem, they discuss how the perception of this issue has evolved in the matter and meaning in light of the increasing understanding of the global environmental crisis. Yet, as they argue, the definitions highlight the scientific interpretation of the plastic problem, distancing from and objectifying the problem and the Arctic regions. Similarly, the chapter by Trubina demonstrates how animals, such as polar bears, are objectified in a nature documentary series in a way that enhances scientific authority and tells a particular climate change narrative about the ongoing global displacement of water and ice. Both chapters showcase the potential influence of media and mediation on the roles of other actors such as NGOs, commercial interests, and local communities amid the Arctic environmental change. Media plays a central role in public perceptions and shaping the narratives of the Arctic region (Nilsson & Christensen, 2019), which will likely play a prominent role in how the Arctic is framed in political arenas, as relatively few people have any firsthand knowledge of the region.

Part II: Hydrological Space and Politics

In the scholarly debates about aquatic space, a wide range of concepts have been employed to grasp its specificity – from "aquatic regimes" (Mustonen, 2014) through "blue space as caring space" (Buser et al., 2020) to an "empire of air and water" (Carroll, 2016). "Aquatic regime" is the notion of grasping the contrast between the conventional uses of lakes and rivers and industrial resource extraction, while "caring blue space" refers to water's capacity to be a material and site of care, that is, ecological concerns, in diverse urban settings. The metaphor of the "empire of air and water" is employed to capture the poles, the Arctic, and the oceans as formerly inaccessible blank spaces, which, for centuries, captivated writers as atopic and otherworldly. However, existing studies are often concerned with finding new ways of understanding our planet's "warmer" oceans and blue spaces. It is no wonder that the most well-known conceptual movement in this regard is called the "blue humanities." In this volume, we also discuss the Arctic's "white spaces"; water in these spaces is often frozen, although, as stated, water's seemingly endless existence morphs into a fluid mode in nearly real time.

How does the Arctic's natural and political space and, more generally, the space of polar waters lend itself to national and international governance? The states and the international bodies, that is, the Arctic Council and the EU, find ways to collaborate on shared issues. The Northern aquatic space comprises a palimpsest of strategies used to conquer and use it, such as early exploration, cooperation, and geopolitical rivalries – mitigating climate change's consequences. These recent and more distant strategies signal the lack of correspondence between the global concerns and the Arctic's division among separate states. The strategic interests of the states appear somewhat disconnected from the global agenda, while the international organizations often lack much-needed leverage. In its multicultural and social meanings, water calls for unified action. It is particularly urgent when the renewed interest in exploiting new trade routes brings more risks, including the possibility of further environmental disasters stemming from oil drilling or shipping. Northern waters are inseparable from the area's wildlife and fisheries, and the habitats in the area are fragile. The Arctic waters are vast but contested. One can confidently predict that more disputes regarding the rights and access to resources are imminent, demanding the concerted effort of all Arctic governments and international bodies to prevent conflicts.

The authors of this part of the book develop different strategies to make sense of the processes unfolding on different scales – to "zoom in" and "zoom out" these processes using ethnographic research, conceptual analysis, and media analysis – spending extended periods of ethnographic fieldwork in their designated areas and in the localities they are intimately familiar with. They focus on the locals living and working on the coast and at different locations on the rivers. They examine the locals' relationships with salmon in the Kemijoki River in Northern Finland and how these relationships are shaped by regulations and restrictions, as well as the corporate interests and international natural resource management institutions. **Monica Tennberg**'s chapter discusses these relationships, demonstrating that river water proves too volatile and fluid to obey the intricately devised multilevel systems of governance. Accordingly, specialized systems of knowing water prove pivotal for retaining power over it. **Anna Stammler-Gossmann** investigates riverine grasslands and villages by the Amma River in Yakutia as the residents suffer from the floods that are happening more often. The locals live off cattle, and cattle depend on natural grazing, so when the river shores are flooded, the locals (animals and people) do not have enough grass. The ethnographically oriented scholars in this part of the volume examine the waterways as the sets of land–water interfaces whose human and non-human inhabitants are subject to different bureaucratic procedures. In the case of the Yakutian Amma River, the locals accustomed to raising cattle are unhappy about the local administration's attempt to shake off its responsibility to care for the local pastures.

Jaana Kouri and **Kirsi Sonck-Rautio** continue reflecting on local ecological knowledge and environmental heritage while building the case of the Finnish archipelago in the Baltic Sea. They introduce the concepts of environmental ethnography and "icegraphy" to investigate the results of their fieldwork in the region and capture the changing attitudes of the locals towards the consequences of climate, namely, the loss of permanent ice cover during the winter months. Locals depend on ice and miss its abundance. This ice is now gone forever. The authors compellingly demonstrate how the rapidly changing ice landscape is captured in the ambivalent narratives of the local dwellers whose nostalgia for former times can be bitter yet retain happier moments. The scholars demonstrate a possibility of tactful field research that is keen to preserve the local, often unspoken ecological knowledge. The drama of those who collected this knowledge for generations is comprised of the fact that they may well be its last bearers: With ice rapidly disappearing, its deep multisensory awareness faces the risk of remaining obsolete without being passed to future generations. However, it is precisely this kind of knowledge obtained through living with ice that may prove pivotal for resolving the strategies for sustainably treating the Arctic.

However serious and important, security interests must closely connect to the environmental changes. In their chapter, **Heather Nicol** and **Barry Scott Zellen** demonstrate how current and increasing environmental change also brings changes to security paradigms. They examine the ongoing problematization of traditional geopolitics in transforming coastal and maritime regions. They investigate shifts in understanding security paradigms caused by melting ice instead of the ice cap deemed sturdy and stable by previous geopolitical assessments. The dynamics of climate change also resulted in the changing importance of islands and archipelagos. Again, conventionally understood in terms of geopolitics as inaccessible barriers and, thus, as crucial natural obstacles to the Soviet and other threats, today, the authors argue, understanding security must be broadened to ensure that aquatic space with related environmental concerns is compellingly incorporated. This is similar to Tennberg's chapter in which the case of the city of Rovaniemi's watershed in the Finnish North is examined to show how the EU's regulations provided local authorities much-needed leverage to deal with the hydropower company whose operations caused sufficient environmental harm. The various intersecting governmental strategies penetrate the watershed's materiality. Numerous social and spatial relations lead to conflicting perceptions of the river flow and disputes from local activists and authorities regarding corporate attempts to control the river floor. Thus, this chapter demonstrates how environmental scientific knowledge, local knowledge of Rovaniemi inhabitants, the research commissioned by the water plant, and the political expertise of the local and regional politicians and administrators intersect to pursue conflicting but overlapping interests.

Part III: Narrating and Visualizing Cold Waters

Part III discusses the ways of knowing cold waters as narrated in literary fiction and depicted in visual arts such as cinema. It focuses on how the social and the aesthetic are linked to the awareness of the cold waters in recent discussions concerning climate change and its consequences in the northern parts of the world. The chapters consider how the knowledge of water embedded in novels and cinematic renderings contributes to mapping spaces marking the history of the northern cartography of power. While human effects, memories, and dreams linked to water have not been central questions in techno-scientific and hydrobiological discourses, the following chapters on literature and visual culture emphasize the dialogue among (hydro)science, social, and cultural aspects. The authors share the aim to think *about* the changing world *with* water. Exemplary cases of northern histories are mapped through real and imaginary waters. The focus is on life in the North and the multiple meanings of water for coastal people; on river pollution in Russian Far East and its consequences to the indigenous peoples; how the environmental imagination of water is linked to climate change and melting polar ice; and how cold water works as a poetic space of mind and memory.

Cold water – its presence and experience – is especially considered in the context of local and indigenous cultures. The cold water of the river is seen as emblematic of the northern indigenous epistemology and way of thinking, as in **Eeva Kuikka**'s chapter. According to her postcolonial and ecocritical reading of Khanty author Eremei Aipin's novel *Khanty, or the Star of the Dawn* (*Khanty, ili zvezda utrennei zari*), the northern river is a watery map connecting the indigenous Khanty people on a geographical and temporal level. Also, the cold river water transmits the indigenous epistemology, underlining the inseparability of people and non-human nature. **Mika Perkiömäki**'s analysis of the work of Boris Shergin and Stepan Pisakhov – two Pomor writers of the early Soviet period – argues that a distinctive feature of Pomor identity is the dependency on Northern waters. His chapter suggests that early Soviet Pomor fiction idealizes and romanticizes the traditional Pomor life that depends on the Northern Sea, while it imagines the arrival of Soviet Arctic modernity as something that unproblematically coexists with the traditional way of life.

As these chapters demonstrate, people's ways of acting concerning cold waters also tell about the relationship of one's physical being in the world and inward experiences with the surrounding water environment. The stories are told in local surroundings and specific cultural contexts. They include the maritime Arctic Ocean and the coastal life between water and Earth, the Northern Russian rivers transmitting the indigenous epistemology, imaginary waters in New England, or even an extended atopic space in the Arctic as a possible future home for the surviving humans after the environmental apocalypse. Thinking with and about water means considering lived contexts connected to the sea or river that, as the chapters show, can be experienced as an individual *Bildungsroman* and defined by a collective way of life with its traditions and a specific relationship with water. The connection to the sea, lakes, and rivers depends on the liquid topology. Each community's language, symbols, and traditions – national, ethnic, gendered, or generational – affect and constitute how memories and traditions are circulated. Experiencing the maritime or riverine environment as a cultural phenomenon brings a specific water-bound identity endowed with cultural and national consciousness emerging in the historical alteration of knowledge, ideology building, and local and global power structures.

As suggested, water is not only a concern for the natural sciences but a cognitive and cultural medium of social imagination and political power. Water is a central part of modern history and a potent symbol of creative thinking and poetic imagination. Thus, the authors focus on the diverse Northern waters in authentic historical and geographical contexts while maintaining the idea that the water motif reflects the flow of the poetic language of art and imagination. In short, water is liquid physicality endowed with symbolic power. The chapters take this symbolic-material entanglement of water as their starting point in examining the exterior and interior water worlds when defining human activities in specific natural and social environments.

In their chapter, **Tatu Laukkanen** and **Arja Rosenholm** write about indigenous communities coping with the changing world on the Arctic coastline, as represented in two Soviet/Russian films about the Chukchi from different eras: *The Most Beautiful Ships* (*Samye krasivye korabli*) and *The Whaler Boy* (*Kitoboi*). In a psychosocially informed analysis of Russian cinema about the Arctic Ocean, the shore's multidimensional qualities are conceptualized through the lenses of liminality, instability, and transitoriness. **Markku Lehtimäki**'s chapter focuses on residents' experiences of cold water and Maine's wintry landscape in the New England region of the United States, as imagined and represented in the fiction of Paul Harding. In Harding's novels, *Tinkers* and *Enon*, cold water functions as a kind of transmission medium, bearing memories of the dead and enabling mystical connections to the historical past. In these two chapters, water is read regarding trauma and memory so that, for example, cold Northern waters have their subconscious and mysterious connections to identity and knowledge. In his classic *Water and Dreams*, Gaston Bachelard sees that water in its various substances and appearances has deep significance as a symbolic element. He suggests that contemplating depth can connect people to their subconscious and the deceased as they grapple with infinity and darkness, whereas the surface of the water forms a boundary between life and death, conscious and subconscious, and reality and imagination (Bachelard, 1983, p. 47). While cold water in its material, physical, and haptic qualities is potentially known to people in these literary and cinematic narratives, its deeper meanings may remain unknown. Here, the capacities of the region's imaginative renderings emerge.

Indeed, the realm of water is both "real and imagined" (Soja, 1996). In Part III, this realm is explicitly situated in the Arctic Ocean and the rivers that flow towards the North, as do the Russian rivers, as well demonstrating the imbrication of nature and culture by the poetic, imaginary, and speculative waters. The common idea, shared by the chapters in this part of the volume, is that human self-representation is strongly intertwined with water. Accordingly, the chapters relate themselves to the concept of *aquagraphy* or "water writing," meaning that water can be read as "writing about water" or as "writing modes produced by water itself" and by the liquid imaginary (Capeloa Gil, 2008, p. 10; Schmitz-Emans, 2008, p. 37). The chapters offer micro-histories of specific places – the Arctic shores of the White Sea and the Barents Sea, the marine shoreline of the Arctic Ocean in the Russian Far East, northern rivers in Western Siberia depicted as a watery map that connects the indigenous Khanty people, and New England in the northeastern area of the United States – as well as analyses of texts situated in those places. The chapters show how the artists in visual culture and literature are working against abstraction that plays a role in scientific and technical discourse.

The chapters correspond to a central premise in the critical maritime history, as well as sea and ocean studies, according to which water bodies (e.g., seas and oceans) must be analyzed as historical locations (Klein & Mackenthun, 2004, p. 2; see also Costlow et al., 2017; Tynkkynen et al., 2021). Indeed, especially the chapters concerning the Arctic Ocean in this volume take issue with "the cultural myth that the ocean is outside and beyond history" (Klein & Mackenthun, 2004, p. 2). The chapters not only discover the modern historical experience of transnational contact zones in the history of seas, rivers, and oceans but suggest that (northern) oceans, seas, and rivers play a conceptual role in cultural and spatial orientation. Telling those histories involves considering social processes occurring in specific water-bound chronotopes that conceptually integrate space and time (Bakhtin, 1981, p. 84). Liquid maps reflect different ways of relating to the surrounding environment with their own historical context. Consequently, the chapters particularly examine the relationships among space, place, and water. Their explorations draw maps to the "real-and-imagined" water-bound areas as intertwined and geographically, physically, and spiritually connected. Thus, it is argued that water metaphors and images are transnational, universal, and contextually characterized by recognizable cultural identities and spatiotemporal mental maps.

While Russian or American fiction and cinema in Part III deals with representations of actual waters in their historical and cultural contexts, **Heidi Hansson** and **Maria Lindgren Leavenworth** argue that speculative fiction also has its origins in real-world concerns. Thus, in their analysis of Julie Bertagna's fantastic atopias concerning the Arctic Ocean, the writers envision those readers imaginatively engaging with contemporary risk scenarios by encountering future worlds devastated by climate change. Here, in the book's conclusion, the previously peripheral and inhospitable Arctic emerges as a new, visionary home.

The importance of literature and the visual arts is clearly increasing in the multidisciplinary Arctic studies published in the last decades (e.g., Ryall et al., 2010; Mackenzie & Westerståhl Stenport, 2016; Hansson & Ryall, 2017; Wood-Donnelly, 2019; Lehtimäki et al., 2021). These studies are based on the premise that nature and culture cannot be considered separate entities. Thus, the multidisciplinary Arctic studies overlap with the recent material ecocriticism (Iovino & Oppermann, 2014), emphasizing a methodological reaction to the environmental crisis, the destruction of habitats of local communities, and the "scrambling for resources" in the Arctic areas. Understanding the complex interaction of nature and culture from a literary and artistic perspective entails paying attention to water as an existential element and resource with its cultural and political implications.

University of Turku, Turku, Finland Markku Lehtimäki

University of Tampere, Tampere, Finland Arja Rosenholm

UNC Global Research Institute, Chapel Hill, NC, USA Elena Trubina

Åbo Akademi University, Turku, Finland Nina Tynkkynen

References

Alaimo, S. (2012). States of suspension: Trans-corporeality at sea. *ISLE* (*Interdisciplinary Studies in Literature and Environment*), *19*(3): 476–493. https://doi.org/10.1093/isle/iss068

Arbo, P., Iversen, A., Knol, M., Ringholm, T., & Sander, G. (2013). Arctic futures: Conceptualizations and images of a changing Arctic. *Polar Geography*, *36*(3): 163–182. https://doi.org/https://doi.org/10.1080/1088937X.2012.724462

Bachelard, G. (1983). *Water and dreams: An essay on the imagination of matter* (trans. E. R. Farrell). The Dallas Institute of Humanities and Culture. (Original work published 1942)

Bakhtin, M. (1981). *The dialogic imagination*. M. Holquist (Ed.). University of Texas Press.

Blum, H. (2010). The prospect of oceanic studies. *PMLA*, *125*(3): 670–677. https://doi.org/10.1632/pmla.2010.125.3.670

Böhme, H. (1988). Umriss einer Kulturgeschichte des Wassers: Eine Einleitung. In H. Böhme (Ed.), *Kulturgeschichte des Wassers* (pp. 7–44). Suhrkamp Verlag.

Braidotti, R. (2013). *The posthuman*. Polity Press.

Burkart, P., & Christensen, M. (2013). Geopolitics and the popular. *Popular Communication*, *11*(1): 3–6. https://doi.org/10.1080/15405702.2013.751853

Buser, M., Payne, T., Edizel, O., & Dudley, L. (2020). Blue space as caring space – Water and the cultivation of care in social and environmental practice. *Social & Cultural Geography*, *21*(8): 1039–1059.

Carroll S. (2016). *An empire of air and water: Uncolonizable space in the British imagination, 1750–1850*. University of Pennsylvania Press.

Capeloa Gil, I. (Ed.) (2008). *Fleeting, floating, flowing: Water writing and modernity*. Königshausen & Neumann.

Chen, C., MacLeod, J., & Neimanis, A. (2013). Introduction. In C. Chen, J. MacLeod & A. Neimanis (Eds.), *Thinking with water* (pp. 3–22). McGill-Queens University Press.

Christensen, M. & Nilsson, A. (2017). Arctic sea ice and the communication of climate change. *Popular Communication*, 15 (4): 249–268. https://doi.org/10.1080/15405702.2017.1376064

Cohen, M. (2006). The chronotopes of the sea. In F. Moretti (Ed.), *The novel*. Volume 2: Forms and themes (pp. 647–667). Princeton University Press.

Costlow, J., & Rosenholm, A. (Eds.) (2017). *Meanings and values of water in Russian culture*. Routledge.

Costlow, J., Haila, Y., & Rosenholm, A. (Eds.) (2017). *Water in social imagination: From technological optimism to contemporary environmentalism*. Brill/Rodopi.

Crutzen P. J. (2006). The "Anthropocene". In E. Ehlers & T. Krafft (Eds.), *Earth system science in the Anthropocene* (pp. 13–18). Springer. https://doi.org/10.1007/3-540-26590-2_3

DeLoughrey, E., Didur, J., & Carrigan, A. (2015). *Global ecologies and the environmental humanities: Postcolonial approaches*. Routledge.

Dittmer, J., Moisio, S., Ingram, A., & Dodds, K. (2011). Have you heard the one about the disappearing ice? Recasting Arctic geopolitics. *Political Geography*, *30*(4): 202–214. https://doi.org/10.1016/j.polgeo.2011.04.002
Dodds, K., Kuus, M., & Sharp, J. P. (Eds.) (2013). *The Ashgate research companion to critical geopolitics*. Ashgate.
Dodds, K., & Wilson Rowe, E. (2021). Red Arctic? Affective geopolitics and the 2007 Russian flag-planting incident in the central Arctic Ocean. In M. Lehtimäki, A. Rosenholm & V. Strukov (Eds.), *Visual representations of the Arctic: Imagining shimmering worlds in culture, literature and politics* (pp. 178–195). Routledge.
Driessen, H. (2004). A Janus-faced sea: Contrasting perceptions and experiences of the Mediterranean. *Mast Maritime Studies*, *3*(1): 41–50. http://hdl.handle.net/2066/64721
Economist (2012). Cosy amid the Thaw: The Arctic Council works well – Because of the region's riches. March 24, 2012. www.economist.com/node/21551029/print.
Gillis, J. R. (2012). *The human shore: Seacoasts in history*. University of Chicago Press.
Gilroy, P. (1993). *The black Atlantic: Modernity and double consciousness*. Harvard University Press.
Goodbody, A., & Wanning, B. (Eds.) (2008). *Wasser, Kultur, Ökologie*. V&R unipress.
Guerrieri, V. (2019). The spatiality of hope: Mapping Canada's northwest energy frontier. *Globalizations*, *16*(5): 678–694.
Hansson, H., & Ryall, A. (Eds.) (2017). *Arctic modernities: The environmental, the exotic and the everyday*. Cambridge Scholars Publishing.
Iovino, S., & Oppermann, S. (Eds.) (2014). *Material ecocriticism*. Indiana University Press.
Keil, K. (2014). The Arctic: A new region of conflict? The case of oil and gas. *Cooperation and Conflict*, *49*(2): 162–190. https://doi.org/https://doi.org/10.1177/0010836713482555
Klein, B., & Mackenthum, G. (2004). Introduction: The sea is history. In B. Klein & G. Mackenthum (Eds.), *Sea changes: Historicizing the ocean* (pp. 1–12). Routledge.
Kouri, J., Räsänen, T., & Tynkkynen, N. (Eds.) (2020). *Muutoksen tyrskyt ja kotirannan mainingit: Kulttuurisia näkökulmia merentutkimukseen*. Suomalaisen Kirjallisuuden Seura.
Latour, B. (2005). *Reassembling the social: An introduction to actor-network-theory*. Oxford University Press.
Lehtimäki, M., Meretoja, H., & Rosenholm, A. (Eds.) (2018). *Veteen kirjoitettu: Veden merkitykset kirjallisuudessa*. Suomalaisen Kirjallisuuden Seura.
Lehtimäki, M., Rosenholm A., & Strukov, V. (2021). Introduction: Visualising the Arctic. In M. Lehtimäki, A. Rosenholm & V. Strukov (Eds.), *Visual representations of the Arctic: Imagining shimmering worlds in culture, literature and politics* (pp. 1–20). Routledge.
Lemke, P. (2011). Arctic processes and global climate. In S. Wasum-Rainer, I. Winkelmann & K. Tiroch (Eds.), *Arctic science, international law and climate change: Legal aspects of marine science in the Arctic Ocean* (pp. 45–54). Springer.
Linton, J. (2010). *What is water? The history of a modern abstraction*. UBS Press.
MacKenzie, S., & Westerståhl Stenport, A. (Eds.) (2016). *Films on ice: Cinemas of the Arctic*. Edinburgh University Press.
Morgunova, M., & Westphal, K. (2016). Offshore hydrocarbon resources in the Arctic: From cooperation to confrontation in an era of geopolitical and economic turbulence? *SWP Research Paper* 3/2016. https://nbn-resolving.org/urn:nbn:de:0168-ssoar-46526-8
Mustonen, T. (2014). Endemic time-spaces of Finland. *Aquatic regimes*. *Fennia*, *192*(2).
National Research Council of the National Academies. (2014). *The Arctic in the Anthropocene: Emerging research questions*. The National Academies Press.
Nielsen K. H., & Nielsen H. (2021). *Camp century: The untold story of America's secret Arctic military base under the Greenland ice*. Columbia University Press.
Nilsson, A., & Christensen, M. (2019). *Arctic geopolitics, media and power*. Routledge.
NSIDC (National Snow & Ice Data Center). (2012). *Arctic sea ice falls below 4 million square kilometres*. Boulder. http://nsidc.org/arcticseaicenews/2012/09/arctic-sea-ice-falls-below-4-million-square-kilometers/

Ojala-Fulwood, M. (2021). Arctic regions in early modern maps. In M. Lehtimäki, A. Rosenholm & V. Strukov (Eds.), *Visual representations of the Arctic: Imagining shimmering worlds in culture, literature and politics* (pp. 23–41). Routledge.

Østerud, Ø., & Hønneland, G. (2017). Geopolitics and international governance in the Arctic. *Arctic Review*, *5*(2). https://arcticreview.no/index.php/arctic/article/view/1044

Palosaari, T. (2012). The amazing race: On resources, conflict, and cooperation in the Arctic. *Nordia Geographical Publications*, *40*(4): 13–30.

Palosaari, T., & Tynkkynen, N. (2015). Arctic securitization and climate change. In G. Hønneland & L. C. Jensen (Eds.), *Handbook of the politics of the Arctic* (pp. 87–104). Edward Elgar.

Peters, K., & Anderson, J. (2014). *Water worlds: Human geographies of the ocean.* Routledge.

Pratt, M. L. (1992). *Imperial eyes: Travel writing and transculturation.* Routledge.

Rediker, M. (1987). *Between the devil and the deep blue sea: Merchant seamen, pirates, and the Anglo-American maritime world, 1700–1750.* Cambridge University Press.

Reusch, T. B. H., Dierking, J., Andersson, H. C., Bonsdorff, E., Carstensen, J., Casini, M., et al. (2018). The Baltic Sea as a time machine for the future coastal ocean. *Science Advances*, *4*(5), [eaar8195]. https://doi.org/10.1126/sciadv.aar8195

Rice, E. (Ed.) (1996). *The sea and history.* Sutton Publishing.

Richter, V. (2016). Where things meet in the world between sea and land: Human-whale encounters in littoral space. In U. Kluwick & V. Richter (Eds.), *The beach in anglophone literatures and cultures: Reading littoral space* (pp. 155–173). Routledge. https://doi-org.libproxy.tuni.fi/10.4324/9781315613932

Ryall, A., Schimanski, J., & Howlid Waerp, H. (Eds.) (2010). *Arctic discourses.* Cambridge Scholars Publishing.

Schmitz-Emans, M. (2008). Water writing: Fluid languages and the voices of the sea. In I. Capeloa Gil (Ed.), *Fleeting, floating, flowing: Water writing and modernity* (pp. 37–64). Königshausen & Neumann.

Secord, J. A. (2004). Knowledge in transit. *Isis*, *95*(4): 654–672.

Smith, J. R. (2021). "Exceeding Beringia": Upending universal human events and wayward transits in Arctic spaces. *EPD: Society and Space*, *39*(1): 158–175. https://doi.org/10.1177/0263775820950745

Soja, E. (1996). *Thirdspace: Journeys to Los Angeles and other real-and-imagined places.* Blackwell.

Steinberg, P. (2001). *The social construction of the ocean.* Cambridge University Press.

Stoddart, M. C. J., & Smith, J. (2016). The endangered Arctic, the Arctic as resource frontier: News media narratives of climate change and the north. *Canadian Review of Sociology*, *53*(3): 316–336.

Strang, V. (2004). *The meaning of water.* Berg.

Tynkkynen, N. (2016). The pollution of the Baltic Sea: A mirror image of modernization. In J. Costlow, Y. Haila & A. Rosenholm (Eds.), *Water in social imagination: From technological optimism to contemporary environmentalism* (pp. 167–192). Brill/Rodopi.

Van Alstine J. & Davies W. (2017). Understanding Arcticness: Comparing resource frontier narratives in the Arctic and East Africa. In I. Kelman (Ed.), *Arcticness: Power and voice from the north* (pp. 89–102). UCL Press.

Wood-Donnelly, C. (2019). *Performing Arctic sovereignty: Policy and visual narratives.* Routledge.

Young, O. (2012a). Arctic politics in an era of global change. *The Brown Journal of World Affairs*, *19*(1): 165–178. https://www.jstor.org/stable/24590936

Young, O. (2012b). Arctic tipping points: Governance in turbulent times. *AMBIO*, *41*: 75–84. https://doi.org/10.1007/s13280-011-0227-4

Zellen, B. S. (2009). *Arctic doom, Arctic boom: The geopolitics of climate change in the Arctic.* Greenwood Publishing Group.

Contents

About the Contributors

Heidi Hansson, Ph.D. is a professor of English literature at Umeå University, Sweden. She has published widely on images of the North and the Arctic in travel writing, fiction, and popular culture, often in relation to gendered visions and accounts of the region. In recent years, she has paid particular attention to representations of the North in popular culture, such as crime fiction, speculative fiction for young adults, and the response to Arctic exploration in comic journals. A recent publication is the collection *The Arctic in Literature for Children and Young Adults*, co-edited with Maria Lindgren Leavenworth and Anka Ryall (2020).

Jaana Kouri, Ph.D., M.Ed. is a researcher of the study of religions under the profiling area The Sea at Åbo Akademi University, Finland. She specializes in the meanings people give to their environments. Her research area is the environmental heritage related to the sea and coastal areas as lived, reminisced, and experienced environment. Her approach is multidisciplinary and methodological, which comes through in her use and development of environmental ethnography, autoethnography, and creative writing as a method for co-creating and verbalizing experience-based knowledge. Her latest publications are (with Savitri Jetoo) "Voicing and Visualizing Change: Perceptions of Environmental Heritage in the Baltic Sea Region," *Heritage*, 4:3 (2021), and (with Elina Hytönen-Ng) "The change in the researcher's position in the study of shamanism," *Approaching Religion*, 12:1 (2022).

Eeva Kuikka, M.A. is a doctoral researcher in Russian language and culture at Tampere University, Finland. Her research interests lie in Russian culture and literature, human–animal studies, posthumanism, postcolonial theory, and LGBTQ+ themes. Her ongoing doctoral research addresses indigenous literature of the Russian Arctic. The study asks how the changes that the Soviet regime brought to indigenous communities in the Arctic region are reflected in their literature through relationships with non-human animals. Her research material consists of literary works by Nenets, Chukchi, and Khanty authors from the postwar Soviet era until the collapse of the Soviet Union.

Otto Latva, Ph.D. is a historian focusing on environmental history as well as human–animal and human–plant studies. He has widely studied the early modern as well as nineteenth- and twentieth-century societies and cultures. At the moment, he works as a researcher at the University of Turku and leads the research project *Disappeared, Endangered and Newly Arrived Species: The Human Relationship with the Changing Biodiversity of the Baltic Sea* (2021–2025), funded by the Academy of Finland.

Tatu Laukkanen, Ph.D. is a film scholar and industry professional affiliated with Tampere University. His current research interests include the gangster and war genres, the screening of economic change in the BRICS countries, cinematic geopolitics, and films about the Arctic. Recent publications include "Neoliberal Capitalism and BRICS on Screen" (with Iiris Ruoho) in *BRICS: Framing a New Global Communication Order?*, Nordenstreng and Thussu (Eds.) (Routledge, 2021), and "A Prophecy of Bolsonaro: Masculinity and Populism in the Elite Squad Films" in *Culture and Politics of Populist Masculinity*, Hakola et al. (Eds.) (Lexington Books, 2021).

Maria Lindgren Leavenworth, Ph.D. is professor of modern English literature at Umeå University, Sweden. Her research connected to the North and the Arctic has addressed nineteenth-century travel writers Selina Bunbury, Bayard Taylor, and S. H. Kent, as well as twentieth-century fiction by Ursula K. Le Guin, Dan Simmons, and Michelle Paver. She is co-editor of *The Arctic in Literature for Children and Young Adults* and is finishing a project on the imagined Arctic in speculative fiction.

Markku Lehtimäki, Ph.D. is a professor of comparative literature at the University of Turku, Finland. His fields of expertise are narrative theory, visual culture, ecocriticism, and American literature. Lehtimäki's recent studies focus on the relationship between nature and narrative, especially in the fiction about the changing North. His research projects include *Natural Narratology, Cognitive Poetics, and Ecocriticism* (2009–2011) and *The Changing Environment of the North: Cultural Representations and Uses of Water* (2017–2021), both funded by the Academy of Finland. He has co-edited several books, including *Narrative, Interrupted: The Plotless, The Trivial and the Disturbing in Literature* (De Gruyter, 2012) and *Visual Representations of the Arctic: Imagining Shimmering Worlds in Culture, Literature and Politics* (Routledge, 2021).

Heather N. Nicol is the Director of the School for the Study of Canada and Director of the Frost Centre for Canadian Studies and Indigenous Studies at Trent University. She is a professor in the School of the Environment and the coordinator of the Trent Circumpolar Studies online diploma. Nicol is also a founding member and co-director of the UArctic Læra Institute of Circumpolar Education. She is the author of many books and articles related to circumpolar policy, borders, and security and was the 2015–16 Fulbright Research Chair in Arctic Studies at the Canadian Studies Center, Henry M. Jackson School of International Studies, University of Washington.

Mika Perkiömäki, Ph.D. works as a postdoctoral researcher at the Aleksanteri Institute, University of Helsinki. His field of expertise covers Russian cultural and media studies, ecocriticism, and the environmental humanities. His dissertation "Imagined Riverography of Late Twentieth-Century Russian Prose" (Tampere University, 2021) is an ecocritical study on meanings of the river in Russian natural-philosophical prose of the late Soviet period. He has also studied representations of nature in early Soviet literature of the Pomors, and he is the editor-in-chief of *Idäntutkimus*, the Finnish review of East European studies. Currently, he studies practices of mediatized knowledge production and distribution on fossil and renewable energy and climate change in Russia as part of the project *FLOWISION: Best from Both Worlds – Enhancing Energy Transition in Russia and Finland by Making Resource Flows Visible*.

Tuomas Räsänen, Ph.D. works as an associate professor of environmental history in the Department of Historical and Geographical Studies at the University of Eastern Finland. His research interests include the history of marine pollution, the history of environmentalism, and the history of human–wildlife relations. He is currently leading the research project *Humans and Ticks in the Anthropocene* and works as a researcher in the projects *Living with the Baltic Sea in a Changing Climate* and *Culture of Unsustainability: Animal Industries and the Exploitation of Animals in Finland since the Late Nineteenth Century*, all of which are funded by the Academy of Finland.

Arja Rosenholm, Ph.D. is professor (emerita) of Russian language and culture at Tampere University, Finland. Her fields of scholarship are Russian literature and culture, gender studies, and ecocriticism, including space studies. Her research projects include *Water as Social and Cultural Space: Changing Values and Representations* (2012–2016) and *The Changing Environment of the North: Cultural Representations and Uses of Water* (2017–2021), both funded by the Academy of Finland. She has co-edited several scholarly volumes, among them *Meanings and Values of Water in Russian Culture* (Routledge 2017), *Water in Social Imagination: From Technological Optimism to Contemporary Environmentalism* (Brill/Rodopi, 2017), and *Visual Representations of the Arctic: Imagining Shimmering Worlds in Culture, Literature and Politics* (Routledge, 2021).

Maria Sakaeva, Ph.D. independent researcher, worked as a fellow at Warsaw University on the project *Native Voices and Oil-Water Relations in the Russian Arctic*, funded by the Kirkland Foundation (2021). She received her Ph.D. in economic sociology in 2015 from the National Research University Higher School of Economics in Moscow. Her current research interests concern local journalism, digital media, anthropology of water, and Arctic indigenous cultures. Her publications have appeared in *Demokratizatsya: The Journal of Post-Soviet Democratization*, *Laboratorium: Russian Review of Social Research*, and *Studies of Transition States and Societies*. From 2010 to 2014, she worked as a journalist in print and online media.

Kirsi Sonck-Rautio, Ph.D. is a researcher at Åbo Akademi University, Finland. She works under the profiling area The Sea, and her research interests focus on global environmental problems especially connected to marine environmental change, political ecology, local ecological knowledge, and consumer cultures regarding plastic packaging. She received her Ph.D. in ethnology from the University of Turku in 2019. Her recent publications include "Power and responsibility in transition to sustainable food packaging" in *Social and Cultural Aspects of the Circular Economy*, V. Pal (Ed.) (Routledge, 2022) and "Nordic and Nature: The Human-Nature Interface in *Ethnologia Scandinavica* 1970–2020," *Ethnologia Fennica* 51 (2021).

Anna Stammler-Gossmann, Ph.D. is senior researcher in social anthropology and coordinator of the Arctic Studies Programme at the Arctic Centre at the University of Lapland in Rovaniemi, Finland. Her research focuses on practical and theoretical developments in community studies, specifically how different forms of knowledge become entangled in negotiations for legitimacy. Her current work draws on theoretical trends linking environmental anthropology of snow, permafrost, and water disaster (flooding) with tourism and food studies. Publications related to human–water relationship include "People of the Permafrost Land" (*Science and Technology*, 2019, in co-authorship, Russian), "The Big Water of a Small River: Flood Experiences and a Community Agenda for Change" (book chapter, Springer, 2012), and "'Translating' Vulnerability at the Community Level: Case Study from the Russian North" (book chapter, Springer, 2010).

Monica Tennberg, Ph.D. is a research professor at the Arctic Centre, University of Lapland, Finland, and leads the University of the Arctic thematic network on Critical Arctic Studies. She is a political scientist and has worked with Arctic topics, such as international environmental cooperation, climate change adaptation, and Arctic urbanism since the early 1990s. She is currently co-editing two books: with Marjo Lindroth and Heidi Sinevaara-Niskanen, *Unravelling the North – Critical Studies of the Arctic* (Routledge), and with Hannah Strauss-Mazzullo, *Living and Working with Snow, Ice and Seasons in the Modern Arctic: Everyday Perspectives* (Palgrave MacMillan).

Elena Trubina, Ph.D. worked as a scholar on the framework of the project *The Changing Environment of the North: Cultural Representations and Uses of Water*, funded by the Academy of Finland. She is currently a lecturer at the Global Research Institute at the University of North Carolina, where she is also a fellow at the Center for Slavic, Eurasian and East European Studies. Her research interests are postsocialist space, urban development, memory, and cultural industries. Selected publications include "Sidewalk fix, elite maneuvering and improvement sensibilities: The urban improvement campaign in Moscow," *Journal of Transport Geography*, 83 (2020), and "Postcolonial Criticism and Urban Theory," in Russian, *Novoe literaturnoe obozrenie*, 161 (2020).

Nina Tynkkynen, Ph.D. works as a professor in environmental governance and policy under the profiling area The Sea and is the head of subject in public administration in the Faculty of Social Sciences, Business and Economics at Åbo Akademi University. She received her Ph.D. in environmental policy in 2008 from the University of Tampere. Her research interests include environmental policy, politics and governance in multilevel governance settings, and the politics of environmental knowledge. Tynkkynen's research is featured in international and domestic high-impact social scientific journals. Selected publications include *Russia and the Politics of International Environmental Regimes: Environmental Encounters or Foreign Policy?* (a monograph by Edward Elgar in 2015; with A. Korppoo & G. Hønneland).

Barry Scott Zellen, Ph.D. is a research scholar specializing in Arctic geopolitics, international relations, and the tribal foundations of world order. In the 2000s, he ran the Arctic Security Project at the Center on Contemporary Conflict at the Naval Postgraduate School (2004–2012), while editing the journal *Strategic Insights*. He completed his doctorate in international relations at the University of Lapland (2015), after which he received postdoctoral research support from the Kone Foundation (2016–2018). He was appointed a visiting scholar in the Department of Geography at the University of Connecticut (2018–2022), Arctic Scholar at the Center for Arctic Study and Policy (2019–2022), and Fulbright Scholar at the Polar Law Institute, University of Akureyri (2020). He has authored, edited, or co-edited over a dozen books on Arctic, indigenous, and strategic issues.

List of Figures

Part I
Mediating the Change

Chapter 1
The Problem of Plastic in the Arctic

Otto Latva and Nina Tynkkynen

Abstract "The plastic problem" was recognised as one of the world's largest growing environmental concerns by the United Nations Environmental Programme (UNEP) in 2016. It is estimated that by 2050, there will be more plastic than fish in the oceans. Scientists have documented a growing amount of plastic in the Arctic Ocean. Plastics travel on ocean currents to the far north and accumulate in deep-sea sediments, water, ice, and animals. With its harsh living conditions and limited food web, the Arctic is particularly vulnerable to the plastic burden exacerbated by climate change. This chapter describes how marine plastic litter in the Arctic has evolved as an environmental problem. We look at the evolution of the Arctic's plastic problem by first describing the issue based on research literature. Second, we examine how newspapers have written about marine plastic in the Arctic since the problem was first publicly noted in the 1980s. The chapter provides a chronology of the news, paying particular attention to transitions in how the problem has been defined. Further, the role of (cold) water in the differing interpretations of this problem is scrutinised. The chapter contributes to the current volume by telling "a plastic story" that links with many binaries and concepts discussed in the volume: matter and meaning, fluid and solid, scarcity and abundance, the liquid and solid forms of water, as well as materiality and transformations by water.

Keywords Marine plastic · Plastic pollution · Socio-material construction of environmental problems · Materiality

O. Latva
University of Turku, Turku, Finland
e-mail: onilat@utu.fi

N. Tynkkynen (✉)
Faculty of Social Sciences, Business and Economics, Åbo Akademi University, Turku, Finland
e-mail: nina.tynkkynen@abo.fi

M. Lehtimäki et al. (eds.), *Cold Waters*, Springer Polar Sciences,
https://doi.org/10.1007/978-3-031-10149-6_1

In the late 1960s, biologists from Newfoundland to Australia studying seabird feeding ecology started noticing plastic in the diets of their research objects (Ryan, 2015; Coleman & Wehle, 1984). Similar news of plastic ingestion and entanglement was soon heard regarding fish off the southern New England coast, sea turtles of Costa Rica and Japan, and whales in the North Atlantic. By the end of the 1970s, plastic particles were found in the planktonic and benthic samples (Feder et al., 1978) and in some beaches in such large quantities that they seemed to be covered with "plastic sand" (Gregory, 1978). A new problem was added to the list of global environmental concerns: plastic pollution at sea.

Plastic pollution is everywhere – from the equator to the Arctic, the sea surface to the seafloor, and from densely populated beaches to remote coastlines. Whilst only 10% of all solid waste is plastic (Heap, 2009), up to 80 per cent of the waste accumulating on land, shorelines, the ocean surface, or seabed is plastic (Barnes et al., 2009). According to estimates, between 4.8 and 12.7 million tons of mismanaged land-based plastic waste entered the oceans in 2010 (Jambek et al., 2015). These plastics come from many sources – on land and at sea. Wind and global ocean currents spread plastic particles in the marine environment around the world. Ocean plastic pollution is alarming due to its persistence, complexity, steady growth, and pervasive impacts on all aspects of ecosystems, including human communities (Villarrubia-Gómez et al., 2018).

The Arctic has some of the greatest amounts of plastics on Earth. Scientists have documented that the Arctic Ocean surface waters hold the most plastics of any ocean basin (Barrows et al., 2018). Plastics travel on ocean currents to the far north and accumulate in the deep-sea sediments, water, ice, and animals. With its harsh living conditions and limited food web, the Arctic is particularly vulnerable to the plastic burden. That burden's effects are further exacerbated by climate change. The prospects that plastic will end up in human food are of major concern to scientists, as people in the north heavily rely on the marine ecosystem for sustenance. Moreover, marine plastic litter poses aesthetic and cultural threats and causes socio-economic losses in marine-based sectors, affecting Arctic human life in various ways.

This chapter describes how marine plastic litter in the Arctic has evolved as an environmental problem. Our examination is inspired by the ontology of new materialism, posthumanism, transcorporeality, the notion of material-discursive relations and the entanglement of the material environment and human reasoning (e.g. Latour, 2004; Barad, 2007; Coole & Frost, 2010; Alaimo & Hekman, 2008). Thus, we approach environmental problems as co-constituted by the interaction between the "matter" – a change in the quality of the environment – and "meaning", i.e. human experience and interpretation of that change, leading to the definition of the problem constantly evolving (see Tynkkynen, 2008; Haila & Levins, 1992). We look at the evolution of the Arctic plastic problem by first describing the issue based on research literature. Second, we examine how newspapers have written about marine plastic in the Arctic since the problem was first publicly noted. For approximately the last decade, the plastic debris plaguing Earth's oceans has evoked a great deal of media attention and made news headlines around the world. Compared to many other environmental problems, plastic debris is easy to visualise to raise awareness. On top of the impressive images, such as those of a sea turtle with a plastic straw stuck in its nostril, or of the gyres of plastic debris, including the Great Pacific Garbage Patch known as the largest accumulation of plastic in the world, the media are increasingly reporting on the problem (see Alaimo, 2012, pp. 487–489). Yet, while collecting material about the Arctic plastic, we found surprisingly little written about the topic in the Arctic context.

The newspaper texts analysed in this chapter contain a selection of the pieces of news describing plastic pollution in the Arctic context.[1] We obtained these texts from the English ProQuest Central – a collection of databases providing access to thousands of journals, magazines, newspapers, dissertations and other international publications. We included only English-language newspapers in our study. The ProQuest Central database contains 2544 newspapers, 2276 in English. There is no geographical framing for these newspapers, but the database includes newspapers published around the world. This extensive and international content makes the database a fruitful and representative source collection, which is mainly why we used it as the primary source for our article.[2]

We examined the evolution of the plastic problem by first broadly searching the database, using search the terms "Arctic Ocean" with the words "plastic" and "trash/waste/litter". With this method, we discovered slightly over one hundred newspaper articles from the 1980s to the 2010s from which we separated 64 original articles describing the existence of plastic in the Arctic waters: 10 articles from the 1980s, six from the 1990s, 24 from the 2000s, and 24 from the 2010s. In some cases, especially after the 2000s, the same news (by a certain news agency) circulated in several newspapers, so we included that news only once. Had duplications been included, the number of hits would have been much higher. As noted, we were surprised by the scant number of hits and articles describing Arctic marine plastic. While we are aware of the limitations of our material in excluding non-English language newspapers, such indicates that the problem is newly emerging in the news. Moreover, the US and Canada are major Arctic countries, and their media are included in this search.[3]

Unfortunately, the ProQuest did not include images for most of the newspapers or articles that were among the hits. There were some images included, mostly depicting the scientists that were cited in the article. This means that our analysis

[1] As the media material was retrieved from ProQuest database, which does not contain full bibliographic information about all the hits, we have decided to refer to the media material with the names of the journals and magazines and dates of publication in the text, and leave this information out of the reference list. Copies of the articles are available upon request.

[2] Our initial intention was to limit the study to the newspapers of the Arctic countries. However, the language barriers and the Nordic source material being mostly copyrighted, making it digitally inaccessible, was challenging.

[3] To further justify the choice and outline of our material, we checked the Finnish national daily paper *Helsingin Sanomat*. The plastic problem in the Arctic context was mentioned only three times between 1990 and 2021, demonstrating that the issue is not yet better covered in regional (Arctic) media.

focused mainly on the text. As the data contains only 64 articles from different newspapers, there is little point in trying to track e.g., the political emphasis behind the newspaper in question and its impact on the problem definition. Moreover, as turned out, most of the articles had a very neutral and scientific tone, which further lessened the need to pay attention to the publication venue and its ideological stance.

The chapter begins with a discussion about the problem of Arctic marine plastic based on the scientific literature. After this, the results of the media analysis are presented. In the analysis, we first created a chronology of the news, paying particular attention to the transitions in the way in which the problem was defined to make sense of the evolution of the problem. Further, we scrutinised the role of (cold) water in the differing interpretations of this problem and in the transitions of its evolution to frame the topic in terms of the current volume. The chapter contributes to the current volume by telling "a plastic story" that links with many binaries and concepts discussed in the volume: matter and meaning, fluid and solid, scarcity and abundance, the liquid and the solid, as well as materiality and transformations by water.

1.1 The Peculiar Arctic Plastic

The bodies of almost all marine species, ranging in size from plankton to marine mammals, contain plastic. Plastics turn up in bird nests, are worn by hermit crabs instead of shells, and are present in sea turtle, whale and albatross stomachs. Ingestion and entanglement result in impaired movement and feeding, reduced reproductive output, lacerations, ulcers, and death (e.g. Derraik, 2002). Microplastics – micro- and nano-sized plastic particles derived from small particles developed for specific applications or produced through the breakdown of larger items – are increasingly consumed by marine life that confuse them with food sources. Plastic debris is persistent: Plastic swallowed by an albatross had originated from a plane shot down 60 years earlier, almost 10,000 km away (Wabnitz & Nichols, 2010).

Marine plastic doesn't just harm the marine environment but human wellbeing and socioeconomic sectors such as tourism, aquaculture and navigation (e.g. Werner et al., 2016). Plastic is so widespread throughout the environment that it can be considered a geological marker of the Anthropocene Epoch – the most recent period in Earth's history when human activity has significantly impacted the planet's climate and ecosystems (Zalasiewicz et al., 2016).

The concentration of plastic within the Arctic is at the same or even higher level than in urban areas, despite the Arctic being far away from industrialised and highly populated areas (Hallanger & Gabrielsen, 2018). Sea-based sources, including commercial fishing, aquaculture, shipping, and the oil and gas sector, are responsible for most of the plastic pollution in the Arctic Ocean and along the coast (Robbins et al., 2020). Also, human settlements within the Arctic contribute to plastic pollution (Hallanger & Gabrielsen, 2018). The greatest and deadliest form of plastic debris in the Arctic is formed by lost, abandoned or derelict fishing gear. The problem is only expected to worsen as fishing moves north due to climate change (Robbins et al., 2020). Moreover, microplastics contaminate the entire marine food chain, including fish species such as Arctic cod, which live exclusively in far northern waters (ibid).

Robbins et al. (2020) give an example of the magnitude of the Arctic plastic problem by presenting the case from the Pribilof Islands in the southeastern Bering Sea off the coast of mainland Alaska. According to them, the islands are among the world's most productive ecosystems regarding fisheries for walleye pollock, king crab, salmon, Pacific cod, and other species responsible for about half the entire fish supply of the United States. The cultural and economic wellbeing of indigenous communities on the islands heavily depends on a healthy marine ecosystem. Moreover, these islands provide vital breeding and feeding areas for more than half the world's population of northern fur seals, endangered Steller sea lions, harbor seals and millions of seabirds. Yet, the same ocean currents bringing fish and other animals to the islands carry massive amounts of plastic to the shores of the Pribilof Islands that stand at the heart of the currents. Hundreds of thousands of kilograms of plastic debris, consisting mostly of fishing gear from faraway places, have been removed from the beaches. One single clean-up event on the beaches in May 2019 removed almost 9100 kg of plastic litter from a little over eight kilometres of shoreline.

How did marine plastic become such a problem? Mass production of plastic took off rapidly in the 1950s, shaping modern society's development (e.g. Thompson et al., 2009). Global production of plastic resin increased from around 1.5 million tons in 1950 (Plastics Europe, 2008, p. 24) to 322 million tons in 2015 (Plastics Europe, 2016). Innovation in the plastic industry led to low-cost, versatile, durable, corrosion-resistant and lightweight plastics. Major end-applications for plastics include packaging; building and construction materials; and electrical, electronic and medical equipment. Almost all aspects of daily life around the world involve plastic. Plastic significantly contributes to our health, safety and sustainability, including clean, safe and equal distribution of water and food, as well as substantial energy savings in transportation (Tynkkynen et al., n.d.). However, plastic production consumes natural resources (oil). Moreover, most plastic is not biodegradable as it does not rot like paper but stays in the environment for centuries. Almost half the plastic produced in the world is single-use (Xanthos & Walker, 2017).

A vast majority of marine plastic derives from land-based sources, and the sources of plastics are diverse. Marine plastics differ from those onshore; most marine plastics cannot be seen because they are tiny and dispersed (Emmelhiez, 2015). Ironically, the same characteristics making plastic so usable lead to problems regarding its occurrence in the marine environment. Also, a major portion of plastic produced is used to make disposable packaging items or other short-lived products that are discarded within a year of manufacture (Hopewell et al., 2009). The rising standard of living has allowed people living in remote communities to buy more single-use plastic items, while waste management is inadequate, and communities are left to shoulder the impacts of the ocean plastic crisis (Phelan et al., 2020). Thus, the problem of marine plastic litter is rooted in unsustainable production and consumption patterns, poor solid waste management, lack of infrastructure and financial resources, lack of adequate legal and policy frameworks and poor enforcement, including international trade of waste.

Ocean plastic pollution has received growing international attention in recent years. Major advances in global initiatives have addressed marine plastic pollution, including high-level statements and programmes like the United Nations Environment Programme (UNEP) on marine litter. International initiatives addressing marine plastic pollution need the support of scientific research, business and community organisations, and engaged and committed government action on different scales addressing the use and disposal of plastics (Haward, 2018). Of course, the best way to address the problem of plastics in the ocean and the Arctic would be to stop them from getting there in the first place. Such would require fundamental changes in consumer behaviour, corporate practices and public policies around the world. The most direct way to fight ocean plastic pollution would be to organise clean-ups for local beaches and waterways – done in many places around the world, including the Arctic region. However, clean-ups do not alleviate the microplastics problem and are mostly an onshore solution without addressing plastic in the water.

1.2 From Funny Plastic Toys on an Arctic Expedition…

Plastic and the Arctic Ocean's shared past occurs over a short period in recent history. According to our data, the words "plastic" and "Arctic Ocean" were not mentioned in connection with each other in newspaper articles until the 1980s. However, plastic was not regarded as harmful waste as, for instance, oil and other industrial pollutants, which in the public discussion appeared to be the main threat to the Arctic environment (e.g. *Toronto Star*, 1988). For example, an article in *Edmonton Journal* (1989) stated that "The Arctic is a sink for industrial pollutants carried north from Europe, the Soviet Union and North America by ocean currents and the northerly winds." In other words, the perception existed in the 1980s that Arctic waters carried pollution, but the contamination was mainly defined regarding toxic substances. This focus resonated with the public's concern regarding the pollution of the oceans at large in the 1970s and 1980s; for instance, Rachel Carson had warned about toxic substances, including radioactive waste, in her famous book *The Sea Around Us*, first published in 1950 (Carson, 1950). At that time, little attention was given to the flows of visible material substances in the oceans.

Although plastic waste was not considered a serious problem in the seas in the 1980s, plastic's presence in Arctic waters emerged in the existing public discussion. *Globe and Mail*, for instance, reported in 1982 how workers of a High Arctic Oil Rig had drilled a hole in the Arctic ice and dumped their wastes through it into the sea. Among the empty oil drums, ash cans and a battered pick-up truck, sheets of plastic were discarded into the sea. Some newspaper articles mentioned that, for example, old plastic containers were found on the coast of the Arctic Ocean in North Canada (*Wall Street Journal*, 1985). This reporting indicates the conception of the oceans as capable of diluting and dispersing harmful items and pollutants (see Laakkonen & Laurila, 2001; Patton, 2006).

Altogether, plastics were rarely mentioned apart from other waste drifting in the Arctic Ocean in the 1980s. In 1988, for example, an article referenced "highways of floating garbage" that "choke an estimated 100,000 whales and a million seabirds every year" (*Toronto Star*, 1988). The garbage mentioned surely consisted mainly of plastic litter. Nevertheless, this was not highlighted in the text. The article was exceptional, as the impact of plastic debris on marine animals was rarely discussed in the 1980s. Overall, plastic waste was dumped into the sea and washed ashore in the 1980s but was not perceived as a major concern. Thus, one can assume that people were not paying attention to marine plastic or even knew how plastic litter impacted marine life. Primarily, the plastic litter in the seas, particularly on its shores, was perceived as a cosmetic problem in the 1980s (see e.g. *Globe and Mail*, 1989).

The understanding that there is a great deal of plastic floating in all the world's major oceans, even in the high Arctic, became more visible in the public discourse during the 1990s. Yet, the danger of plastic litter in the Arctic was still not widely understood. For instance, when *The Spectator* (1998) published one of the first articles of our dataset that was very concerned about how the North Pole was becoming increasingly more polluted, chemicals were considered the leading cause for the contamination. Plastic was mentioned only once, even though it had been found in Arctic waters a few decades earlier.

The relationship between plastic waste and the Arctic Ocean in the 1990s was well-crystallised by newspaper articles describing the movements of plastic tub toys that had spilt into the Pacific Ocean from a cargo ship in 1992. Many of these toys drifted into the Arctic Ocean and were predicted to drift through Arctic waters with the northern ocean currents into the North Atlantic until around the year 2000 (*Wall Street Journal*, 1994). Numerous newspapers wrote about these 28,880 tub toys. Instead of explaining the accident as an environmental disaster, it was enthusiastically welcomed in the public discussion. *Wall Street Journal* (1994) highlighted the accident as a "golden opportunity" to learn about the ocean and its currents and "as cheap data as you can get." *The Times* (1994) even compared the plastic toys crossing the Arctic Ocean as "following in the wakes of such legendary explorers as Fridtjof Nansen", who made an expedition to the Arctic Ocean between 1893 and 1896. In other words, these plastic toys were perceived almost entirely as hilarious tools for ocean science, not as a threat for aquatic life – indicative, eventually, of the practice of scientific stabilizing, distancing and detachment (see Pickering, 2015). Seen from a posthuman viewpoint that encourages us to look beyond the nature/culture divide, this perception highlights the (falsified) detachment of sea life from the toys and nature and culture from each other – as if they existed in two separate realms.

Ultimately, only a few newspaper articles in the 1990s highlighted the dangers of marine plastic litter, with only a small proportion paying attention to the plastic garbage in the Arctic Ocean. One such article was "Eternal voyage across a sea of plastic," published in *The Guardian* (1990), which effectively described the dangers of plastic in the sea by stating that "plastic chokes dolphins, kills turtles and poisons coral, but it will never die". Moreover, the author stated he had witnessed "a parade of plastic waste" in the high Arctic, and along the shores of Novaya Zemlya, he had seen the "bright and brazen colours of plastic". Contrary to the news admiring the thousands of plastic toys drifting in the Arctic Ocean, this piece did not go viral in the press.

1.3 … to Worried Oceanographers

The dangers of plastic garbage in the Arctic Ocean and all oceans slowly became a focus in the 2000s. The journey of the tub toys across the Arctic was mentioned in newspapers almost yearly during this decade. The toys were still mainly described in a funny context. *Daily Mail* (2003), for instance, wrote that "they've braved tempests and icebergs". *Sunday Times* (2003) explained that "the toys are thought to have spent about 5 years encrusted with snow and ice as they made their way past polar bears and seals and eventually round the tip of Greenland". Nevertheless, in some of the articles describing these toys' adventures, a faint message began to appear: that these tub toys ending up in the ocean was not a good thing. For instance, *The Times* (2005) mentioned that these toys were "merely the tip of the mountains of rubbish drifting through the world's seas". During the 2000s, the fact that marine plastic litter in the Arctic was indestructible became a more significant part of the public discussion (see e.g. *Telegram and Gazette*, 2007), reflecting the growing global concern over plastic debris at large.

Remarkable about the debate on plastic and the Arctic Ocean in the 2000s was the understanding that the rubbish drifting in the ocean was above all plastic and not just some vague "garbage". Newspapers reported of garbage on the shores of the Arctic that would neither decompose nor be transported away (see e.g. *Toronto Star*, 2005). At the end of the 2000s, a more detailed debate emerged on the significant amount of plastic garbage in the uninhabited areas of the Arctic Ocean (see e.g. *New York Times*, 2008) and that plastic litter was discovered 6770 meters deep in the Arctic Ocean's Frahm Strait (*Dominion Post*, 2009). By the 2010s, there was an awareness that plastic garbage was present in all the world's seas, even in the remote Arctic Ocean – not only on its shores or the water's surface but its depths. In 2010, the conservationist Phillippe Costeau highlighted the beach sandals washed up in the Norwegian Arctic, symbolising the global nature of marine litter (*The Providence*, 2010).

In the 2010s, significantly more press articles on the Arctic Ocean's plastic problem appeared than in previous decades. The articles no longer only stated there was plastic in the Arctic but paid growing attention to the problem's seriousness and causes. Scientists specialising in studying the Arctic environment took central stage in the news. For example, an interview with biologist Jennifer Provencher in 2011 concerning the shocking amounts of plastic rubbish discovered in the intestines of fulmars gained a great deal of media attention. In her interview, Provencher reported that "84 per cent of the [fulmars] the researchers examined from two Arctic colonies had plastics in their guts". However, a study of 181 fulmars in the 1970s did not find any plastic from these seabirds (see e.g. *Calgary Herald*, 2011). Notably, these new observations of embodied plastic spoke of "animal bodies invaded by human consumerism" (cf. Alaimo, 2012, p. 488), even if the connection to and danger that plastic contamination caused human bodies was still largely unnoticed.

Interviews with marine biologist Melanie Bergmann also received widespread attention in the international press during the 2010s. Her studies revealed that plastic trash litters the seabed of the Arctic Ocean. Moreover, deep-sea observations on the northeastern coast of Greenland showed the amount of plastic garbage on the floor of the Arctic had doubled since 2002. Bergmann also pointed out that, in addition to birds, plastic garbage had been discovered in the stomachs of Greenland sharks, indicating the existence of plastic in the depths of the Arctic (see e.g. *Toronto Star*, 2012; *The Nation*, 2015). Bergmann speculated that the large quantities of plastic in the Arctic might be "a trash vortex forming in the Barents Sea north of Norway and Russia, which is thought to be fed by the densely populated coastal regions of Northern Europe." She also mentioned that "another possible cause of the Arctic trash is receding Arctic Sea ice due to global warming, allowing fishing trawlers and cruise liners to operate – and leave litter – farther north" (*The Nation*, 2015). Furthermore, there were newspaper articles stating that "a large fraction of the plastic waste from the United Kingdom gets into the ocean drifts northward and ends up in the Arctic Ocean" (*Hindu*, 2016).

In 2017, a research team led by Dr. Andres Cozar determined that the Arctic Ocean contained 300 billion pieces of plastic. This study, which received widespread attention in the public discussion, discovered "two major 'dead ends' – in the Barents and Greenland seas – where plastic transported over thousands of miles by ocean currents has been gathering" (see e.g. *The Independent*, 2017; *Los Angeles Times* [online], 2017; *Citizens' Voice*, 2017; *Washington Post*, 2017).

That same year, the news reported on a research team that found large pieces of "polystyrene lying on remote ice floes only 1000 miles from the north pole, which have been previously inaccessible to scientists because of sea ice" (*The Observer*, 2017; *Telegraph* [online], 2017; *The London Free Press*, 2017; *The New Zealand Herald*, 2017; *The Western Morning News*, 2017). Accordingly, the link between the plastic problem and climate change started gaining visibility in the media. For example, *The Times* reported in April 2018 that Arctic ice contains many various plastics, noting that part of the trash was "probably the result of increased industrial activity in the Arctic enabled by shrinking sea ice".

An important phase in the discussion of plastic garbage in the Arctic Ocean was the emergence of the microplastics concept in the mid-2010s. Studies showed that "polar sea ice is becoming a major sink for microplastic contamination". At the time, however, scientists could not determine how much microplastics were in the Arctic Ocean and where they came from (*The Vancouver Sun*, 2016). In 2018, a news report of a study suggested that microplastics were frozen in the Arctic in much larger quantities than previously thought. According to this study, 1 litre of sea ice contained about 12,000 particles of plastic (*Irish Independent*, 2018; *The Times*, 2018).

All these comments by researchers in newspapers in the 2010s added depth to the debate on the Arctic Ocean's plastic problem. These scholars sought answers to this problem, explained the scale of the problem, and disseminated knowledge on plastic in the Arctic to a larger audience. Notably, there was a clear reaction to what researchers had said. For example, newspapers reported how Starbucks and many other companies stopped using plastic straws in the late 2010s (*The Daily Telegraph*, 2018). Another example of the change in perceptions was the article concerning the *Blue Planet II* documentary, which appeared in many newspapers and mentioned the plastic bath toys that drifted into the Arctic Ocean in the 1990s. However, their drifting into the sea was no longer referenced as a fun occasion. Instead, the context in which the toys were mentioned dealt with how plastic destroys marine life (*Daily Star*, 2017; *Daily Mail*, 2017), as it becomes entangled with animal bodies.

Interestingly, the news on Arctic plastic still mainly emphasised the harm plastic debris causes "wildlife" rather than humans (e.g. *Irish Independent*, 2018; *The Western Morning News*, 2017; *The London Free Press*, 2017). In other words, the news often frames the Arctic plastic problem as impacting nature and animals rather than the people living there. Other concerns, such as plastic ending up in human bodies due to diets heavily reliant on seafood, have hardly been discussed in the news.

1.4 The Plastic Debris and the Cold Water

If anything, plastic symbolises progress and modernity, promising sealed, perfected, clean and smooth abundance. As Westermann (2013, p. 69) argues, "vinyl's plasticity and its chemical creation captured what high modernity expected from technology at large: a world freed from the material restrictions that nature traditionally imposed on humanity. By implication, we would also have a world freed of scarcity, a world of plenty". As our chronology in the above section illustrates, the early ways the media interpreted the plastic in the oceans, particularly in the Arctic, reflect this promise. Marine plastic was seen as a sign of modernity and an opportunity for science to follow and study the ocean currents and marine life "as cheaply as it gets" (*Wall Street Journal*, 1994), with the help of tub toys symbolising small "Fridtjof Nansens" in their joyful adventures. The relationship between the plastic material and water (in all forms) was presented as straightforward and unproblematic, even positive. Water was interpreted as a carrier of plastic to new places yet to be discovered. The plastic in the water was seen as a medium helping scientists study the oceans or the Arctic, giving joy to the spectators following the journey of funny plastic ducks. This perception greatly underlines the distinction between matter and meaning, typical for the ontological nature-culture dualism, yet gives little to understanding the problem in its essence, which lies in the entanglement of human life and lifestyles, including consumption patterns and their consequences for nature.

In later years' interpretations, the flow of water and the plastic drifting with it turned more problematic, reacting with destructive aspects related to the rise of technological civilisation and the demolition of nature (cf. Böhme, 1988). As Alaimo writes, "everyday, ostensibly benign, human stuff becomes nightmarish as it floats forever in the sea" (2012, p. 487). The same ocean currents bringing fish and other seafood from the southern waters to the north carried plastic to the Arctic, causing pollution far away from where the plastic was produced and used. "The trash vortex", as well as the flip-flops (beach sandals) brought by the sea currents from warm places to the cold Arctic, indicate the transboundary and global nature of the problem and environmental problems at large. In this way, these items tie the Northern waters as part of modern history in an era of lavish consumerism. However, the Arctic also represents a dead end: Global currents like the Gulf Stream carry the plastics all the way from the Americas and Europe to the Arctic, where they stop and stay out of sight – outside the realm of humans.

The relationship between polar sea ice and plastics also evolves in media's interpretations. Initial reports presented sea ice as a major sink of waste, for example, workers of a High Arctic Oil Rig who drilled a hole in the ice and dumped their plastics there. The media labelled the hole as a storage space or even as a museum of tub toys. Again, the dualism of modernisation prevails positive technological development on land (the invention of plastic as a saviour of humankind), bringing negative occurrences to the sea, as the unwanted side effects of material prosperity on land were pushed from sight (see Tynkkynen, 2017). Later, the plastic in the ice starts being seen as a time bomb that may be fired by global warming and the receding Arctic Sea ice, especially when the attention turns to microplastics from the late 2000s on. This symbolism has a double effect, linking two major environmental problems of the Arctic: the melting permafrost and sea ice as an effect of climate change and the release of plastic pollution as an implication of the melt.

Similarly, climate change and its complex relationship with ice and water is a dilemma for Arctic communities. Palosaari (2012) discusses the 'Arctic Paradox' in the context of climate change, permafrost, Arctic oil and gas resources: Hydrocarbon use contributes to climate warming, making the Arctic sea ice melt, so new oil and gas resources become available. Using those resources further accelerates climate change while potentially increasing local people's social and economic wellbeing. Many international actors question the sustainability and ethics of Arctic oil and gas exploration. Some Arctic indigenous communities have perceived these concerns as neocolonialism: Oil is found on their territories, but they are prohibited from benefiting from it (Palosaari & Tynkkynen, 2015). Plastic usage can raise similar reactions because of its symbolism and preservative functions, which can help remote Arctic communities. Why should using plastic be avoided now that the rising standard of living finally allows people in remote communities to buy more single-use plastic items, as others have benefitted from such usage for long?

1.5 Conclusion

Global discourses on the Arctic have, in recent years, highlighted melting glaciers, rising sea levels, environmental problems and visions of conflicts between states on resources. In contrast, positive visions of the Arctic's economic prospects, based on using natural resources and opening sea routes, have been introduced and linked to the increased wellbeing of local populations and indigenous peoples (Palosaari & Tynkkynen, 2015). The plastic pollution in the Arctic, related to globalisation's negative and positive impacts, has not yet received major attention as the global plastic concern typically focuses on the more southern seas. Yet, the "Arctic turn" may bring soon more attention to this problem. Small signs of this attention already exist, as scientific reporting regarding the Arctic marine plastic has proliferated over the last decade.

In this chapter, we demonstrated how the mass production of plastic, which started in the 1950s, became a growing problem in the oceans and Arctic waters. We examined how the public discussion in the international press defines the Arctic Ocean's plastic problem and how the perceptions of these plastics whirling in the cold Arctic waters and their effects have evolved since the first newspaper articles published in the 1980s. Examining the four decades of public discussion concerning the plastic problem in the Arctic Sea reveals several details about the human relationship with the environment, the sea and consumption. This scrutiny also highlights the need for a long-term understanding of the Arctic environment and the linkages among various problems, such as plastic debris and climate change in the Arctic.

As our analysis indicates, the Arctic marine plastic problem has evolved from the first observations where plastic was seen as an interesting object used for studying ocean currents to a more multifaceted problem regarding the marine life that consumes plastic. To a certain extent, the current definition of the problem focuses on "the unsettling and unintended consequences of substances and things" (Alaimo, 2012, p. 487). However, this does not (yet) seem to extend to the human realm as much as marine wildlife. Of course, it is important that plastic's impact on marine animals in the Arctic hemisphere has been noticed and emphasised. However, humans' experience of the problem, or its cultural impacts, are not discussed.

The global nature of the problem and its linkages to other major global concerns, such as climate change, is well-covered in the media. In all its changing forms and movements, water plays a central role in this interpretation. At large, the Arctic has long been "global" in the sense that Northern fishing grounds, whaling, the fur trade, and mining have connected the Arctic to markets around the world (Heininen & Southcott, 2010, p. 1). Today, the forces of globalisation are boosted by climate change, and the Arctic is becoming closer to being integrated into the global economy. The melting sea ice opens new Arctic sea routes, which may increase local plastic pollution. Though the problem is global, news articles often frame the Arctic as a remote place – a realm of nature rather than humans. Such reflects the Western habit of seeing places dualistically – as civilised or uncivilised.

Throughout the years, the news has taken an utmost scientific stance of detachment and distancing, underlining the divide between nature and culture as well as the matter and meaning (cf. Pickering, 2015) – distinctive for discourses of modernity and unsurprising as such. In the Arctic context, the scientific way of reporting on the problem prevails and has a very "hard science" tone. As a realm of nature, the Arctic is an object rather than a place where nature and humans meet. Further, the Arctic region's geopolitical sensitivity certainly plays a prominent role: Scientific distancing keeps environmental issues non-politicised – a familiar strategy from the Soviet era (see Darst, 2001). The role of science and the scientific exploration of the Arctic has been emphasised since the 'Murmansk Initiative' by Soviet state secretary Mikhail Gorbachev in 1989. This initiative aimed to transform the Arctic from a sensitive military theatre to an international zone of peace (see Åtland, 2008, pp. 289–290), which may impact the tone regarding how Arctic plastic is reported.

To conclude, marine plastic is a major human problem with its roots in our daily consumption patterns, which can have severe consequences for all life on Earth, including human life and culture. As Alaimo (2012, p. 476) describes with the concept of transcorporeality, flows of substances, such as plastic, are substantially and perpetually interconnected with humans. The Arctic marine plastic is no exception. Consequently, the transcorporeality should, to a growing extent, be addressed in public discussion, research and practice concerning marine plastic debris. The question then is this: What kind of knowing, being and acting would be needed to bring this issue to the fore? Globally, climate change has already made its way from the scientific agenda to the political one. Climate change's impacts are now more visible and better documented. With rising public awareness, issues on the scientific agenda have become recognised by policymakers. The same is happening concerning marine plastics, even if not yet in the Arctic context.

Undoubtedly, the scientific distancing of environmental issues must be supplemented by more complicated models of entanglement and emergence. Here, "blue humanities" plays a role. Scientific captures are inevitable to create a complete understanding of the marine plastic problem or any transcorporeal problem (cf. Latour, 2010). The co-composing of problem definitions where matter and meaning link, highlighting the need for interdisciplinary knowledge. Such is the only way to genuinely understand and solve the wicked, global problems of our time – on seas and on land.

References

Alaimo, S. (2012). States of suspension: Trans-corporeality at sea. *ISLE (Interdisciplinary Studies in Literature and Environment), 19*(3), 476–493. https://doi.org/10.1093/isle/iss068

Alaimo, S., & Hekman, S. (Eds.). (2008). *Material feminisms*. Indiana University Press.

Åtland, K. (2008). Mikhail Gorbachev, the Murmansk initiative, and the desecuritization of interstate relations in the Arctic. *Cooperation and Conflict, 43*(3), 289–311. https://doi.org/10.1177/0010836708092838

Barad, K. (2007). *Meeting the universe halfway: Quantum physics and the entanglement of matter and meaning*. Duke University Press.
Barnes, D. K. A., Galgani, F., Thompson, R. C., & Barlaz, M. (2009). Accumulation and fragmentation of plastic debris in global environments. *Philosophical Transactions of the Royal Society B: Biological Sciences, 364*, 1985–1998. https://doi.org/10.1098/rstb.2008.0205
Barrows, A. P. W., Cathey, S. E., & Petersen, C. W. (2018). Marine environment microfiber contamination: Global patterns and the diversity of microparticle origins. *Environmental Pollution, 237*, 275–284. https://doi.org/10.1016/j.envpol.2018.02.062
Böhme, H. (1988). Umriss einer Kulturgeschichte des Wassers: Eine Einleitung. In H. Böhme (Ed.), *Kulturgeschichte des Wassers* (pp. 7–44) Suhrkamp Verlag.
Carson, R. (1950). *The sea around us*. Oxford University Press.
Coleman, F., & Wehle, D. H. S. (1984). Plastic pollution: A worldwide oceanic problem. *Parks, 9*(1) https://marinelab.fsu.edu/media/3069/1984_coleman_wehle_plasticpollution-parks.pdf
Coole, D., & Frost, S. (Eds.). (2010). *New materialisms: Ontology, agency, and politics*. Duke University Press.
Darst, R. (2001). *Smokestack diplomacy: Cooperation and conflict in East-West environmental politics*. MIT Press.
Derraik, J. G. B. (2002). The pollution of the marine environment by plastic debris: A review. *Marine Pollution Bulletin, 44*, 842–852. https://doi.org/10.1016/S0025-326X(02)00220-5
Emmelhiez, I. (2015). Images do not show: The desire to see in the Anthropocene. In H. Davis & E. Turpin (Eds.), *Art in the Anthropocene* (pp. 131–154). Open Humanities Press.
Feder, H. M., Jewett, S. C., & Hilsinger, J. R. (1978). Man-made debris on the Bering Sea floor. *Marine Pollution Bulletin, 9*, 52–53. https://doi.org/10.1016/0025-326X(78)90534-9
Gregory, M. R. (1978). Accumulation and distribution of virgin plastic granules on New Zealand beaches. *New Zealand Journal of Marine and Freshwater Research, 12*(4), 399–414. https://doi.org/10.1080/00288330.1978.9515768
Haila, Y., & Levins, R. (1992). *Humanity and nature: Ecology, science and society*. Pluto Press.
Hallanger, I., & Gabrielsen, G. (2018). *Plastic in the European Arctic*. Brief Report No. 045. Norwegian Polar Institute. https://brage.npolar.no/npolar-xmlui/bitstream/handle/11250/2478285/Kortrapport45.pdf
Haward, M. (2018). Plastic pollution of the world's seas and oceans as a contemporary challenge in ocean governance. *Nature Communications, 9*, 667. https://doi.org/10.1038/s41467-018-03104-3
Heap, B. (2009). Preface. *Philosophical Transactions of the Royal Society B: Biological Sciences, 364*, 1971. https://doi.org/10.1098/rstb.2009.0030
Heininen, L., & Southcott, C. (2010). Globalization and the circumpolar North: An introduction. In L. Heininen & C. Southcott (Eds.), *Globalization and the circumpolar North* (pp. 1–21). University of Alaska Press.
Hopewell, J., Dvorak, R., & Kosior, E. (2009). Plastics recycling: Challenges and opportunities. *Philosophical Transactions of the Royal Society B: Biological Sciences, 364*, 2115–2126. https://doi.org/10.1098/rstb.2008.0311
Jambeck, J. R., Geyer, R., Wilcox, C., Siegler, T. R., Perryman, M., Andrady, A., Narrayan, R., Law, L. K., Narayan, R., & Law, K. L. (2015). Plastic waste inputs from land into the ocean. *Science, 347*, 768–771. https://doi.org/10.1126/science.1260352
Laakkonen, S., & Laurila, S. (2001). The sea and the cities. *AMBIO: A Journal of the Human Environment, 30*(4), 263. https://doi.org/10.1579/0044-7447-30.4.263
Latour, B. (2004). *Politics of nature: How to bring the sciences into democracy* (trans: C. Porter). Harvard University Press.
Latour, B. (2010). An attempt at a compositionist manifesto. *New Literary History, 41*(3), 371–490. https://doi.org/10.1353/nlh.2010.0022
Palosaari, T. (2012). The amazing race: On resources, cooperation and conflict in the Arctic. *Nordia Geographical Publications, 40*(4), 13–30. https://nordia.journal.fi/article/view/75946

Palosaari, T., & Tynkkynen, N. (2015). Arctic securitization and climate change. In G. Hønneland & L. C. Jensen (Eds.), *Handbook of the politics of the Arctic* (pp. 87–104). Edward Elgar.
Patton, K. C. (2006). *The sea can wash away all evils: Modern marine pollution and the ancient cathartic ocean*. Columbia University Press.
Phelan, A., Ross, H., Setianto, N. A., Fielding, K., & Pradipta, L. (2020). Ocean plastic crisis – Mental models of plastic pollution from remote Indonesian coastal communities. *PLoS One, 15*(7), e0236149. https://doi.org/10.1371/journal.pone.0236149
Pickering, A. (2015). Science, contingency, and ontology. In L. Soler, E. Trizio, & A. Pickering (Eds.), *Science as it could have been: Discussing the contingency/inevitability problem* (pp. 117–128). University of Pittsburgh Press.
Plastics Europe. (2008). *The compelling facts about plastics: An analysis of plastics productions, demand and recovery for 2006 in Europe*. https://plasticseurope.org/wp-content/uploads/2021/10/2006-Compelling-facts.pdf
Plastics Europe. (2016). *Plastics – the facts 2016*. https://plasticseurope.org/knowledge-hub/plastics-the-facts-2016/.
Robbins, G., Mallos, B., LeVine, M., & Huntington, H. P. (2020). Reducing plastics and other waste in the Arctic Ocean. *Environment: Science and Policy for Sustainable Development, 62*(5), 27–30. https://doi.org/10.1080/00139157.2020.1791638
Ryan, P. G. (2015). A brief history of marine litter research. In M. Bergmann, L. Gutow, & M. Klages (Eds.), *Marine anthropogenic litter*. Springer. https://doi.org/10.1007/978-3-319-16510-3_1
Thompson, R. C., Swan, S. H., Moore, C. J., & vom Saal, F. S. (2009). Our plastic age. *Philosophical Transactions of the Royal Society B, 364*, 1973–1976. https://doi.org/10.1098/rstb.2009.0054
Tynkkynen, N. (2008). *Constructing the environmental regime between Russia and Europe: Conditions for social learning*. Acta Universitatis Tamperensis 1301. Tampere University Press.
Tynkkynen, N. (2017). The pollution of the Baltic Sea: A mirror image of modernization. In J. Costlow, Y. Haila, & A. Rosenholm (Eds.), *Water in social imagination: From technological optimism to contemporary environmentalism* (pp. 167–192) Brill/Rodopi.
Tynkkynen, N., Hakala, H., Soini K., Bor, S., Katajajuuri, J. M., Sonck-Rautio, K., & Sundqvist Andberg, H. (n.d.) *Sustainability trade-offs in food packaging: Introducing the functional sustainability perspective*. Unpublished manuscript.
Villarrubia-Gómez, P., Cornell, S. E., & Fabres, J. (2018). Marine plastic pollution as a planetary boundary threat – The drifting piece in the sustainability puzzle. *Marine Policy, 96*, 213–220. https://doi.org/10.1016/j.marpol.2017.11.035
Wabnitz, C., & Nichols, W. J. (2010). Editorial. Plastic pollution: An ocean emergency. *Marine Turtle Newsletter, 129*, 1–4.
Werner, S., Budziak, A., Van Franeker, J. A., Galgani, F., Hanke, G., Maes, T., Matiddi, M., Nilsson, P., Oosterbaan, L., Priestland, E., Thompson, R., Veiga Joana, M., & Vlachogianni, T. (2016). Harm caused by marine litter. *European Commission*. https://doi.org/10.2788/690366
Westermann, A. (2013). The material politics of vinyl: How the state, industry and citizens created and transformed West Germany's consumer democracy. In J. Gabrys, G. Hawkins, & M. Michael (Eds.), *Accumulation: The material politics of plastic* (pp. 68–86). Routledge.
Xanthos, D., & Walker, T. R. (2017). International policies to reduce plastic marine pollution from single-use plastics (plastic bags and microbeads): A review. *Marine Pollution Bulletin, 118*(1–2), 17–26. https://doi.org/10.1016/j.marpolbul.2017.02.048
Zalasiewicz, J., Waters, C. N., Ivar do Sul, J. A., Corcoran, P. L., Barnosky, A. D., Cearreta, A., Edgeworth, M., Gałuszka, A., Jeandel, C., Leinfelder, R., McNeill, J. R., Steffen, W., Summerhayes, C., Wagreich, M., Williams, M., Wolfe, A. P., & Yonan, Y. (2016). The geological cycle of plastics and their use as a stratigraphic indicator of the Anthropocene. *Anthropocene, 13*, 4–17. https://doi.org/10.1016/j.ancene.2016.01.002

Chapter 2
Rivers through the Prism of Oil Spills: Native Voices from the Russian Arctic

Maria Sakaeva

Abstract Relying on the case of the Komi Republic – the Russian oil-producing region – I examine the perceptions and interpretations of the Pechora River by the local citizens following damages caused by oil spills. I use an anthropological approach to analyze fresh waters and rivers and their role in the socio-cultural world of the locals of the Komi Republic. I demonstrate that the connection between the lives of the rivers and the people living nearby are deeply connected with and threatened by spills and other petroleum-driven damages. This chapter is based on the suggestion that oil's long-term presence is the crucial factor, as it shapes the inhabitants' perceptions of the Pechora River. I analyze the materials published by the regional environmental NG *Save Pechora*, to examine what the river means to the inhabitants who live near it. These publications serve as the sources through which Pechora River's inhabitants construct, develop and frame the discourses about their river environment. The key finding of my research is that the residents of the Pechora River blame oil as a river killer. Continuous presence of oil-driven damages in the lives of Komi locals dramatically influences the perceptions of their river homeland, which is mainly associated with multi-faceted dearth.

Keywords Oil spills · Pechora River · Russian Arctic · Native people · Environmental NGO · Water anthropology

2.1 Introduction

During the last decade, as Wagner and Jacka (2018) noted, the relations of local inhabitants with fresh waters and rivers became more prominent in the anthropological research (p. 3). In particular, the aspect of petroleum-driven river devastation has been introduced into an investigation of broader environmental challenge

M. Sakaeva (✉)
Independent Scholar, Syktyvkar, Komi Republic, Russia
e-mail: eu.sakaeva@gmail.com

M. Lehtimäki et al. (eds.), *Cold Waters*, Springer Polar Sciences,
https://doi.org/10.1007/978-3-031-10149-6_2

provoked by oil spills (Gill et al., 2011; Smith, 1993). The existing literature addresses the risk of river degradation caused by natural resource extraction in some Western states (Sylves & Comfort, 2012; Wickman, 2013). Scholars mention that the worries of indigenous people regarding their rivers link to recent hydrocarbon reactivation, such as the development of old mining facilities in Norway (Reinert, 2016).

In underdeveloped countries like Russia, water devastation appears in another form: it often stems from oil extraction in the Arctic.[1] For decades, Russian policy and practice of oil production were characterized by highly irresponsible land use and ignoring of many environmental risks. As a result, by the end of Soviet rule, the Russian North had turned into an industrial desert (Roginko, 1992; Stammler, 2011). Since the Komi Republic contains the largest oil supply and the greatest volume of freshwaters within the European North, investigating the damage caused by oil extraction and its impact on the lives of the residents is highly relevant.

Through the prism of oil spills, I explore a series of perceptions from the Komi Republic's residents regarding the Pechora River. Environmentalist David Orr (2004) suggested exploring the globe through the dialogue between oil and water (p. 17). My contribution indicates that "conflict" rather than dialogue seems a more appropriate definition if we analyze water-oil relations from the perspective of the Komi natives who suffered from oil spills. Here is why: As Deutsche Welle (DW) reported, Russia tops the global list for oil spills. An oil spill happens somewhere in the Russian Arctic almost every half hour (Kondratenko, 2021). Since the early-2000s Greenpeace has documented a stable increase in the number of oil spills in the Komi Republic (Bachman, 2010). The most severe environmental devastation concerns the Pechora River because the main flow of spilt oil has been entering it. The Pechora River is the sixth-longest river in Europe and carries its waters to the Barents Sea and covers more than 2/3 of the Komi Republic.

One of my key findings is that Pechora River's residents blame oil as a river killer. Such necessitates saying a few words about what rivers mean to the local population. Rivers have occupied an important place in the historical and social lives of people in the Russian European North. In this swampy and densely forested part of Russia, permanent settlements were possible only near rivers, which networked different communities into the interdependent system (Shabaev & Istomin, 2017, p. 100). The Pechora River with its many tributaries constructs such a system with villages and rural communities inhabited by the Komi locals. Since the late sixteenth century, the Pechora River basin has been an important network of waterways supporting fishing and commerce (Wilson & Istomin, 2019, p. 1292). The Pechora River created the soul of the native inhabitants of the region and determined their lifestyle's core characteristics.

I rely on an anthropological approach when investigating water-oil relations, focusing on a set of existential, cultural, and social meanings regarding water among

---

[1] 70 percent of the Russian lands are in the Arctic. 70 percent of the oil and 90 per cent of the gas is extracted in the Arctic part of the country.

locals. I follow the finding of Stephanie LeMenager (2016): oil spill stories swarm with agents – human and nonhuman – involving human activists and nonhuman victims (p. 23). She argues that oil spills connect oil and water, as well as people and nature, closely and more deeply than any technological disaster.

I base my research on a bottom-up approach. In indigenous studies, this approach means taking the side of the natives – looking at their perceptions and understandings of wildlife (Hindman, 1998; Widener & Gunter, 2007). According to Wagner and Jacka's (2018) literature review on water anthropology, the researchers have been chiefly interested in the role of technologies and geopolitics rather than what rivers contribute to the socio-cultural lives of the natives (p. 5). Marama Muru-Lanning (2018) claims that visualizing a river through the natives' perspective organizes our research optic around the notion that rivers have social lives (p. 140).

Some scholars noted that the public and scientists had ignored the voices of the Russian Northerners regarding their environmental visions (Roginko, 1992; Slezkine, 1994; Stammler, 2011, 2020). Under this light, bottom-up approach helps me introduce the voices of the residents of the Pechora River basin into the debate about the oil-driven devastation of the Russian Arctic. I argue that their non-scientific, non-academic knowledge about oil's impact on river life is vital for discussion. Thus, my research illustrates an underdeveloped aspect of the global discourse on the Arctic environment, showing that the livelihood of Northern residents cannot be separated from river life and its sustainability.

Residents of the Pechora River area have mainly manifested their visions and demands through the non-governmental organization titled NGO *Save Pechora*, founded in the Komi Republic in 1989 amid the wave of environmental protests across the Soviet Union (Henry, 2010). Pechora River's inhabitants initiated this organization in response to the frequent, long-term and massive oil-driven devastation of their river homeland. *Save Pechora* defines its profile as socio-ecological because its activity includes protecting the people and the nature they inhabit. NGO positions itself as the organization speaking for the people and acting for the future of the Pechora River (Komitet spaseniia Pechory, 2013). *Save Pechora* unites activists and enthusiasts from several settlements of the Komi Republic, who experience significant suffering from oil damages. I define *Save Pechora's* website as the leading public platform as the main public platform where these issues are raised.

In the following section, I will describe my theoretical background and data collection. Then I will provide an overview of oil spills in the Russian Arctic and the Komi Republic in recent times. In the first empirical part – relying on the case of the Usinsk oil disaster – I explain how oil-driven devastation transforms the capacity of Pechora residents to speak about their river. The last section illustrates my main research finding: continuous presence of oil-driven damages in the lives of Pechora River locals dramatically influences their perception of their river homeland, which is mainly associated with multi-faceted dearth.

2.2 Introducing the Voices of Pechora Residents into the Oil-Water Debate: Theory and Method

Recently, there has been a rising wave of critical remarks: that academic literature on the oil industry's environmental impacts has ignored the voices of the local population, who experience major suffering from oil spills (LeMenager, 2016; Widener & Gunter, 2007). Indeed, scholars primarily discovered environmental state or corporate responses to spills (Gill et al., 2011; Smith, 1993; Sylves & Comfort, 2012). Scholars on indigenous people and oil pollution in the Russian Arctic have also been preoccupied with the historical analysis of state policy (Roginko, 1992) or corporate responsibility in oil provinces (Wilson & Istomin, 2019). LeMenager (2016) defines interdependency between water and oil as the structural foundation of the livelihood of local residents in oil-producing countries. She insisted that giving these residents a voice meant shifting from a large-scale perspective provided by state and business agents to a bottom-up vision. That approach requires paying attention to how the inhabitants construct, develop and frame the images of an oil-damaged water environment, which is embedded in and shaped by their everyday socio-cultural life (LeMenager, 2016, p. 23). I employ her anthropological mindset about the connection between water and the people in oil-producing areas. However, a significant difference exists between the accidental marine oil disaster in Santa Barbara, explored by LeMenager, and numerous spills in Komi. As I will show, Pechora River's residents and members of the NGO *Save Pechora* define the oil boom of the 1970s as the starting point of river devastation. My research is based on the suggestion that the long-term total presence of oil is the key factor shaping the perceptions of the Pechora River among its inhabitants.

Save Pechora proclaimed its primary purpose as follows: "Protecting nature as a traditional habitat for the local population – native residents of the Pechora River basin" (Komitet spaseniia Pechory, 2013). "Native residents of the Pechora River" are those living in the settlements near the Pechora River and its many tributaries. The question about native and non-native residents of the Komi Republic remains very debatable, as its population comprises over 100 nationalities and ethnic groups (Smetanin, 2004). Ethnic Komi make up only 24 percent of its population.[2] Identifying as indigenous residents of the republic, ethnic Komi do not form a homogeneous group. The Pechora River's inhabitants belong to the northern group of Komi, known as the Izhma Komi, Komi-Izhemtsy or Izvatas. The Izhma Komi population is about 40,000, and there has been much discussion regarding achieving a separate indigenous status, with little progress (Kim et al., 2015).

Save Pechora unites activists and enthusiasts from rural areas inhabited by Komi-Izhemtsy. Supporters and members of *Save Pechora* also live in the cities of Usinsk, Pechora, Ukhta and Sosnogorsk, which flourished during the oil boom of the 1970s.

[2] The Komi people, who number 300,000 overall, are not recognized by Russian Federation law as one of Russia's "small-numbered indigenous peoples of the North" (korennye malochislennye narody Severa).

Consequently, many people from surrounding rural districts migrated there. Many ethnic Komi in urban areas continue participating in civic activities of *Save Pechora*.

I analyzed a series of materials published in 2016–2020 on *Save Pechora*'s website: press releases, resolutions, open letters, reports, and news. I paid attention to the publications covering the oil industry and spill issues in detail. My data collection also includes several publications from *Ekologicheskii Vestnik* (*Ecological Herald*), a print newspaper published by *Save Pechora* since 1989 and uploaded to its website. The newspaper is sent to all the villages of the Pechora River basin. In the rural districts where the internet is expensive and slow, these publications are an essential source of information (Wilson & Istomin, 2019, p. 1295).

I base my tools of empirical data collection on the methodology suggested by Widener and Gunter (2007), who investigated the recovery of Alaska natives after the 1989 Exxon Valdez oil disaster in Prince William Sound. Widener and Gunter's understanding of local communities is very similar to LeMenager's (2016) because the scholars put the voices of the residents, their petroleum-driven worries and their perceptions and knowledge into the center of analysis. Widener and Gunter (2007), to unpack the box of indigenous knowledge about oil's impact on native life, shifted their attention from mainstream media to the newspapers of the Alaska natives. The researchers argue that alternative media should occupy a more prominent place in studies on oil spills because they provide a venue for the marginalized voices of the natives and "signal the collective, though not unanimous interpretations of the spill" (p. 769).

In contrast to their study, I focus not on the natives' recovery from oil spills but the images of the Pechora River as portrayed in *Save Pechora*'s publications. I limit my research to data sources from *Save Pechora*'s website for several reasons. All materials in this source are prepared, edited, and published by ordinary people from the above mentioned settlements. I refer to BBC, DW and other media outlets to include the locals' comments in their reports about spills in the Komi Republic. However, professional journalists speak for the people – this is the nature of that field (Roudakova, 2017). The *Save Pechora* website functions as an alternative media source, providing a public platform and allowing people to speak for themselves. Finally, journalists cover a limited number of spill problems: health impact, corporate accountability, or ecological damages. *Save Pechora* members have never limited the debate to such terms. Instead, they focus on the river's cultural, symbolic and existential meanings.

I emphasize that analyzing journalistic pieces have a sense regarding the studies concerned with media coverage of oil spill recovery, as noted in the research by Widener and Gunter. I'm not interested in the coverage of spills in their variable interpretations and focus on a single interpretation: the vision of Pechora residents regarding the role of oil spills in river life. I argue that materials published by *Save Pechora* present multiple voices from the ground – unchanged by a journalist's interpretations, editor's revisions, or editorial policies. I am inspired here by Elizabeth Hindman's (1998) claim that in community outlets, where issues override events due to a publication's time delay, residents become active producers of "news" and definers of issues important to them (p. 29).

2.3 Oil Spills in the Komi Republic: An Overview

Oil production, providing more than half of the annual regional budget, exacts a considerable price on the environment and residents in the Komi Republic. Lukoil-Komi, the region's dominant oil-gas company, is unofficially known as the owner and operator of a facility from which oil is discharged. Lukoil-Komi was established in 1999 when the Russian corporation Lukoil bought the old state enterprise Komineft. My Google search query "Lukoil-Komi history" resulted in a long list of information and articles about oil spills.

The scope of oil spills in Komi continues attracting the attention of world media. For example, in 2016, *The Guardian* published a special report with the following notion: "These spills are relatively small and rarely garner media attention. But they add up quickly, threatening fish stocks, pasture lands and drinking water" (Luhn, 2016). In May 2021, a message appeared in news feeds about another massive oil spill on the Kolva River – the tributary of the Pechora River (Hundred tons, 2021). In 2014, during a ten-day "oil patrol" Greenpeace reported the discovery of 201 contaminated sites (mostly from pipeline breaks) in the Usinsk oilfield (Sredi apokalipsisa, 2017).

While working on the present contribution in April 2021, I made another Google search query – "oil and Arctic Rivers" – garnering over 13 million sources on the last significant oil accident: June 2020, near Norilsk. Greenpeace claimed it was the worst accident of its kind in modern times (Greenpeace, 2020). The accident provoked a series of journalistic investigations on oil pollution in Russia in national and foreign media. According to the report by independent media, Vazhnye istorii (Important stories), the media reports on less than 1 per cent of oil leaks in Russia every year (Vazhnye istorii, 2020). Leaks most often occur in three major regions with high output production of oil and oil loading facilities in the Arctic: the Komi Republic, Khanti-Mansi Autonomous Okrug and Yamalo-Nenets Autonomous Okrug.

The case of the Komi Republic illustrates why distinguishing between oil spills and environmental disasters is crucial in debates concerning the relations between indigenous people and the environment. In contrast to Alaskans, who are still recovering from the only Exxon Valdez disaster, Pechora residents, as I described, have lived through endless oil spills for decades. However, the literature on spills is dominated by American cases and concentrated on Alaskan natives' recovery following the 1989 Exxon Valdez spill (Dyer, 1993; Gill & Picou, 2001; Smith, 1993; Widener & Gunter, 2007). Chad Wickman (2013) added that only catastrophes like the Exxon Valdez or the 2000 Deepwater Horizon spill raised public and academic debate about a detrimental impact of oil extraction on the environment (p. 5). However, the Komi Republic tops the list of the world's oil disasters. Usinsk, with a population of 39,000 – in the northern part of Komi – became famous in 1994 when the pipeline had been leaking for seven months. Though the full scale of the accident is unknown, the amount of spill ranged from 100,000 to 200,000 tons (Kondratenko, 2021). This accident is still estimated as the world's largest land oil disaster (Luhn, 2016). In my empirical analysis, I rely on that case not because of its scope, ecological or social consequences but as the starting point for Pechora River inhabitants to speak for the river. In the next section, I analyze their worries, feelings and expressions towards the Usinsk spill to illustrate how and why this technological disaster irreversibly changes people's perceptions of the river and their understanding in speaking for it.

2.4 The Transformative Power of an Oil Spill: The Usinsk Disaster

The literature on the social impact of technological disasters explores them through the terms of loss and dearth because these disasters left human victims helpless (Erikson, 1998, p. 154) and destroyed the lives of nonhumans (LeMenager, 2016). Scholars define oil spills in the Arctic as world-changing because of their multiple social effects on the natives: community erosion (Picou, 2000; Picou et al., 2004) or social anomie (Dyer, 1993). My empirical analysis shows it was a world-changing event for Pechora River residents who lived through the Usinsk spill. However, in contrast to the studies mentioned, the list of spill consequences is not limited by emotional suffering or social erosion. The transformative power of the Usinsk disaster was implied in three main dimensions: the capacity of *Save Pechora* to speak for the river, the nature of people's loss and human attachment to the river. These dimensions were closely related through the interdependency between the lives of the Komi natives and the Pechora River.

As mentioned, since the 1970s, Pechora residents have suffered from small and large-scale spills. In 1989, *Save Pechora* founders listed speaking for the river as the organization's main purpose (Komitet spaseniia Pechory, 2013). Meanwhile, five years later, the narrative of worrying about the river was transformed into the willingness of acting for its sake. I found that this shift from waiting to acting, from asking to claiming, was the first feature of transformative power, significantly influencing *Save Pechora*'s profile. The *Save Pechora* member A. Gabov manifested this shift in his speech at the fourth Annual Conference, which took place in August 1994 after the Usinsk disaster. Gabov presented a collective opinion of Mutnyi Materik village's inhabitants. Here is the fragment of his speech published by the *Ekologicheskii Vestnik*:

> Can we say that we achieved our goals and Save Pechora finally saved Pechora from death? I would not answer in the affirmative, although I have been a permanent member of the organization all these five years. I would say more: today the Pechora River needs urgent salvation more than 5 years ago. Great river is really dying today (Gabov, 1994, p. 4).

He added:

> Our silence can lead the indigenous population of the Pechora River and the entire Republic to a sad ending. Save Pechora must not allow this to happen. The time has come for action (Gabov, 1994, p. 4).

In other words, fighting against oil pollution transformed into fighting for the river's life. Protests, public discussions, demands to several federal and regional state bodies, open letters in a dozen villages and cities of the Komi Republic attracted the attention of Greenpeace and other global organizations, international media, and foreign governments. The scale of the protests and the high level of their consolidation were unexpected, even for *Save Pechora* members. "Perhaps for the first time, the members of *Save Pechora* from the city of Ukhta voiced themselves so loudly. The protests were also attended by the residents who did not own formal membership. High school students also responded to the Usinsk tragedy," reported *Ekologicheskii Vestnik* in October 1994 (Po trubam ili po reke, 1994, p. 1).

The tension surrounding the river's recovery after the Usinsk spill proved the possibility of collective self-defense of the Pechora River residents facing a large problem. This finding confounds the research on the Russian Northern locals, including the indigenous people, that traditionally characterizes them with low levels of organized resistance (Stammler, 2011, p. 247). Conversely, the experience of *Save Pechora* fits the global debate about oil disasters as the issue driving social mobilization and civic activism in local communities (LeMenager, 2016).

The Usinsk spill has changed the very perception of the Pechora River among its residents, who started imparting a deep sense of loss into their understanding of it. This is the second dimension of transformative power I investigated in my research. Under technological or ecological damages, rivers can be a source of pain because they reveal what "homeland" means to indigenous people (Schwartz, 2006; Silverman, 2018). Analyzing the narratives about the Pechora River I found that the Komi citizens define it mostly as "homeland" and "motherland". Unsurprisingly, in 1994 it was something of a shock to observe the river – and its flora and fauna – only to find it totally, finally, and irretrievably destroyed:

> The inhabitants of the Kushora village remember that after the oil spill of 1994, when oil flowed through the Kolva River, their horses and cows died. Drank water and died. They could not give birth to offspring. Now there are no cows in the village, only goats, sheep and eight horses (Save Pechora, 2019, April 7).

This news report was published in 2019 and is related to the main tributary of the Pechora River – the Kolva River. It is still famous as one of the most oil-polluted rivers in the world (The 10 biggest oil spills, 2013). However, the expressions of the shock over the 1994 spill seem like the Usinsk disaster happened yesterday. Narratives like this reveal a much more important challenge regarding oil spill recovery. Over 25 years have passed since the tragedy, but it is not forgotten. In this sense, my research on the Komi Republic supports a few bottom-up studies with the evidence that local residents view technological disasters as a specific type of catastrophe. Such a disaster divides history into two totally disconnected periods – before the disaster and after it (Erikson, 1998; LeMenager, 2016).

My finding goes further, showing that feelings of pain and loss have enforced human attachment to the river as a homeland. River attachment is not unique to the Komi, as rivers have cultivated the national soul and homeland for other Northern nations, such as the Baltic ethnic group. In Katrina Schwartz's (2006) study on the transformation of national identity among rural Latvians, she highlighted their attachment to rivers and forests as connected to their collective and self-identification. According to my study on the Komi Republic, when we look at a vision of wildlife among Russia's Northern natives, we should define their environmental imagination as river-centric.

Moreover, I argue that oil spills make this river attachment developed from generation to generation visible. Homeland narratives in their connections with the Usinsk spill allowed Pechora residents to construct the link between the emotional and ethical meanings of human-nature relations. This link was made through understanding the river as a bridge connecting the past, present and future. The understanding of the river as the time bridge is evident, for example, in the report published in *Ekologicheskii Vestnik* after a protest in the city of Ukhta on 6 October 1994:

> The action began with picketing of the Komineft building, the square was full of posters. Here are some of them: "Oil poisoning of the Usinsk land is an environmental crime", "We and our children live on the Komi land. This is our land. We must protect it from environmental disaster", "Komi Republic – is our home and no one is allowed to destroy it", "Who are responsible for the crimes of the oil workers?" (Po trubam ili po reke, 1994, p. 1).

Unlike several studies focusing on the social or health dimensions of spills (Picou et al., 2004; Rodin et al., 1992), in this section, I was interested in an alternative aspect of petroleum-driven impact rooted in the life of the river rather than the people. I found that oil is perceived as the source of pain because spills have led to river dearth. In other words, Pechora residents construct their demands and express their feelings mainly through victimizing the river, in which the river is viewed as a suffering subject. However, it does not mean that oil spills don't destroy the everyday lives of Pechora residents but that oil pollution and land-water devastation construct an essential part of their social reality. Following this idea in the next section, I will explain how and why Pechora residents focus on river suffering more than on their own sufferings.

2.5 Death, Loss and Emptiness: Natives' Representations of Pechora Homeland

Since the mid-1990s, the metaphors of damage and dearth became the core characteristic of how the Komi natives understood, felt, and perceived their river in light of oil production. People reflect the river death not only as a metaphor but understand and perceive it as reality – an essential part of their daily lives. In this section, I will show that according to the voices on the ground, death has at least three faces: dearth as damage to people's lives, dearth as a loss of wildlife area and dearth as a war.

First, the inhabitants of the Pechora area have faced multiple health problems. People expressed strong emotions about the dying river and its destroyed surroundings, which became dangerous to people's well-being. "Children die, young people are dying. Oil and refined products are among the most severe carcinogens. We are simply afraid for ourselves, for the health of our children. People go to the forest, pick berries, and feed them to children. When and where *Lukoil* becomes accountable for the spills?". These words came from L. Redkina, a resident of the village of Shchelyayur, as quoted in the *Save Pechora* news report on 7 March 2017 (Save Pechora, 2017a). The report concerned a meeting of the inhabitants of two villages on oil spills. This fragment is just one of many narratives illustrating the interdependency between human suffering and nonhuman dearth. It means that Pechora's residents base health vulnerability on environmental determinism.

Another dimension of death is associated with the loss of river wildlife. Because of the 1994 Usinsk spill, locals described their river as "empty", "dying" and "killed". According to my analysis, over 20 years later, nothing has changed. "Dead" and "empty" are the central definitions in people's descriptions regarding the water and river. These descriptions are based on their victimization of the river. Narrating a river-oil link in their reports, letters, and speeches, Pechora residents primarily attribute pollution to river dearth, the emptiness of freshwaters, the disappearance of fish and killed nature. On May 28, , 2016, protesters from the villages of Ust'-Usa and Novikbozh voiced their worries about forest sustainability. *Save Pechora* referred to their expressions in the report, published a day later:

> Lilia Petrovna Smirnova from the village of Novikbozh said that the locals already had no fresh air to breathe – there were towers all around, and next to it are "smoking lamps" – torches, and even worse are the so-called "technological processes" of burning who knows what: – Smoke spreads all over the district, the sun became invisible! The sky is painted in incredible acid colors. What will happen to our forests – where we pick mushrooms and berries?! There are places for fishing and hunting, and all this... for oil?! (Save Pechora, 2016a, May 28).

As we can see, when Pechora inhabitants speak their fears concerning river sustainability, they speak of their fear for forest survival and vice versa. As mentioned, forests are crucial for understanding the Pechora River environment. Historically, for the Komi people, the forest was a mosaic of numerous flora and fauna, which made taiga alive and provided the foundation for human living (Shabaev & Istomin, 2017; Smetanin, 2004). However, my research reveals that during the last decade, this is not the case. *Save Pechora* activists regularly report on the total, multidimensional and continuous erosion of the forest ecosystem, which is "turning into an industrial zone" and characterized by the "disappearance of many fauna species", "dying trees" and "empty land not able to give a birth".

Traditionally, the Komi developed intimate relationships with rivers and forests by balancing exploitation and cherishing them as closely linked to their identity. Residents of the Pechora communities strongly and unconditionally associated the destruction of this balance with Lukoil-Komi. On 10 December 2017b, a former forest guard reported against further corporate expanding:

> No one ever dared to even think of cutting down or damaging a tree in the floodplain of the river. Why can Lukoil destroy everything?! For centuries, all generations have kept, protected forests, boreal rivers, Mother Pechora! We always ate, breathed, rested and lived in

> the forest and the river. What kind of war has begun that everything needs to be cut down? (Save Pechora, 2017b, December 10).

The metaphor of war seems one of the most central expressions among the voices related to Lukoil-Komi. In different materials, Pechora residents define company activity as a war on human and nonhuman lives. One of the most unexpected findings in my research was that *Save Pechora*'s long-term opposition to oil was provoked by devastating and totally irresponsible oil extraction practices rather than people's willingness to stop any oil activities in the region. A list of critical comments is very much a rejection of barbarian land-use practices exposed by Lukoil-Komi for decades: oil leaks, garbage piles, rusty pipelines, and declarative land recultivation.

People don't just express criticism but hold the company responsible. In May 2016, after another spill, the inhabitants of the villages of Tom, Mutnyi Materik and the town of Ukhta brought to the collective protests a series of handmade posters with the following demands: "Lukoil, you are ruining our rivers!", "Lukoil, the Izhma River – not a toilet!", "Who will be responsible for the Yarega disaster?", "We are against the tyranny of the oil workers!", "We stand for a clean Pechora" (Save Pechora, 2016a).

A double-coined consciousness of suffering drives such a struggle against the oil industry: When the inhabitants spoke of river and wildlife suffering, they spoke of communities suffering. Moreover, ethical evaluation is an essential element in their understanding of protecting nature. I want to illustrate this statement referring to the resolution of the inhabitants of the village of Izhma and Izhemskii's rural districts.[3] On 6 May 2016, over 150 people protested in Izhma after a massive oil spill that polluted rivers, streams, swamps, and forests:

> For centuries, we, Izhemtsy, have been trying to live with dignity in our native land – in the basins of Izhma and Pechora rivers. The forest and the river are the sources of our livelihood, the basis of our culture. But in recent decades, activities of oil-producing companies, and above all Lukoil-Komi, year after year destroys our habitat, depriving us of the possibility of traditional healthy life in our native land. Our demands are being ignored. The policy of the company operating on our ancestral lands is fundamentally contrary to the interests of the local population (Save Pechora, 2016b).

I refer to these voices because they perfectly illustrate the main conclusion of this empirical section. The natural environment, community, wildlife, and homeland appear irreversibly linked in the river-oil imaginations of Pechora natives. Such warm, intimate attitudes illustrate the essential and unquestionable interdependency between human wellbeing and river life. The central tragedy is that the oil industry's long-term presence in the Pechora area with its continuous, multiple, and devastating effects on river life made petroleum the core aspect of natives' perceptions of the river. When people wrote of their losses, they mostly spoke of despair and

[3] Izhma is a rural locality and the administrative centre of Izhemskii district. Izhma is located nearby the Izhma River. This village is one of the most ancient and large rural community in the Northern part of the Komi Republic. In 2010 its population was estimated about 3750 people.

sadness on losing their living river. In this sense, my study echoes Widener and Gunter's (2007) idea: that the residents' narratives give us more ideas of what the sources of their pain, loss and worry might be (p. 771). The Komi indigenous inhabitants claim the Pechora River should be indicated as the main victim, not communities or individuals. By victimizing the river, the Komi natives not only manifest their losses and the river's suffering but map their communities and connect the past with the future. Perhaps the most significant finding of my research is that oil production anywhere manifests the end of all living things, human and non-human beings. Residents of the Pechora River frame this vision of oil through desperate cultural, symbolic and natural losses, as well as personal and collective drama.

2.6 Conclusions

My initial goal was to explore the interpretations and perceptions of the Komi natives towards the Pechora River in their connections to oil damages. As my analysis of *Save Pechora* publications progressed, I realized the interdependence between oil and river life is essential, natural, and unquestionable for the inhabitants. So, I viewed the materials prepared by the non-governmental organization *Save Pechora* as the sources that people used to construct, develop, and frame the discourse about their river environment. Following their central definition of the river as homeland, I explored the natives' perspectives on river life through the prism of oil spills.

At first glance, any debate on oil spills provokes a battle among corporations corporate, state and public actors regarding who must take responsibility. Over the decades, the typical outcome of the discussion moved to the final stage, where nobody takes responsibility (Dyer, 1993; LeMenager, 2016; Picou et al., 2004). Within the long-term academic and public debate about balancing water sustainability and oil production, my research supports a non-mainstream position. Pechora residents' experiences proclaim the meaninglessness of further searching for balance and stable connection.

The environmental degradation of the Arctic demonstrates the rising relevance of David Orr's (2004) opinion: "We are smart enough to build technologies, but not smart enough to use it safely" (p. 35). Relying on reckless, aggressive exploration, oil companies have transformed the territory of the Komi Republic into oil flows contradicting river flows. Conflicting flows are not limited to rivers and natural resources. As Jean Féaux de la Croix (2014) claims, many contradicted flows – material and cultural, natural and industrial – precisely provide promising discussion about disconnection, inequality, and injustice (p. 98).

Despite disconnection and injustice, Pechora inhabitants have put great effort into making their voices heard. Even now, I cannot answer what is the main driver of their persistence, power, and capability to defend their river homeland for decades: hope or despair. At least, my study provides the other vision of the Russian Arctic natives. Most literature gives the impression that the Russian North is inhabited mostly by passive people who flee to the tundra in case of danger (Stammler,

2020). For example, Stammler and Peskov (2008) described the lifestyles of the Khanty and Nenets as their main reason for avoiding civic activism (p. 247). My research concerns the most famous and once the most active environmental NGO in the Russian European North: *Save Pechora,* established by ordinary rural people who settled near the river and stayed from generation to generation, never reaping the benefits from the oil industry. Speaking for themselves and the Pechora River, they have aimed to improve river-oil coexistence going forward.

Oil-driven environmental damages don't often make the news unless they are serious oil spills, like the 1994 Usinsk disaster in the Komi Republic. What lessons could we learn here? We can hardly find a strong reason for optimism. While working on this contribution, in May 2021, a message appeared in news feeds about another massive oil spill on the Kolva River (Britskaia, 2021). As before, the public and government's disclosure of the incident occurred when the volumes of the spilt oil could no longer be hidden. Such was not the first, and neither would it be the last large oil spill. Again, this development didn't raise a serious debate about the price of oil extracted in Russia.

The Pechora River case shows that stories about oil spill prevention and recovery switch our attention from the core issue of injustice to secondary meanings. According to the rising voices of the natives, this debate could be ongoing unless the powerful actors and politicians proclaim sustainability's impossibility regarding oil industries. The question is not about the struggle against oil-gas pipelines or corporations, given that Russian Arctic natives don't own sufficient power capacity to stand equally in that battle. Rephrasing the words of LeMenager (2016), another question remains open – whether oil or water, black flow or blue flow will determine the landscapes in the Komi Republic.

References

Bachman, J. (2010, November 8). Oil and ice: Worse than the Gulf spill?. *Reuters*. https://www.reuters.com/article/us-russia-oil-idUSTRE6A71IL20101108

Britskaia, T. (2021, June 1). A very dirty lie [Ochen' griaznaia lozh']. *Novaya Gazeta*. https://novayagazeta.ru/articles/2021/06/01/ochen-griaznaia-lozh

Dyer, C. L. (1993). Tradition loss as secondary disaster: Long-term cultural impacts of the Exxon Valdez oil spill. *Sociological Spectrum, 13*(1), 65–88. https://doi.org/10.1080/02732173.1993.9982017

Erikson, K. T. (1998). *Trauma at Buffalo Creek. Society, 35*, 153–161. https://doi.org/10.1007/BF02838138

Féaux de la Croix, J. (2014). 'Everybody loves flow': On the art of rescuing a word from its users. *Suomen Antropologi, 39*(2).

Gabov, A. (1994, September). My ne dolzhny molchat'. *Ekologicheskii Vestnik*, 9(4). https://savepechora.ru/page.php?p=349

Gill, D., & Picou, S. (2001). The day the water died: The Exxon Valdez disaster and indigenous culture. In S. Biel (Ed.), *American disasters* (pp. 277–299). New York University Press. https://crrc.unh.edu/sites/crrc.unh.edu/files/media/docs/Workshops/human_dimensions/reading_materials/day_the_water_died_gillpicou_chapter.pdf

Gill, D. A., Picou, S., & Ritchie, L. A. (2011). The Exxon Valdez and BP oil spills. *American Behavioral Scientist, 56*(1), 3–23. https://doi.org/10.1177/0002764211408585

Greenpeace. (2020, June 5). *The river runs red – catastrophic oil accident in the Russian Arctic*. https://www.greenpeace.org/international/story/43553/oil-spill-russian-arctic/

Henry, L. (2010). *Red to green: Environmental activism in post-soviet Russia*. Cornell University Press.

Hindman, E. (1998). Community, democracy, and neighborhood news. *Journal of Communication, 48*(2), 27–39. https://doi.org/10.1111/j.1460-2466.1998.tb02746.x

Hundred tons of oil leaked from the pipeline in Komi. (2021, May 17). *BBC News*. https://www.bbc.com/russian/news-57151151

Kim, H. J., Shabaev, Y., & Istomin, K. (2015). Lokal'naia gruppa v poiske identichnosti. *Sotsiologicheskie issledovaniia, 8*, 85–92.

Komitet spaseniia Pechory. (2013, May 8). *Kratkaia informatsiia o rabote obshshestvennoi organizatsii s 1989 po 2013 gody*. Doklad. [Save Pechora. Short information on activity of the NGO since 1989 till 2013]. www.savepechora.ru

Kondratenko, T. (2021, March 29). Why Russia has so many oil leaks?. *Deutsche Welle*. https://www.dw.com/en/russia-oil-spills-far-north/a-56916148

LeMenager, S. (2016). *Living oil: Petroleum culture in the American century*. Oxford University Press.

Luhn, A. (2016, August 5). The town that reveals how Russia spills two *Deepwater horizons* of oil each year. *The Guardian*. https://www.theguardian.com/environment/2016/aug/05/the-town-that-reveals-how-russia-spills-two-deepwater-horizons-of-oil-each-year

Muru-Lanning, M. (2018). At Every Bend a Chief, At Every Bend a Chief, Waikato of One Hundred Chiefs': Mapping the Socio-Political Life of the Waikato River. In J. Wagner and J. Jacka (Eds.), *Island Rivers: Fresh Water and Place in Oceania* (pp. 137–164). ANU Press.

Orr, D. (2004). *Reflections on water and oil. Moebius, 2*(2), 17–21.

Picou, S. (2000). Talking circle as sociological practice: Cultural transformation of chronic disaster impacts. *Sociological Practice, 2*(2), 77–97. https://doi.org/10.1023/A:1010184717005

Picou, J. S., Marshall, B. K., & Gill, D. A. (2004). Disaster, litigation, and the corrosive community. *Social Forces, 82*(4), 1493–1522. https://doi.org/10.1353/sof.2004.0091

Po trubam ili po reke?. (1994, October). *Ekologicheskii Vestnik, 10*, 1–3. https://savepechora.ru/page.php?p=349

Reinert, H. (2016). On the shore: Thinking water at a prospective mining site in northern Norway. *Society & Natural Resources, 29*(6), 711–724. https://doi.org/10.1080/08941920.2015.1132352

Rodin, M., Downs, M., Petterson, J., & Russell, J. (1992). Community impacts resulting from the Exxon Valdez oil spill. *Industrial Crisis Quarterly, 6*(3), 219–234. https://doi.org/10.1177/108602669200600304

Roginko, A. (1992). Environmental issues in the soviet Arctic and the fate of northern natives. In J. M. Stewart (Ed.), *The soviet environment: Problem, policies and politics* (pp. 213–222). Cambridge University Press.

Roudakova, N. (2017). *Loosing Pravda: Ethics and the Press in Post-Truth Russia*. Cambridge University Press.

Save Pechora. (2016, May 28). Zhiteli Ust'-Usy i Novikbozha prisoedinilis' k protestam protiv neftezagriazneniia basseina Pechory [Residents of Ust'-Usa and Novikbozh joined the protests against oil pollution in the Pechora River basin]. https://savepechora.ru/page.php?p=160

Save Pechora. (2016a, May 6). K protestuiushchim zhiteliam Izhmy segodnia prisoedinilis' poselok tom, gorod Ukhta i selo Mutnyi Materik [Today, the village of tom, the city of Ukhta and the village of Mutnyi mainland joined the protesting residents of Izhma]. https://savepechora.ru/page.php?p=155

Save Pechora. (2016b, May 6). Izhemtsy – Za chistuiu Izhmu [Izhemtsy - for clean Izhma]. https://savepechora.ru/page.php?p=154

Save Pechora. (2017a, March 7). V Shchelyayure podtverdili svoiu pozitsiiu po nedopushcheniiu stroitel'stva vblizi naselennykh punktov. [Shchelyayur reaffirmed its position on preventing construction near settlements]. https://savepechora.ru/page.php?p=214

Save Pechora. (2017b, December 10). Nedelia protesta nachalas' mitingami v Ust'-Use i Novikbozhe. [The week of protest began with rallies in Ust'-Us and Novikbozh]. https://savepechora.ru/page.php?p=262

Save Pechora. (2019, April 7). Izhemtsy otvergli ocherednoi proekt Lukoil-Komi [Izhemtsy rejected another project of Lukoil-Komi]. https://savepechora.ru/page.php?p=345

Schwartz, K. Z. S. (2006). *Nature and national identity after communism: Globalizing the ethnoscape*. University of Pittsburgh Press.

Shabaev, Y., & Istomin, K. (2017). Territorial'nost', ètnichnost', administrativnye i kul'turnye granitsy: komi-izhemtsy (iz'vatas) i komi-permjaki kak "drugie" komi. *Ètnograficheskoe obozrenie, 4*, 99–114.

Silverman, E. (2018). The Sepik River, Papua New Guinea: Nourishing tradition and modern catastrophe. In J. Wagner & J. Jacka (Eds.), *Island rivers: Fresh water and place in Oceania* (pp. 187–221). Australian National University Press. https://library.oapen.org/bitstream/handle/20.500.12657/29543/anubook.pdf?seque#page=201

Slezkine, Y. (1994). *Arctic mirrors: Russia and the small peoples of the North*. Cornell University Press.

Smetanin, A. (2004). *Istoriia Komi* (Vol. 1). Komi Publishing House.

Smith, C. (1993). News sources and power elites in news coverage of the Exxon Valdez oil spill. *Journalism Quarterly, 70*(2), 393–403. https://doi.org/10.1177/107769909307000214

Sredi apokalipsisa. Usinsk. Liudi. Neft'. (2017, September 25). 7 x 7. https://lr.7x7-journal.ru/usinskneft/?fbclid=IwAR03GeuHwBDCQaY2wDSl2o5I1Mz4G117GSJNqBwKvXlKHfUlLsKHORVR-iA

Stammler, F. (2011). Oil without conflict? The anthropology of industrialization in Northern Russia. In A. Behrends, S. P. Reyna, & G. Schlee (Eds.), *Crude domination: An anthropology of oil* (pp. 243–269). Berghahn Books.

Stammler, F. (2020). Animal diversity and its social significance among Arctic pastoralists. In F. Stammler & H. Takakura (Eds.), *Good to eat, good to live with: Nomads and animals in Northern Eurasia and Africa* (pp. 215–243). University of Lapland.

Stammler, F., & Peskov, V. (2008). Building a 'culture of dialogue' among stakeholders in north-west Russian oil extraction. *Europe-Asia Studies, 60*(5), 831–849. https://doi.org/10.1080/09668130802085182

Sylves, R. T., & Comfort, L. K. (2012). The Exxon Valdez and BP *Deepwater horizon* oil spills. *American Behavioral Scientist, 56*(1), 76–103. https://doi.org/10.1177/0002764211413116

The 10 biggest oil spills – The Kolva River. (2013, April 19). *Oil and Gas*. https://www.oilandgasiq.com

Vazhnye istorii. (2020, October 15). Neftianye avarii sluchaiutsia kazhdye polchasa: issledovanie real'nykh masshtabov zagriaznenii prirody. https://istories.media/investigations/2020/10/15/neftyanie-avarii-sluchayutsya-kazhdie-polchasa-issledovanie-realnikh-masshtabov-zagryaznenii-prirodi/

Wagner, J. R., & Jacka, J. K. (2018). *Island rivers: Fresh water and place in Oceania*. Australian National University Press.

Wickman, C. (2013). Rhetorical framing in corporate press releases: The case of British petroleum and the Gulf oil spill. *Environmental Communication, 8*(1), 3–20. https://doi.org/10.1080/17524032.2013.816329

Widener, P., & Gunter, V. J. (2007). Oil spill recovery in the media: Missing an Alaska native perspective. *Society & Natural Resources, 20*(9), 767–783. https://doi.org/10.1080/08941920701460325

Wilson, E., & Istomin, K. (2019). Beads and trinkets? Stakeholder perspectives on benefit-sharing and corporate responsibility in a Russian oil province. *Europe–Asia Studies*, 71(8), 1285–1313. https://doi.org/10.1080/09668136.2019.1641585.

Chapter 3
Baltic Seals and Changing Marine Frontiers in the Twentieth Century

Tuomas Räsänen

Abstract The chapter investigates the environmental history of Baltic seals from the Finnish perspective. The northern Baltic Sea, which is home to two seal species, the Baltic ringed seal and the grey seal, has been treated by humans as a frontier environment with no other purpose for society than to provide resources. In these processes of frontierisation the Baltic seals have been among the worst victims. During the first half of the twentieth century humans tried to eradicate seals because they preyed on fish. When the hunting pressure eased, the diminished seal populations were almost wiped out by toxic chemicals, as the sea was used as a sink for human waste. In the most recent decades the seals have suffered especially from the warming climate and iceless winters. Seals are thus tragic examples of the changes happening in the northern seas. The ringed seal, which previously lived all along the Finnish coast, is currently acutely threatened by climate change everywhere but the northernmost and coldest tip of the Baltic Sea.

Keywords Baltic Sea · Seals · Environmental history · Frontier · Conservation

3.1 Introduction

The Northern Baltic Sea is a habitat for two seal species, the Baltic ringed seal (*Pusa hispida botnica*) and the grey seal (*Halichoerus grypus*). Until the 1970s because of the harm they caused to fisheries, the Baltic seals were killed on sight. This was encouraged by bounties paid by the government. When seal hunts began to reduce, populations were severely hit by toxic chemicals, which made female seals sterile. The Baltic seals were finally protected in the 1980s, when only a handful of individuals were left, and many scientists and environmentalists predicted that seals would disappear from the Baltic Sea. When this did not happen, and the

T. Räsänen (✉)
University of Eastern Finland, Joensuu, Kuopio, Finland
e-mail: tuomas.rasanen@uef.fi

M. Lehtimäki et al. (eds.), *Cold Waters*, Springer Polar Sciences,
https://doi.org/10.1007/978-3-031-10149-6_3

number of seals surprisingly started to grow in the 1990s, seals were again met with scorn, if not hostility. Today, both species of northern Baltic seals are facing not only by the growing threat of direct human aggression but also climate change. Although grey seals in the Baltic Sea have adapted to cold climes in the winter and may face problems with shorter and warmer winters, climate change is a much more severe problem for the Baltic ringed seal, which in its current habitat is a relic of the last glacial period and a colder environment. The Baltic ringed seal gives birth in the late winter on the sea ice and therefore depends on ice for its survival (Ukkonen et al., 2014, p. 1694).

The purpose of this article is to examine the environmental history of Baltic seals from the perspective of marine environmental changes and the knowledge of them. I will argue that the northern, semi-arctic nature of the Baltic Sea has played and continues to play a special role in the human relationship with seals. In the early twentieth century seals were seen as important for the survival of humans in the harsh northern environment, being both a threat to the subsistence and the source of income and nutrition, with grave consequences for seal populations. In more recent years the Baltic Sea's coldness and northernness has meant it has been prone to anthropogenic environmental changes, while seals have been particularly vulnerable to these changes. Scientists have therefore used them as an indicator species. The study of seals has revealed the true and terrifying scope and speed of the environmental changes of which humans are also potential victims. The seals can therefore be seen as junctures where environmental changes at different times of the twentieth century have converged and materialised in local settings. The main research question can therefore be formulated as follows: how were seals connected to, and how did they embody, the environmental changes in this northern marine environment throughout the twentieth century?

The research has been conceptualised around the notion of the frontier. In environmental history the frontier concept originates among American historians, who have seen it as a crucial element in the process of conquering the American west and exploiting its natural resources. The frontier was a space, as much a state of mind as a geographical place, in which nature was claimed by civilisation by taming indigenous communities, as well as beasts of nature (Cronon, 1995, p. 7–9). In other words, the frontier was a space in which nature – as well as traditional cultures – was sacrificed on the altar of the human development.

In recent decades the frontier concept has spread from its American origins to include myriad themes that share the idea of being on the edge of something unknown. Frontiers have been found on every continent, where they have arguably shaped people's mindset (e.g. Gammone, 2015; Urbansky, 2020), as well as in areas as diverse as science[1] and gastronomy (for gastronomy see Peyton & Dyce, 2017). In this article the concept is used in its somewhat more traditional meaning, in terms

[1] As for science, there are of course numerous esteemed journals that are named frontiers of psychology, education, and neurology, to name just three of many, indicating the quest for uncharted knowledge.

of environmental history, albeit without a connotation with settlers or permanent settlements.

To put it simply, I use the frontier concept as spaces of conquest on the borders of civilisation, which are encroached and exploited for their resources and possibilities. By the early twentieth century terrestrial frontiers in America and elsewhere had all but disappeared as the continents were mapped, and their resources were exploited by the global economy. However, continents only constitute some 30 percent of the earth's surface, and humanity has since found new frontiers to discover and research in open seas. It is a sea that during the twentieth century has been an empty space belonging to no one, according to international law, not unlike the nineteenth-century American west. Obviously, there have been no settlers in the sea, but from the perspective of civilisation the result resembles that of terrestrial frontiers. This 'lawlessness', the absence of sovereignty regarding the sea and emptiness in cultural constructions, has allowed industrial nations to treat the sea as their new frontier, where the most powerful are ready to claim and excavate all the resources from the underwater world and the seabed. Left behind are the ecological impacts of conquest, such as pollution, habitat loss, and declining biodiversity (see e.g. Kroll, 2008; Steinberg, 2001). It is no coincidence that one of the biggest riches of the sea, even the relatively small Baltic Sea, was the capacity of its huge water masses to dissolve the waste rapidly industrialising nation states were producing (e.g. Hela, 1966). Necessarily, just as the taming of the frontier in the American west meant the death knell of the bison and the wolf, the conquest of the sea has had negative consequences for marine life, from the smallest plankton to fish and marine mammals. Obviously, humans have impacted Baltic marine fauna unevenly. Some species have done reasonably well; others have been hit hard by multiple changes in marine ecosystems. Among the worst victims have been large predators like seals.

The article is divided into three sections, indicating different stages and scopes of marine frontiers, in which I will sketch the three distinctive yet partly overlapping processes of marine change, and how seals have related to and embodied them. In the first section. I will examine the period from the early twentieth century until the 1970s, when seals were primarily seen as targets of the hunt, whether because of their villainousness or for their flesh and blubber. The second section focuses on the period from the 1960s until the 1990s, when the hunting gradually ended, but seals were ravaged by toxic chemicals. The third section covers the period from the 1990s until the present, which has been characterised by the warming climate, and seals have tangibly embodied the global climate crisis in the local context.

The main sources used for this research are the documents produced by the World Wide Fund for Nature[2] Finland's Marine Seals Working Group (originally Baltic Seals working group).[3] This collection consists of scientific reports on seals, records of the working group meetings, and memos, notes, and letters written by the

[2] Originally World Wildlife Fund.

[3] The name of the group was originally the Baltic Sea Seals Working Group and was later changed to Marine Seals Working Group. In sake of clarity, I will mostly use the latter name throughout the chapter.

working group's members. Some documents have been published in scientific journals, popular magazines, and newspapers. However, most consist of unpublished materials that are not properly archived or organised but are stored in the personal collections of Dr. Antti Halkka, the long-time chair of the WWF Marine Seals Working Group.[4]

There is a growing body of literature concerning the present-day human–seal relationship, most of which focuses on the conflicts between fishermen and seals (e.g. Reunanen & Mellanoura, 2013; Sonck-Rautio, 2019). As for historical studies of Baltic marine seals, the only existing study, by Juha Ylimaunu, examines the human–seal relationship from the perspective of the last few thousand years. It focuses on the development of hunting practices and the importance of seal hunting for coastal communities (Ylimaunu, 2000). The more generic histories of hunting also address the cultures of seal hunting and their changes over time (e.g. Nieminen, 2020, pp. 266–284).

3.2 The Baltic Sea as the Hunting Frontier

World-system theorists have divided the world into different spheres according to their place in the power networks of commerce and exploitation. At its heart are the centres that run the system and where capital accumulates. These centres are supported by semi-peripheries and peripheries, which supply the centres with valued commodities and manage to withhold some surplus for themselves. For example, these are regional hubs and provincial towns to which materials flow from hinterlands. Finally, these peripheral commercial hubs are surrounded by commodity frontiers, whose only purpose for the system is to provide raw material for the capitalist machine. When the resources from one place are depleted, the exploiters, masterminded by the capital flowing from the centre, move to another place, where the land (or the sea) remains rich in resources (Moore, 2010; Wallerstein, 2004, pp. 28–30). Whether in the early modern Baltic Sea, with the Netherlands as the centre and the Baltic cities as the (semi-)periphery, or colonial America with London as the centre and coastal cities in confederate states as the periphery, in my opinion this world-system analysis neatly simplifies the fate of different geographies and their relations to global networks.

Interestingly but unsurprisingly, these analyses have tended to focus on dry lands.[5] Regarding these terrestrial commodity frontiers, whether in the remote Finnish wilderness or American west, there was always somebody who benefited from the system by supervising the extraction of materials, organising transport, and

[4] I am grateful to Antti Halkka, who generously gave me access to these materials and commented the final draft.

[5] It is telling that even when pretending to write about the sea, whether the North Atlantic or the Mediterranean, they only speak about the land and people surrounding it (e.g. Braudel, 1995; Moore, 2010).

selling the products. In this sense, the commodity frontier was still part of the functioning machine. However, terracentrism conceals the fact that there are also vast spaces on earth that can be called ultimate frontiers, namely, the sea. The sea was a commodity frontier par excellence, for it was never anything other than a fuel for the machine. As the geographer Philip Steinberg has put it, the sea was a great void from a cultural perspective. It was occasionally crossed or exploited, but it was never a human environment (Steinberg, 2001). There were no oppressors and oppressed in the sense that there were at dry land commodity frontiers, but simply robbery that left behind only losses of lives and destructed habitats.

If the sea was the ultimate commodity frontier, the cold seas have been its even more extreme examples. For centuries the polar seas were spaces of slaughter and death in which Europeans killed whales and seals in their millions. There was no aim to civilise or Westernise the sea as was always the case in overseas colonial enterprises. The sole aim was to kill as much as possible. When the supply dried out, the hunters moved further north until there were no targets left to hunt. A famous example, along with whaling, of this frontier attitude of slaughter and habitat destruction in northern waters was the fate of Steller's sea cow, whose population was hunted to extinction in just a few decades in the eighteenth century (e.g. Estes et al., 2016). Interestingly, while attitudes towards marine mammals started to gradually change in the colonising nations (e.g. Lambert, 2002), in the cold seas of the periphery the killing only intensified, and with few dissenting voices. The Baltic Sea, where the purpose of the seal hunt was not to feed the populous industrial centres of the world with food or raw materials as was the case in the polar seas, was a prime example of this. On the shores of the Baltic Sea it was coastal fishing and the needs of coastal communities that formed the rationale for the hunt. Yet the prime motor behind the intensified killing was the state and local governments.

Local people in the Finnish archipelago had traditionally hunted seals for their meat and blubber. When the sea was covered with ice, it was relatively easy for humans to club the seal cubs lying on the ice before their fur had matured. As Olavi Stenman from the Finnish Game and Fisheries Research Institute put it, "[the seal] is pathetically defenceless against killing in both breeding ice and in its traditional flocking islets" (Stenman, 1991). More importantly, however, there was an interspecific conflict over the fish, and the Finnish government pursued ever tighter control of its natural resources, including those living in the underwater world. Throughout the first half of the twentieth century and until the 1970s it was one of the principal goals of Finnish fishery policy to circumscribe fish catches for humans and humans alone. Animals that contested for fish with humans were enemies of civilisation and were thus met with hatred and elimination. "The grey seal is among the fisherman's worst enemies," as an article in the fishing journal *Fiskeritidskrift* put it in 1907 (Anon, 1907, p. 213). It is therefore no exaggeration to claim that what the wolf was on land – a despicable mistake of nature looked down on even by some famed conservationists (Ilvesviita, 2005, p. 145) – seals were in the sea. They therefore merited only extermination.

For the Baltic Sea this entailed a transformation of largely forgotten space in which local communities stood alone into a commodity frontier that supported the progress of the Finnish nation. From the perspective of the human–seal relationship this simultaneously meant the opening of the hunting frontier, which enabled the commodification of the marine resources, mainly fish. For seals this transformation entailed relentless slaughter, which was supported by government bounties. Bounties for marine seals were enacted in 1909, and they were paid until 1975 (Lampio & Stenman, 1975). Bounties provided the ends for hunters; guns provided the means. By the early twentieth century the hunt was made easier by the development of more efficient weapons that were used to kill adult seals. In addition to guns and clubbing, the seals were hunted using nets, hooks, and traps (Ylimaunu, 2000; see also Gottberg, 1921, p. 17).

Every year the fishery journal pieced together the annual success in the "extirpation war against seals", as an author in the fishing magazine called it, and calculated the annual number of kills: 5458 dead ringed seals and 1674 dead grey seals in 1924; 6368 dead ringed seals and 2774 dead grey seals in 1933; 2238 dead ringed seals and 1650 dead grey seals in 1938 (cf. Gottberg, 1921, p. 17, 1925, p. 52, 1934, p. 128, 1939, p. 170). The bodies piled up.

As the war against seals continued for decades, it necessarily had a devastating effect on the seal populations, which is evident in the hunting statistics. During 1909 and 1918 some 10,000 ringed seals were killed annually in Finland alone. In the 1920s and 1930s the annual number of hunted ringed seals averaged 4000 individuals. The hunting continued unabated; there were just fewer seals left to hunt. In the 1950s and 1960s the number had fallen to an average of 1400 (Härkönen et al., 1998). A rough estimate is that at the beginning of the twentieth century there were perhaps 300,000 seals in the Baltic Sea, some 200,000 ringed seals and approximately 100,000 grey seals. When the populations of seals sank to their lowest in 1970s and 1980s, the number of ringed seals in the entire Baltic Sea was approximately 3000 individuals, and of grey seals 1500 (Härkönen et al., 1998). Martti Soikkeli, the long-time chair of the WWF marine seals working group, estimated at the turn of the 1980s and 1990s that during the twentieth century a total of 600,000 seals had been killed in the Baltic Sea (Soikkeli, n.d.). The scale of the massacre can therefore easily be compared with the destruction of much better-known frontier killing fields such as the hunting of whales in the Arctic seas or bison in the American prairies.

3.3 Seals at the Chemical Frontier

During the first half of the twentieth century the scientific knowledge of marine seals in Finland was closely related to the "fight against seals", as fishing advocates continually referred to the seal hunt. It was only in the 1950s that seals were approached as a justified object of study of their own. The first studies, undertaken by Göran Bergman, a noted biologist of marine and archipelagic fauna, focused on measuring the number and distribution of the Baltic seals.[6] It was then thought that the decline of the seal population was caused only by the overkill (Bergman, 1958). However, from the late 1960s it became increasingly clear that it was not just hunting but also the toxic chemicals that ravaged seals.

In the late 1960s Swedish chemists discovered that the Baltic Sea was the most toxic sea area in the world. Fish caught from the sea contained five to ten times more DDT and PCBs than fish in any other researched marine ecosystem (Jensen et al., 1969). These findings were followed by intensive discussion about the chemicals and their impact on the marine ecosystem. Concerning individual species, the main interest was in the white-tailed eagle, which faced huge reproduction problems because of chemicals (Räsänen, 2020; for a discussion on chemicals in the marine ecosystem see also Räsänen, 2012).

While the attention was primarily on the white-tailed eagle, seals were facing similar difficulties. Although less high than in eagles, Baltic seals contained up to ten to twenty times more toxins in their tissues than seals in other sea areas (Hook & Johnels, 1972; Jensen et al., 1969; Karppanen & Henriksson, 1974). The consequences were similarly alarming. Swedish scientists reported that only one out of eight seal females in the Bay of Bothnia shot by locals during the breeding period was pregnant. Closer scrutiny in joint research by Swedish and Finnish researchers showed that the local observations had not been too far off the mark. Less than a third of females were pregnant; in normal conditions this should have been around three quarters (Helle, 1976; Jensen & Olsson, 1976). In the following years the sterility of Baltic ringed seals was identified as caused by uterine occlusion, the result of PCBs and possibly also DDT, which prevented them giving birth. This disability was permanent, which made the survival of the ringed seals in the Baltic Sea uncertain. The problem was most acute in the 1970s, when the levels of toxic chemicals in seawater were at their height. Nevertheless, more than 20 years later, at the end of the 1990s, a quarter of seal females were sterile. Uterine occlusion seemed to affect ringed seals in particular. Grey seals had their own problems with toxic chemicals. It was discovered that grey seal females had a lot of tumours in their uterus, which similarly lessened their reproductive capacity (Helle, 1980; Helle, 1999).

Decades of mass slaughter had reduced the seal populations to tiny fragments of their past abundance of hundreds of thousands of individuals. The great majority of the Baltic ringed seals, some 3000 individuals in the early 1990s, lived in the Bay of Bothnia. There were also small populations in the eastern Gulf of Finland, the Gulf of Riga, and the Archipelago Sea in southwestern Finland. As noted earlier, ringed seals live only in places that freeze during the winter. There was a total of some 5000 ringed seals in the entire Baltic Sea. The population of grey seals was more widespread, but it also collapsed. There were approximately 4000 grey seal individuals in the Baltic Sea, of which there were less than 1000 in Finland and more than 2000 in Sweden (Helle, n.d.; Marine Seals Working Group, 1993a).

[6] Bergman was also the first scientist in Finland to raise concerns about the decline of populations of the white-tailed eagle.

PCBs had been used in various industrial processes, and DDT was mainly used in agri- and silviculture to improve production. In agriculture the introduction of pesticides and artificial fertilisers has been referred to as the Green Revolution. However, since the roots of both developments lie in the progress of industrial chemistry, it would be better called the second chemical revolution,[7] characterised by the mass production of chemically engineered substances for every imaginable human means. Of course, the origins of the second chemical revolution date much further back, but it took off with a vengeance after the Second World War and is therefore an element of the so-called 50s syndrome, when all the environmental stress indices leapt to a new trajectory (Pfister, 2010).

This revolution, whether green or chemical, was turning the sea into a blue desert and wiping out the remnants of the seal populations. The role of the sea in this development was the role of the frontier, which was sacrificed to human progress. The sea was simultaneously a dumping place for human waste – municipal wastewater, radioactive waste from nuclear facilities, scrap vehicles, and so on – and a place which could be consigned to the shadows of environmental awareness. The fact that it was also a home for seals, eagles, fish, and myriad other living beings meant that marine life was doomed to a spiral of overuse and overload. The Baltic Sea was a special case among world seas. As a small and almost landlocked sea area with poor water exchange, it was a trap for toxic chemicals. The cold climate and coldness of the water through most of the year ensured there was less microbial activity in the Baltic Sea than in warmer climes, and the decaying process of xenobiotics was slower. This, along with slow water renewal in this semi-enclosed sea, explains the world record content of DDT and PCBs in the marine organisms, and hence into the 1990s "the contents of biocides measured from seals [were] still ten times higher than those measured from oceans" (Martinson, 1997a).

By the 1970s it had already become clear to researchers that both seal species were endangered because of hunting and toxic chemicals. This population decline became evident when scientists surveyed the traditional resting places for seals. They realised that where there were once seals in their thousands, they could now find only hundreds, while places that used to have seals in their hundreds had now become entirely desolate (Lampio & Stenman, 1975). Against this background it may sound surprising that no conservation projects targeted seals as they did many other formerly persecuted animal species such as the white-tailed eagle, the peregrine falcon, the close cousin of the Baltic ringed seals, the Saimaa ringed seal, and even the feared wolf, all of which had been subjects of conservation projects since the early 1970s (Räsänen, 2020, p. 297). While the grey seal was protected in Finland in 1982, the hunting of the ringed seal continued until 1989, whereas in other Baltic Sea countries hunting had already ceased in the 1970s. For example, the grey seal was declared critically endangered in Sweden in the 1980s, while its protection status in Finland remained merely endangered. According to one Finnish

[7] The term chemical revolution refers to the birth of modern chemistry, which began with a series of discoveries after the seventeenth century.

advocate of seal protection Swedish conservationists were angry with Finland for allowing hunting to continue (Helle, 1988).

On the other hand, in terms of the number of killed seals, fishing was a bigger problem than hunting. For example, it was estimated that roughly a third of seal cubs born in a year drowned when they got tangled in fishing nets. This meant hundreds of dead seals per year (Marine Seals Working Group, 1990). One fishing official from southwestern Finland noted that "[p]eople would be amazed if they knew how many seals are dying in the nets. During the busiest fishing period five out of six fishing boats may have a dead seal on its deck" (Anon, 1987; see also WWF, 1991). This differed strikingly from species like the white-tailed eagle or Saimaa ringed seal, which is basically the same species as the Baltic ringed seal. It is easy to hear echoes of past attitudes towards the sea, held even by marine scientists, which postulated that the sea was eternally unchangeable and its resources inexhaustible, and therefore in no need of care and protection (Rozwadowski, 2002).

However, there were others who did care, and who tried to raise concern about seals and their difficulties at the marine chemical frontier. Isolated voices of concern had been raised since Göran Bergman's time in the 1950s and 1960s, and in the 1970s Finnish conservationists occasionally called for the protection of seals (Helle et al., 1976; Niemelä, 1974). However, it was only in 1986 that the protection was institutionalised under the umbrella of WWF Finland with the establishment of the Baltic Seals Working Group.[8] This group consisted mainly of professional scientists working at the Institution for Game and Fishery Research, the Finnish Museum of Natural History, and the University of Turku. Several individual conservationists also dedicated their activities to marine animals.

Due to the ban on and restrictions to the use of DDT and PCBs, their content in the Baltic Sea and its organisms decreased substantially by the late 1980s (e.g. Helle, n.d.). Seal conservationists could only hope that these levels would continue to decrease, and seals could recover from poisoning and improve their reproduction rate. The main emphasis in conservation work was therefore placed on three different aims. First, conservation was to be justified by scientific research. In this respect, one of the tasks was to monitor the content of toxic substances in seals, and how chemicals affected their reproduction rates. Second, when nothing could be done to remove chemicals from the seawater, conservationists aspired to secure protected areas for seals where they could give birth and gather in colonies without human disturbance. If the weak reproduction rate was a bottleneck for the growth of seal populations, it was important that as many new-borns as possible survived. Third, conservationists sought to educate the public through press releases and education campaigns about the problems seals faced (e.g. Marine Seals Working Group, 1993b; Stenman, 1991). Regarding education, conservationists tried to win public support by pointing out the significance of seals in assessing the toxicity of the marine environment by studying seals. After all, toxic chemicals were then considered the most severe environmental problem in the Baltic Sea (Martinson, 1997b;

[8] The Swedish project on the conservation of marine seals was launched in the same year.

Stenman & Pöyhönen, 2000). In other words, seals were an indicator species, and knowledge of them was used to indicate the health of the entire Baltic marine environment.

The hunting of grey seals was banned in 1982 and of ringed seals in 1989. The hunting bans and the gradually declining toxicity levels gave seal populations an opportunity to recover. The number of grey seals therefore grew throughout the 1980s. Yet for the ringed seal especially the problems were not over. Even after the hunting ban the recovery was very slow, mainly because of uterine occlusion and a low reproduction rate. When the trend finally turned upwards in the 1990s, the increase only occurred in the Bay of Bothnia colony, not in the Gulf of Finland or the Archipelago Sea. Besides ongoing reproduction problems, they were also losing the ice on which they depended.

3.4 Victims of the Iceless Climate Frontier

The fight against seals in the early twentieth century relied heavily on winter ice cover. It was much easier for seal hunters to kill seals when they were lying on the ice. Annual hunting statistics can therefore also tell us about the ice cover each year. In years of cold winters with extensive and long-lasting ice cover more seals tended to be hunted than in milder years. The interconnection between the coldness of winters and seal numbers also worked without human involvement. Göran Bergman speculated as early as the 1950s that one reason for declining seal populations in the 1930s was a succession of mild winters (Bergman, 1958).

The basic mechanisms and dynamics of human-induced climate change have been understood by Finnish scientists studying the marine environment since at least the early 1960s (e.g. Hela, 1961). In the 1980s climate change was already widely discussed and accepted by natural scientists, and the issue increasingly permeated the consciousnesses of politicians and the wider public (Weart, 2008, pp. 143–152). However, the connection of the global change to the local circumstances was not clearly articulated – and often not even researched – at least in Finland. For example, in the scientific and conservational discussion of seals references to climate change began to gain ground only in the early years of the twenty-first century. Yet the documents on the conservation of marine seals, which never aspired to study the changing climate, reveal the gradual awakening to the changing weather and its impact on seals. These assessments provide an interesting testimony to the environmental changes that have already taken place and that will probably intensify in the years to come in the contexts of northern marine ecosystems, where the increase in temperature is more pronounced than perhaps anywhere else in the world. Again, the northern marine environment was sacrificed for the progress and affluence of human societies. It was a climate frontier, the first and most severely affected by a changing climate, and seals were among the species most adversely affected by this change.

The first isolated reference to poor ice conditions in the documents can already be found in 1975, when the winter came unusually late. In those days it was still hunting that worried researchers most, and they feared the lack of ice would strain the declining grey seal population and make it even more vulnerable to hunting (Lampio & Stenman, 1975). However, the problem of an iceless sea became truly evident in the early years of the 1990s, when there was a succession of mild winters with abnormally little ice cover. In the spring of 1989 it was said that such poor ice conditions had not been seen since 1975. There was another mild winter in 1991, and the winter of 1992 was even worse. Olavi Stenman, one of the leading experts on seals and a member of the Marine Seals Working Group, described 1992 as follows: 'the worst winter ice in several decades has caused problems for the Baltic seals, as familiar and safe ice for giving birth has been either entirely missing, as in the Gulf of Riga, or is abnormally low in extent, as in the Gulf of Finland and the Bay of Bothnia' (Stenman, n.d.-a, Hyljekuolemat itäisellä Suomenlahdella vähäjäisen talven 1988–89 seurauksena, see also -b, Ajatus hyljehoitolasta syntyy).

When surveying the populations, researchers quickly realised that seals' success in reproduction always declined when the ice was missing. When there was no ice, grey seals were forced to give birth on land. In such conditions surveyors found plenty of ringed and grey seal cubs and young seals that had lost their mothers and were therefore in very poor health (e.g. Anon, 1989a; Marine Seals Working Group, 1991, 1992). Conservationists could do nothing about the weather, but they tried to invent methods to help seals survive iceless winters. One option was to try to get fishermen to avoid fishing with nets. When the ice melted early, seal cubs were forced into the water too quickly after they were born, when they "plough into and get stuck in the nets because of their inexperience" (Stenman, 2000. See also Anon, 1989b, 1989c; Marine Seals Working Group, 2000). Some conservationists also pondered the possibility of building nursing homes for seals and their cubs. The nursing home would help seals that were wounded by fishing gear, for example, but it would also act as a preparation for the mild winters that would inevitably follow (Stenman, n.d.-b, Ajatus hyljehoitolasta syntyy; Marine Seals Working Group, 2001).

There was nothing new about nursing individuals of endangered wild animal species. Some members of the Marine Seals Working Group had been active since the 1970s in a conservation project for the white-tailed eagle, which involved activities I have elsewhere called care protection, such as feeding eagles with pig carcasses, building them artificial nests, and nursing wounded individuals (Räsänen, 2020). Zoos around the world had been nursing endangered animals since the 1970s. Nursing homes for seals were not a unique concept; the Finnish seal conservationists visited nursing homes for seals in Germany, Denmark, the Netherlands, and France, for example. In Finland seals had previously been taken care of at Korkeasaari zoo in Helsinki, but seals were not its only "customers", and the zoo was unable to host many seals at once (Stenman, n.d., Ajatus hyljehoitolasta syntyy; Stenman & Pöyhönen, 2000). Although nursing homes for seals were never established in Finland, the fact that it was seriously considered demonstrates the deep concern conservationists felt for seals, and how important it was to save even individual seals.

The problem of iceless seas affected both seal species. Scientists estimated that up to 20 percent of grey seal cubs born on hard rocks in islets died. For ringed seals the problem was even worse: "the ringed seal obviously has suffered even more from almost iceless winters, that is, three out of the last four winters. The ringed seal is adapted to giving birth on snowdrifts on ice." Since they were already troubled by toxic chemicals, "icelessness may have reduced decisively the reproduction rate" (WWF, 1992). For some terrestrial mammals mild winters could be a problem; for many it was only a relief. For ringed seals especially nothing less than the survival of the species in the Baltic Sea was at stake.

Against the opinions of many conservationists the grey seal recovered quite well from the trough of the 1970s and early 1980s. By the early 1990s their population in the entire Baltic Sea had grown to 4000 individuals, and it seemed it could adapt to iceless conditions surprisingly well. This meant fishermen started again to complain about grey seals stealing their catch and ripping their nets and hoped the population would be thinned (see e.g. Hellsten, 1998; Kaarto, 1999; Marine Seals Working Group, 1993b; WWF, 1994). On the other hand, the ringed seal, which historically had always been the more numerous of the two Baltic seals, was now "numerically more endangered than the grey seal" (Marine Seals Working Group, 1993a). It is therefore unsurprising that the target species for the most urgent conservation changed from the grey seal to the ringed seal.

The springs of 1997, 2000, and 2001 came early. Mild winters followed one after another. By the turn of the millennium it was clear that the lack of winter ice was not only a fleeting anomaly but was related to global climate change: "At the end of the meeting devoted to the weather the chair presented results from recent studies about the warming of the climate" (Marine Seals Working Group, 2001). Concerning the Baltic ringed seal, it was already known that "[t]he population in the Bay of Bothnia [...] may have substantially grown because of the years of little ice, when suitable ice for giving birth in the southern Baltic Sea was almost completely missing" (Institute for Game and Fishery Research, 2000). Many seals from the southern subpopulations had left their traditional habitats and migrated northwards. The subpopulation in the Stockholm archipelago was gone; the Archipelago Sea was empty except for a handful of survivors; only at the eastern end of the Bay of Finland a vastly diminished population struggled to cope with the new circumstances (Halkka, 2001). Although a relatively strong population of some 1000 individuals remained in the Bay of Riga (Helle, 1999), it was a matter of time before the ice would occur too rarely for ringed seals. Unsurprisingly, researchers predicted that the Baltic ringed seal would be unable to survive the iceless winters of the twenty-first century in its previous southern habitats and would be forced to retreat to the last remaining ice in the Bay of Bothnia (Meier et al., 2004, see also Halkka, 2005; Meier et al., 2021), its last refuge in the world.

3.5 Conclusions

Every year perhaps hundreds of Baltic seals drown in fishing nets, unnoticed by the public and undiscussed. The hunting frontier, as I have dubbed the Baltic Sea of the early twentieth century, never really disappeared, as it has allegedly been quite a common, if secret, practice among fishermen to carry guns in their boats and kill seals when they saw one (Marine Seals Working Group, 1998). This is in striking contrast with the treatment of the Saimaa ringed seal, the close cousin of the Baltic ringed seal. Fishing restrictions have been put in place on Lake Saimaa to secure the wellbeing of every seal individual, while the inevitable deaths of seals and their cubs are met with sorrow and protest (Jaakkola et al., 2018). The plight of Baltic ringed seals prompts few headlines, although its southern populations in the Gulf of Finland and the Archipelago Sea are now more endangered than the Saimaa population. As for the other marine seal, the grey seal, the general attitude among the Finnish public ranges from indifference to hostility, and calls to cull its population have grown louder as the population has recovered. Our era celebrates individuals – even animal individuals – which is clearly seen in the online cameras that follow the daily life of certain animal individuals, one of which is dedicated to Saimaa ringed seals (WWF, 2021, norppalive). No cameras target Baltic seals. Evidently, some individuals are more valuable than others.

There are without doubt practical and economic reasons for the indifference to the Baltic seals, such as their role as a competitor with humans for food. However, this does not explain the near total absence of marine seals from the Finnish ecological imagination and the canon of iconic animals. The mixture of hostility and indifference towards marine seals has a long history tied to the mistreatment of its habitat, the Baltic Sea, as a frontier-like space. In the early twentieth century the Baltic Sea was seen as a hunting frontier not unlike the other cold seas in northern latitudes. Their only purpose was to provide newly found marine resources to the emerging nation states. In this process seals were enemies of human progress and were therefore eliminated.

When the hunting of seals petered out in the latter part of the twentieth century because so few were left, the construction of the Baltic Sea as an externalised frontier continued. The sea was now transformed into a wasteland of the Great Acceleration.[9] Since the 1960s the Baltic Sea has been a chemical frontier burdened by toxic substances. The sea has been the ultimate sink for the chemicals produced and spread on the land, and the northern seas have taken the hardest hit. Chemicals have sterilised the Baltic seals and almost driven their populations to extinction.

In recent decades the industrialised world has exploited the seas and their capacity to absorb both heat and carbon dioxide from the air. This has balanced the warming of the land and allowed societies to continue to spew heat-trapping gases into the atmosphere. Temperatures are rising fastest in the highest latitudes, including

[9] For a concise introduction to the history of the Great Acceleration of environmental changes in the latter part of the twentieth century see (McNeill, 2014).

the northern seas. In other words, the Baltic and other northern seas function as a climate frontier where societies have externalised their environmental costs with little concern for the ecosystemic consequences. The results are clear in the Baltic Sea, where seals, especially the ringed seal, accustomed to cold winters are forced to cope with iceless conditions.

The killing and agony of seals amidst environmental changes may resemble countless other tragedies in humanity's relationship with nature in the modern era. Yet its extremity sprang from a terracentrism in which northern marine environments have been and are seen as something extraneous, and their biota as of less value than that of terrestrial ecosystems.[10]

References

Anon. (1989a). *Suomenlahden harmaahylkeet ovat poikineet N-liiton vesillä*. Helsingin Sanomat.
Anon. (1989b, April 15). *Hylkeen poikaset pahoissa vaikeuksissa*. Kotkan Sanomat.
Anon. (1989c, April 15). *Verkkokalastusta olisi rajoitettava* (p. 15). Kymen Sanomat.
Anon. (1907). Gråsälen såsom skadedjur. *Fiskeritidskrift för Finland, 16*(7–9), 213–215.
Anon. (1987). Verkkokalastus niittää norppaa. *Eläinten ystävä, 82*(4), 15.
Bergman, G. (1958). Suomen hyljekannoista. *Suomen Riista, 12*, 110–124.
Braudel, F. (1995). *The Mediterranean and the Mediterranean world in the age of Philip II*. University of California Press.
Cronon, W. (1995). The trouble with wilderness; or, getting back to the wrong nature. In W. Cronon (Ed.), *Uncommon ground: Rethinking the human place in nature* (pp. 69–90). W. W. Norton.
Estes, J. A., Burdin, A., & Doak, D. F. (2016). Sea otters, kelp forests, and the extinction of Steller's sea cow. *PNAS, 113*(4), 880–885.
Gammone, M. (2015). The European dream: The frontier in European history. *Politeja, 12*(37), 55–73.
Gottberg, G. (1921). Sälfånsten år 1909–1918. *Fiskeritidskrift för Finland, 28*(1–2), 16–18.
Gottberg, G. (1925). Sälfångsten år 1924. *Fiskeritidskrift för Finland, 32*(4), 51–56.
Gottberg, G. (1934). Sälfångsten år 1933. *Fiskeritidskrift för Finland, 41*(7), 128–131.
Gottberg, G. (1939). Sälfångsten år 1938. *Fiskeritidskrift för Finland, 46*(12), 170–173.
Halkka, A. (2001). Itämeren norppa tarvitsee apua. *Pandan polku, 11*, 4–5.
Halkka, A. (2005, March 14). *An email to Mikael Nordström*. Antti Halkka's Collection.
Härkönen, T., Stenman, O., Jüssi, M., Jüssi, I., Sagitov, R., & Verevkin, M. (1998). Population size and distribution of the Baltic ringed seal (*Phoca hispida botnica*). *NAMMCO Scientific Publications, 1*, 167–180. https://doi.org/10.7557/3.2986
Hela, I. (1961). Meritieteen kehitys ja nykyiset tehtävät. *Terrain, 73*(2), 51–67.
Hela, I. (1966). Meritieteen nykyisistä tehtävistä. *Natura, 3*(1), 10–16.
Helle, E. (1976). Perämeren norpat ja ympäristömyrkyt. *Suomen Luonto, 35*(1), 9–11.
Helle, E. (1980). Lowered reproductive capacity in female ringed seals (*Pusa hispida*) in the bay of Bothnia, northern Baltic Sea, with special reference to uterine occlusions. *Annales Zoologici Fennici, 17*(3), 147–158.
Helle, E. (1988, July 6). *A letter to Antti Halkka*. Antti Halkka's Collection.
Helle, E. (1999, December 2). *Hylkeet – Ongelma vai ei? Article manuscript*. Antti Halkka's Collection.

[10] The research has been funded by the Academy of Finland (project 330,762).

Helle, E. (n.d.). Undated memorandum from 1991 Or 1992. In *The ringed seal – resident of northern waters*. Antti Halkka's Collection.
Helle, E., Olsson, M., & Jensen, S. (1976). DDT and PCB levels and reproduction in ringed seal from the Bothnian Bay. *Ambio, 5*, 188–189.
Hellsten, S. (1998, March 3). *Fiskare vill minska sälstammen*. Åbo underrättelse.
Hook, O., & Johnels, A. G. (1972). The breeding and distribution of the grey seal (*Halichoerus grypus fab.*) in the Baltic Sea, with observations on other seals of the area. *Proceedings of the Royal Society of London. B. Biological Sciences, 182*(1066), 37–58.
Ilvesviita, P. (2005). *Paaluraudoista kotkansuojeluun: Suomalainen metsästyspolitiikka 1865–1993*. University of Lapland.
Institute for Game and Fishery Research. (2000, June 27). *A press release*. Antti Halkka's Collection.
Jaakkola, M., Vuorisalo, T., & Peltonen, L. (2018). *Saimaannorppa ja ihminen: Sosiaalisesti kestävän luonnonsuojelun haaste*. Into.
Jensen, S., & Olsson, M. (1976). DDT and PCB in the Baltic. In *2nd soviet-Swedish symposium on the pollution of the Baltic, Riga, September 17–24, 1973. Ambio Special Report No. 4*.
Jensen, S., Johnels, A. G., Olsson, M., & Otterlind, G. (1969). DDT and PCB in marine animals from Swedish waters. *Nature, 224*, 247–250. https://doi.org/10.1038/224247a0
Kaarto, L. (1999). Hylkeet kalastajan näkökulmasta. In *A manuscript*. Halkka's Collections.
Karppanen, E., & Henriksson, K. (1974). Kvicksilver- och klorerade kolvätehalter i havs- och insjösälar. *Suomen eläinlääkärilehti, 80*, 378–391.
Kroll, G. (2008). *America's ocean wilderness: A cultural history of twentieth century exploration*. University of Kansas Press.
Lambert, R. A. (2002). The grey seal in Britain: A twentieth century history of a nature conservation success. *Environment and History, 8*(4), 449–474.
Lampio, T., & Stenman, O. (1975, January 21). *A letter to the Ministry of Agriculture and Forestry and attached memorandum*. Antti Halkka's Collection.
Marine Seals Working Group. (1990, June 11). A *letter to the Turku and Pori County administrative board*. Antti Halkka's Collection.
Marine Seals Working Group. (1991, May 6). *A press release*. Antti Halkka's Collection.
Marine Seals Working Group. (1992, April 14). *A press release*. Antti Halkka's Collection.
Marine Seals Working Group. (1993a, February 8). *A record of the seminar between Finnish and Swedish seal researchers*. Antti Halkka's Collection.
Marine Seals Working Group. (1993b, December 14). *A record from the meeting*. Antti Halkka's Collection.
Marine Seals Working Group. (1998, November 6). *News by the WWF*. Antti Halkka's Collection.
Marine Seals Working Group. (2000, April 13). *A record from the meeting*. Antti Halkka's Collection.
Marine Seals Working Group. (2001, January 30). *A record from the meeting*. Antti Halkka's Collection.
Martinson, K. (1997a). Hylkeet ja Itämeren tila. *Apaja, 3*(2), 2–3.
Martinson, K. (1997b). Ympäristömyrkkyjen vaikutus. *Apaja, 3*(2), 3.
McNeill, J. R. (2014). *The great acceleration: An environmental history of the Anthropocene since 1945*. Harvard University Press.
Meier, H. E. M., Dötscher, R., & Halkka, A. (2004). Simulated distributions of Baltic Sea-ice in warming climate and consequences for the winter habitat of the Baltic ringed seal. *Ambio, 33*(4–5), 249–256.
Meier, H. E. M., Kniebusch, M., Dieterich, C., Gröger, M., Zorita, E., Elmgren, R., Myrberg, K., Ahola, M., Bartosova, A., Bonsdorff, E., Börgel, F., Capell, R., Carlén, I., Carlund, T., Carstensen, J., Christensen, O. B., Dierschke, V., Frauen, C., Frederiksen, M., … Zhang, W. (2021). Climate change in the Baltic Sea region: A summary. *Earth System Dynamics. Discussion [preprint]*. https://doi.org/10.5194/esd-2021-67. in review.

Moore, J. W. (2010). 'Amsterdam is standing on Norway' part II: The global North Atlantic in the ecological revolution of the long seventeenth century. *Journal of Agrarian Change, 10*(2), 188–227.
Niemelä, E. (1974). Hallin asiat ovat huonosti. *Suomen Luonto, 33*(5), 241–246.
Nieminen, M. (2020). *Jääkauden voittajat: Nuotiolla villipeuraa, hirveä, norppaa, lohta....* BoD – Books on demand.
Peyton, J., & Dyce, M. (2017). Colony on main: History and the ruins of imperialism in Vancouver's restaurant frontier. *Cultural Geographies, 24*(4), 589–609.
Pfister, C. (2010). The '1950s syndrome' and the transition from a slow-going to a rapid loss of global sustainability. In F. Uekoetter (Ed.), *The turning points of environmental history* (pp. 90–118). University of Pittsburgh Press.
Räsänen, T. (2012). Converging environmental knowledge: Re-evaluating the birth of modern environmentalism in Finland. *Environment and History, 18*(2), 159–181.
Räsänen, T. (2020). Merikotkan ahdinko ja hoivasuojelun synty. In T. Räsänen & N. Schuurman (Eds.), *Kanssakulkijat: Monilajisten kohtaamisten jäljillä* (pp. 285–312). Suomalaisen Kirjallisuuden Seura.
Reunanen, S., & Mellanoura, J. (2013). *Hylje: Vahinkoeläin vai luontoelämys?* Riista- ja kalatalouden tutkimuslaitos.
Rozwadowski, H. M. (2002). *The sea knows no boundaries: A century of marine science under ICES.* International Council for the Exploration of the Sea.
Soikkeli, M. (n.d.). *Memorandum.* Antti Halkka's Collection.
Sonck-Rautio, K. (2019). *The fishers of the Archipelago Sea: Resilience, sustainability, knowledge and agency.* University of Turku.
Steinberg, P. E. (2001). *The social construction of the ocean.* Cambridge University Press.
Stenman, O. (1991). Harmaahylkeet tarvitsevat suojelualueita. *Metsästys ja kalastus, 80*(12).
Stenman, O. (2000, April 14). *A letter to Antti Halkka.* Antti Halkka's Collection.
Stenman, O. (n.d.-a). Hyljekuolemat itäisellä Suomenlahdella vähäjäisen talven 1988–89 seurauksena. An undated report by WWF Finland Marine Seals Working Group. Antti Halkka's Documents.
Stenman, O. (n.d.-b). *Ajatus hyljehoitolasta kypsyy. Undated memorandum from the year 1992.* Antti Halkka's Documents.
Stenman, O., & Pöyhönen, O. (2000, August 2). *Hylkeitä ei tapeta joukoittain.* Helsingin Sanomat.
Ukkonen, P., Aaris-Sørensen, K., Arppe, L., Daugnora, L., Halkka, A., Lõugas, L., Oinonen, M. J., Pilot, M., & Storå, J. (2014). An arctic seal in temperate waters: History of the ringed seal (*Pusa hispida*) in the Baltic Sea and its adaptation to the changing environment. *Holocene, 24*(12), 1694–1706.
Urbansky, S. (2020). *Beyond the steppe frontier: A history of the Sino-Russian border.* Princeton University Press.
Wallerstein, I. (2004). *World-systems analysis: An introduction.* Duke University Press.
Weart, S. R. (2008). *The discovery of global warming.* Harvard University Press.
WWF. (1991, May 6). *Press release.* Antti Halkka's Collection.
WWF. (1992, April 14). *Press release.* Antti Halkka's Collection.
WWF. (1994, July 15). *White paper for the meeting on seals in Åland.* Antti Halkka's Collection.
WWF. (2021). *Norppalive.* Online camera. Retrieved May 12, 2021, from https://wwf.fi/luontolive/norppalive/
Ylimaunu, J. (2000). *Itämeren hylkeenpyyntikulttuurit ja ihminen-hylje -suhde.* Suomalaisen Kirjallisuuden Seura.

Chapter 4
Conceptualizing Arctic Documentary: Combining Scientific Authority and the Interests of Broadcasters in BBC's *Frozen Planet*

Elena Trubina

Abstract This chapter investigates the mediated Arctic experience made possible by the BBC's nature documentary series, *Frozen Planet* (2011). Extraordinary footage shot by BBCs Natural History Unit, David Attenborough's on-site narration, and Discovery Channel-facilitated streaming should help raise awareness about the need to manage the world's remaining resources better than we have thus far. Throughout this expensive production, stunning depictions establish the Arctic as a subject of fascination, skilfully introducing the theme of the endangered Arctic. This chapter explores the connections between aesthetics and politics, touching upon the political-economic implications of this production: Which artistic strategies are involved and to what ends? What kind of climate change narrative does the remarkable adaptations of animals, the close-ups of polar bears and the drone-shot majestic yet desolate landscapes facilitate or hinder? What are the central tropes and facts comprising the film's narrative, and to what extent are they attuned to the ongoing global displacement of water and ice? This chapter critically explores these questions using the arguments of media, film and cultural studies.

Keywords Nature documentary · The Arctic · Climate change · BBC · Sir David Attenborough

4.1 Introduction

In October 2011, a new seven-part nature documentary epitomizing sublime depictions of the Arctic and Antarctic environment and wildlife was broadcasted on BBC1. The series garnered about nine million viewers and was sold to over 30 countries. The final episode, "On Thin Ice", clearly illustrated the new strategies the

E. Trubina (✉)
UNC Global Research Institute, Chapel Hill, NC, USA

M. Lehtimäki et al. (eds.), *Cold Waters*, Springer Polar Sciences,
https://doi.org/10.1007/978-3-031-10149-6_4

global broadcasters had devised to raise awareness of climate change. The images featuring the vast expanse of the ice in the North Pole, the enormous glaciers separating for the first time in human history, the camera crews descending the remaining glaciers and Sir David Attenborough's dignified warnings were joined with a symphonic orchestra's pounding chords and rising harmonies. The episode culminated in footage of the dramatic ice run in the gorge downstream of Louise Falls on Canada's Hay River and in a story about the town of Hay River – spared from flooding in the spring of 2009 (the time of shooting). One local man – famous for his precise predictions of the river condition – was interviewed and stated the exact number of days "before she breaks". Like the rest of the series, the footage in this episode was impeccable and unique. At one point, satellite shots were combined with rare footage of a submarine breaking ice to reach the surface. The cameras zoomed in and out as they tried capturing the scientists' actions amidst the vast expanse of ice. At other points – while depicting the hunting festival in Barrow, Alaska – the shots were vivid and full of saturated colours. Viewers were offered a vision of the Arctic and Antarctic environment that was majestic and alarming, global and personal. The celebrity presenter-led series aimed to raise awareness about the precarious state of the regions whose impact on the world is immense.

The back cover of the *Frozen Planet* series [Blu-ray] DVD release says:

> From the award-winning team behind *Planet Earth* comes the intimate portrait of Earth's polar regions [...]. The scale and beauty of the scenery and the sheer power of the elements – the weather, the ocean and the ice – is unmatched anywhere else on our planet. Yet these harsh environments are teeming with life: home to iconic animals from polar bears to emperor penguins and from killer whales to wandering albatrosses [...]. With spectacular polar landscapes and amazing animal behavior, often filmed for the first time, it captures the drama of an extreme natural world. With both Poles under grave threat from climate change, this extraordinary series may be the last chance to witness these great wildernesses before they change forever.

The discourse advertised includes uniqueness, grandeur, resilience and the promise of splendor. The DVD's marketing team draws a viewer's attention to "the last chance to witness". I argue that by emphasizing "the last chance" the team skillfully uses the threat of global climate change. What is implied is that the change is quickly happening, so better enjoy the Arctic wilderness while it lasts. The series' aesthetics are the spectacles of these remote regions. For the purposes of these essays, that "cold" (and frozen) waters are being aestheticized and that the Arctic is mediated by current conditions is particularly important. For my analysis, I draw on the concepts of cultural studies, film studies and studies of the North to connect the components of documentary production and dissemination – tropes, conventions, messages, technologies, and audiences – with the larger context of climate change, related knowledge and politics. In this chapter, I first consider how science is used for film production and how what I call the hybridization of science and entertainment has been achieved. Then I examine ice, snow and bears – the last being one of the series' main characters. The next part of this chapter is devoted to the economics of nature documentary production, including celebrity engagement. In my conclusion, I will summarize how the cultural and the economic, the natural and the social, are intertwined in making and disseminating the popular genre of wildlife documentaries.

4.2 Selecting Knowledge for Entertainment Through Nature Documentaries

Sir David Attenborough opens the series with this powerful statement:

> Over one third of our planet is frozen, and yet, the icy worlds of the Arctic and Antarctic are as alien to most of us as the surface of another planet. They are places of superlatives. From ice caps that hold nearly 80% of our planet's fresh water to frozen forests that encircle the entire globe. These are places that feed our imaginations, places that seem to be borrowed from fairy tales. They're dominated and shaped by the ice... both by its coming and by its going ("To the Ends of the Earth," *Frozen Planet*; BBC One , 2011a).

An impressive team of camera crew and scientists was created to depict these icy worlds (BBC One, 2011b). The scientists included Dr. Pascal Lee (a research scientist with the NASA Haughton Mars project), Professor Lloyd Peck (a leading polar marine biologist), and Bernard Buigues (a French explorer), while some series-makers combined producing, directing and conducting research (Fredi Devas).

The hybridization of science and entertainment is also seen in several more examples. The coffee table companion book to the TV series (Fothergill et al., 2011–2012) was published to please Attenborough fans and supply geography students with visual material about rarely seen wildlife and landscapes. ("The previously restricted Russian Arctic" is mentioned in the book advertisement on the Amazon website.) More significant is that Open University and BBC co-produced the series. Open University Professor and the Lead Academic Advisor of the series Mark Brandon claims, "Through *Frozen Planet* we were able to bring front-line research into people's homes and actually show them the effects that climate change is having on our world. Giving people that understanding is the first step in encouraging them to change behaviors which may allow us to slow the inevitable changes and the impact these will have" (Hartje, 2020). Brandon had also written the same-name free Open University course, distributed by Amazon (The Open University, 2016). *Frozen Planet* can be watched via Amazon Prime or DVD (also distributed by Amazon). Particularly inquisitive readers and parents of schoolchildren can "top up" these mediated experiences with the "analogue" one: They can download the book for free and study this free university course, learning, for instance, that "the polar regions are cold because only a fraction of the solar energy received at the surface of Earth actually heats them. The remaining energy is reflected out into space" or pondering questions like "What is your personal view of the frozen planet? What three images or impressions come to mind? What three words, phrases or names do you think of first? What is the last news story you remember about each region?"

Suppose I were to answer the last question. In that case, I'd cite the news about the new BBC project in which the cold regions are represented by whales who started communicating differently and penguins walking sidewalks in the cities. The documentary *The Year Earth Changed* (2021), directed by Tom Beard and narrated by David Attenborough, continues with the strategy of prompting people to change their lives by showing how quickly nature can restore itself when left alone, as happened during the pandemic. Attenborough's "velvety, statesmanlike voice" is mentioned as an important component of the comforting story about the partial recovery of nature made possible by the halt of global economic activity caused by the pandemic (Brady-Brown, 2021). Such indicates how nature documentaries may positively influence citizens' attitudes: providing needed visual and aural information, allowing viewers to experience nature through the lenses of increasingly popular discourses associated with concerns for humanity's future. Science is assumed to offer concrete facts and representations of nature. It is salient to examine how the relevant scientific data are arranged and interpreted according to the need to impact viewers.

The epistemological significance of nature documentaries has been investigated for several decades – from Fortner's (1985) comparison of the efficacy of classroom instruction to the study of how environmental sensitivities are linked to the pedagogic use of nature documentaries by Barbas et al. (2009). Awakening and sustaining viewers' ecological concerns in institutionalized settings or during leisure viewing has been achieved with the help of narrative conventions: How to animate nature crystallized in the early 1950s when Disney studios launched the first generation of wildlife films (Bouse, 1995). The growing body of work examining documentary visual representations of the natural environment in general and, under conditions of climate change, evolves from several fields and disciplines across the humanities and social sciences. To date, research has focused on analysing nature documentaries using predominantly interpretive approaches for content and distribution. Craig Mitman (1999) showed how "reel nature", safely distanced from urbanized America and pleasantly anthropomorphized, not only substituted the "real" nature, which has been destroyed, but proved a strong link of using nature as a commodity. Derek Bousé's (2000) comprehensive history of wildlife films found that the narratives of these films stem from the animal fables that reflect human society. These constructed narrative models amplified the idea of the substantial unity of nature and humans while misrepresenting the animals. Cynthia Chris (2006) examined commercial interests behind the producers' weaselling off discussions of environmental politics. Some studies focus on audience responses (Chapman et al., 2016; Perrott, 2002). For this book's purposes (addressing "flows of knowledge"), the issues of generating and transmitting knowledge are fundamental. For the viewers within the spatial, cultural and ideological contours of their respective countries, significant epistemological challenges lie in the need to meaningfully connect to these global and catastrophic environmental challenges. Regarding the Arctic, these challenges are characterized as follows:

> The Arctic is incorrectly thought of as one of the last pristine places on Earth; but Arctic ecosystems are just as spoiled by environmental contaminants as heavily industrialized areas, if not worse. [...] Arctic ecosystems, despite their remoteness and distance to emission sources, act as a sink for legacy and emerging POPs and, on a global scale, are among the most polluted areas (Bourgeon, 2018, pp. 277–278).

These connections can be made more meaningful by providing authoritative knowledge on the subject. Regarding the demand to share responsibility for the impact humanity makes on the environment, to act and connect with an international community, the question I pose is this: How can nature documentaries participate in distributing much needed environmental knowledge while retaining fidelity to the commercial priorities of the producing companies? This question combines the aesthetic-political implications and political-economic implications of wildlife documentary production.

The regional specificity is also significant for this analysis. The collection this chapter is part of is devoted to the "cold waters" of the North. The documentary I analyzed was filmed in the Arctic and Antarctic, raising the question about how the documentaries explicitly produced about the North are significant for developing the genre of nature documentary and creating and transmitting the knowledge and discourses about the North as a transnational region. In the innovative collection *Arctic Cinemas and the Documentary Ethos* the editors state in the introduction that the history of the documentary genre is centered around the Arctic, that the films produced in the 1920s proved significant for the theory and practice of this genre and that in the last two decades, there appeared the films that can be compared to "agents provocateurs addressing climate change, from *An Inconvenient Truth* (Davis Guggenheim, United States, 2006) to *Chasing Ice* (Jeff Orlowski, United States, 2012) and the Inuit Indigenous works of the Isuma and Arnait collectives in Nunavut" (Kaganovsky et al., 2019, p. 2).

The series I analyze in this chapter was produced by the BBC. I examine how this corporation helps share the knowledge generated by scholars. While reviewing the BBC's science documentary strand, *Horizon*, Kantor (2020), drawing on Silverstone's (1999) typology of sensory categories of film, uses four components of documentary rhetoric: image, look, voice and sound. These rhetorical devices prompt audiences to view and respond to a film in a particular way. These devices are employed to achieve the tasks of documentaries. The flair of rationality this rhetoric creates engages viewers in responsible reading strategies, inviting them to use their logic and common sense.

Documentaries present intriguing ways of emotionally charging their logical and objective narratives. Scholars disagree on the role of negative and positive texts and images. Some scholars show that, in contrast to neutral stimuli, negative stimuli have a more serious effect on viewers' attention and actions (Baumeister et al., 2001; Estes & Adelman, 2008). However, others are keen to claim that images connected to "individuals' everyday emotions and concerns [...] tend to be the most engaging" (O'Neill & Nicholson-Cole, 2009, p. 355). *Frozen Planet* presents an impressive example of integrity by skillfully conducting the conversation about the real issues the Arctic and Antarctic regions currently face and making alarming claims:

> In this series, we'll be travelling to all parts of these lonely lands, both north and south, to witness its wonders perhaps for the last time and to discover some extraordinary examples of survival against all the odds, as can be found anywhere on the planet ("To the Ends of the Earth," *Frozen Planet*; BBC One, 2011a).

However, the series also offers what can be called contemplative and reassuring stories about the "ever harsh" regions, where the sun visits only briefly, and "brings a brief opportunity for life" (ibid.).

The series' narrative is hardly marked by the implied possibility of the inevitable dramatic change and possible extinction of some species. Lauren C. Stephen (2010, p. 106), while describing the blockbuster *March of the Penguins*, writes that the film's narrative is "haunted" ("because the possibility of extinction is never mentioned explicitly") and asks, "[i]f part of the commercial success of the film is a sense of pathos at the knowledge the penguins' habitat may be threatened, by global warming, pollution or declining fish stocks. The penguins are 'history' in more ways than one, and *March of the Penguins* fulfils a desire to capture and see the penguins on film before they disappear". *Frozen Planet* also invites the viewer to admire the majestic spectacle of Arctic nature while they still can.

4.3 Ice, Snow and Bears

The Arctic and Antarctic snow cover is essential to the planet's climate because it "acts as a layer which modifies energy exchange between the surface and atmosphere, and as the frozen storage term in the water balance, affecting runoff and streamflow" (Derksen & LeDrew, 2000, p. 469). In *Frozen Planet*, viewers are invited to appreciate the beauty of snow-covered transformations: Viewers watch how snowflakes form as Arctic air cools humid southern air and fall to become the storage of water or, more poetically, "the lifeblood of these silent forests" ("To the Ends of the Earth", Frozen Planet, 2011–2012). A picture of the endless ice fields decorates the DVD's front cover. I argue that, as well as the many living creatures the series depicts (including, again, penguins), the water's transformations in the film play quite an active role in that glaciers can be interpreted as series' heroes.

The authors of the monograph *The White Planet* proclaim ice as an agent and indicator of climate change:

> Ice cover, through its extent and thickness, is both an agent in the climatic system and an indicator of any change: first, on a seasonal scale when the expanses of snow cover or sea ice follow the changes of the thermometer; in the longer term, on the scale of decades, when the change involves the advance or the retreat of mountain glaciers and the expansion or the reduction of their volume; and finally, on the level of centuries and millennia, with respect to the ice shelves, ice caps and ice sheets of the polar regions (Jouzel et al., 2013, p. 14).

This series generously evokes images of ice in its agent-like capacity while leaving some questions open-ended, which will be answered in the last episode – the only episode where climate change is addressed. Glaciology currently comes to the fore of public interest because of the environment's ongoing transformations. Ice floes and melt are carefully monitored; climate science and measurement focus on the polar regions because changes are the most extreme there (Scambos, 2011, p. 28). Yet in the series, the frequency that ice is mentioned paints a picture of eternity and grandeur:

> Millions of tonnes of ice grind their way downstream, driven by the unstoppable force of the meltwater. Boulders and trees are plucked from the bank side. Within just a few days, the rivers of the North are all running. The Arctic's fresh water is flowing again.

The degree to which viewers desire clarity and certainty regarding the state of the Arctic environment remains unknown. They likely welcome ambiguity because they prefer staying optimistic about climate change when documentaries send complex messages. In some parts, the series' narrative cleverly plays with ambiguity: When "vanishing ice world" is mentioned, an alarmed viewer would expect a reference to climate change, but the series addresses the seasonal, recurrent changes:

> The summer is just beginning, and the increasing power of the sun is creating a spectacular new landscape. This melting ice sheet in the Arctic is 150 miles long, with 1000 waterfalls. Vast expanses of ice that were once stone still, have come to life. The polar bear's world is melting away under the constant summer sun […]. The polar bear family must adapt to their rapidly changing home. For the two cubs, this is a whole new watery world. It's the first time they have seen the sea ice break up.

A viewer is invited to share a perspective of bear cubs. The polar bear, justifiably proclaimed as the "unquestionable king of the new non-human streamline polar imaginary" (Dahl, 2015, p. 40), looms large in the series. This figure, well-familiar to audiences from ubiquitous visual representations, is anthropomorphized to prompt viewers to empathize and, consequently, better connect to the visuals. A polar bear is shown searching for a partner, spending winter amidst polar ice, catching a seal to feed herself and her cubs and being attacked by daring ducks. Numerous facts are included in the narratives to demonstrate how polar bears are very well-adapted to life in some of the coldest areas on Earth. Viewers learn about this animal's weight, its fur's excellent qualities, and its habits and predicaments (which sometimes lead male polar bears to hunt for bear cubs). Visual and factual narrative conventions were followed to depict nature and "agents'" behaviour as straightforward and coherent. The BBC corporation is well-positioned to offer such comprehension to different segments of the audience. In any event, this corporation remains an important arbiter of judgment regarding nature documentaries with the highest circulation and popularity. To understand what changed about this genre's visual representation and narrative conventions, one must remember that since the 1960s, nature documentary makers rapidly worked out their specific narrative conventions. These conventions, according to Bagust (2008, p. 219), included (1) depicting mega-fauna, especially enormous predators, (2) visual splendour and spectacular scenery, (3) dramatic narrative – the animals are often anthropomorphized, (4) the absence of history and politics, (5) the absence of people (except occasionally tribal, pre-industrial or 'natural' people, including park rangers), (6) the absence of explicit references to the scientific method (although its application is implicit). Half the components of "classical" nature documentary texts can be found in *Frozen Planet*. Large animals – along with polar bears, whales, bison, wolves, walruses, and muskox – figure prominently. The spectacle of nature gives viewers a chance to connect to hard-to-reach places, people and animals. The intense experience of connecting with nature through spectacle prevents viewers from the more complex and problematic ways of responding to the ongoing processes threatening to radically change the life of humanity. The drama of talking about the North is undoubtedly there and achieved with the help of numerous stories about animals, many of which fall prey in the struggle for survival: Young walruses are crushed to death because of overcrowding on the coastline next to Barrow, a young bison falls behind the pack and is killed by wolves, a seal is hunted down and killed by a herd of killer whales, and so on.

At the same time, the changes in how the narrative concerning climate change is constructed can be found in this series. If the absence of history and politics was an unfortunate feature helping to tell timeless stories about nature, in *Frozen Planet*, we hear a few mentions of history. First, "human recorded history" is evoked to speak about "open water," which will soon appear in the Arctic ("To the Ends of the Earth," *Frozen Planet*; BBC One, 2011a), which the series will repeat in the final episode. Third and fourth mentions can be heard in another episode ("Spring", *Frozen Planet*; BBC One, 2011a). In this episode, a difference is drawn between Antarctica and the rest of the world: "Here, humanity's history has been very different", followed by the explanation about humans' late arrival to this continent due to its remoteness. "Polar history" is brought up regarding Antarctica when the hut – the base for Captain Scott and his team's expedition to the South Pole – is mentioned. Politics remain entirely omitted, apparently, to facilitate viewers' immersion into awe-evoking scenes. The absence of people is also fixed by showing scientists at work and including footage about local life. The series touches upon the everyday life of a group of Inuit men in Chukotka in Russia ("Spring", *Frozen Planet*; BBC One, 2011a). Ice floes are shown as part of "a dangerous maze" the nomads must travel while hunting walrus. Humans are shown busy hunting, uniting them with their animal counterparts in the series.

The shift in narrative conventions allowing one to accommodate new precarious realities of the Arctic is particularly noticeable in the last episode – the only one in which climate change is explicitly addressed. In the script of the last episode in the series, the scientists' efforts and animals' challenges are intertwined to stage an encounter with a realistic representation evoking fear, awe and the potential to produce multiple meanings. How can one respond to the prediction that a new ocean will soon open in the Arctic? The ice-covered surface, which remained in place for millennia, will recede. The optimists will surely consider the possibilities for scientific exploration. Simultaneously, the sceptics are already confident that open water will invite further exploitation of the Arctic as new routes become available. The episode generously shows vast icy spaces while locating the narration in *long duree*: The emergence of open water in this region will happen, we are told, "For the first time in human recorded history". An average viewer's cognitive resources may prove insufficient to understand the changes in question. Indeed, as Chakrabarty (2009, p. 197) warns, "The current crisis can precipitate a sense of the present that disconnects the future from the past by putting such a future beyond the grasp of historical sensibility". Similarly, geographical sensibilities must be awakened to productively respond to the series' claims about the possible repercussions that changes in the Arctic and Antarctic will cause globally. This awakening is done, I argue, through what may be named the personification of the concern about the

Northern waters, including turning climatic worries into something far more evocative than their purely scholarly rendering would suggest. Sir David Attenborough announces his mission "to understand what these changes mean, not just to the wildlife and people that live around the Poles, but for the whole planet" and proclaims that part of the series' agenda is examining the consequences that climate change has on animals. The following scene was shot in April in Svalbard where the presenter joined forces with the team of Norwegian scientists whose task was checking whether the bears were healthy: "By darting them [...] and keeping check on them year after year, that we can be sure we know what is happening to them and the population of polar bears as a whole". The presenter makes another optimistic statement, namely, that while many scientists during recent decades registered the bears' worsened conditions, "not every bear is suffering". The scientific concern about bears' weight (if a mother bear is malnourished, cubs will die) is juxtaposed with the description of the risks that ringed seals – typical bear prey – face: "It's very understandable that this mother ringed seal [...] who's looking at me now, should be a little apprehensive". A touch of humour softens a generally harsh and cruel survival game that polar animals endlessly tackle. The focus on spring's changes and animals' safety, I believe, places *Frozen Planet*'s narrative in the set of wildlife films' traditional discourses. Having considered the cultural construction of nature, Wilson (1992, p. 118) notes that many Disney movies "began at the beginning of the spring, the birth of a bear cub or otter. They end at the beginning too, with words like new life, rebirth, hope. These were old 'eternal' stories about the land". There is a sense that *Frozen Planet's* creators capitalized on these narrative conventions or, perhaps, the achievements of the earlier era. However, these creators got caught in "constricting logic" (ibid.) regarding these earlier narratives. To give just one example, wanting to please a viewer with intimate scenes of cubs being born, the team filmed bear cubs in an animal park rather than in the wild, engaging in a skillful montage of genuine scenes and staged ones. The presenter, Sir David Attenborough, supported this, saying: "But on lee-side slopes, beneath the snow, new lives are beginning" (Walker, 2011). While agreeing with the justifications the corporation and presenter provided, one can't help but think that nature documentaries should have a specific ethos. Without this ethos, the audience may feel somewhat cheated (particularly given that the scenes at the animal park were not included in the "How we made it" ten minutes of the fifth episode). Despite this embarrassing detail, the series otherwise engaged numerous scientists and adventurers and managed to put their efforts in a global spotlight.

4.4 Economies of Celebrity Endorsement

The meditated experience of the polar regions' remote natures is the only possible way to interact with their ecosystems. The uniqueness of screen representations of difficult-to-reach places is hard to underestimate. These representations become even more critical when the obvious need to raise global environmental awareness exists. This awareness should stem from educating viewers. However, international broadcasters (and BBC is surely one of the leading ones) must first entertain: "The first task of any prime-time television show is to entertain: If it founders there, it will fail any other mission" (Levenson et al., 1988, p. 40). BBC is a leader of wildlife film production (along with Discovery, Animal Planet and National Geographic). The ongoing competition among major producers creates new imperatives, namely rankings and profits. One of many paradoxes marking today's cultural production can be identified here: the contrast between the global reach of BBC filming and the selectiveness of its broadcasting. One reason that prompted this author to write this chapter has been that *Frozen Planet* is not shown in Russia. Although *Frozen Planet* is available on Netflix in the UK, Discovery in the US, and on BBC iPlayer (which only works in the UK), it cannot be streamed in my home country. I find it symptomatic that searching for the word "frozen" on Netflix generates only the following titles: *Frozen Dead* and *Frozen Hope* (both Netflix productions). Distribution networks, of course, depend on the power struggles within the global TV industry. The joint operations among the BBC and Discovery Communications started in the 1990s. While their composition is changing, it is still expanding, including a £300 m natural history streaming deal (the "largest content deal the BBC has ever done"), which will last for 10 years (BBC.com). It has been emphasized that "BBC Studios' involvement will be restricted to supplying content. It will not be a shareholder in what will strictly be Discovery's show" (Thomson, 2019). *Blue Planet* was a great success and sold to 234 territories within one year. "Soft power" agenda and creative economy go here hand in hand:

> The documentaries add to the UK's 'soft power' by projecting a compelling image of the UK as a global leader in the production of high-quality, high-impact TV and in environmental science that delivers profound real-world benefits. *Blue Planet II* became the most watched advertising-supported nature documentary in nearly eight years in the US. This same series amassed over 250 million views on China's Tencent video-on-demand service.

This praise is found on the Natural Environment Research Council's website, which boasts the financial support from three academic consultants working on *Blue Planet II*. This site claims that "[i]conic series such as *Blue Planet II, Planet Earth II* and *Frozen Planet* have delivered a dramatic wake-up call to the world on key environmental issues, from climate change to plastic pollution of the oceans" (NECR, 2021). A joint venture with Discovery put serious constraints on the BBC production. Thus, not only "spaces of production" but "networks of distribution" are affected by economic priorities of broadcasting corporations: "We do not all receive – and thus have the potential to consume – the same bundles of televised natures. Invariably, in any individual country, the greatest range of 'natures' is available to those who receive (and typically pay for) some form of multichannel television: be it analogue or digital, cable or satellite" (Christophers, 2006, p. 982). My surprise stems from realizing that viewers in the cold regions of Russia, whose natures are depicted in *Frozen Planet,* have limited or no access to this important cultural production. The confident statements of producers about this series' global impact on the environmental consciousness seem partially overshadowed by money-making priorities.

Thus, BBC wildlife programs could be criticized for pursuing primarily commercial priorities. However, they certainly manage to continue with the "hybridization" of science and entertainment by Sir David Attenborough's presence. Although Attenborough determinedly denies being a celebrity (Bonner, 2020, p. 16), he certainly is one. Recognized as "an avatar of scientific authority within natural history broadcasting" (Jeffries, 2003, p. 529), Attenborough exemplifies the British tradition of natural history film, job longevity and scientific authority. In the book somewhat ironically titled *Nature Saviours* (Huggan, 2013), Graham Huggan places Attenborough in the same category as other celebrity conservationists such as the French ocean explorer Jacques-Yves Cousteau; the American primatologist Dian Fossey; the Canadian academic, science broadcaster and environmental activist David Suzuki; and the Australian educator and host of "The Crocodile Hunter" Steve Irwin. Huggan claims Attenborough's style is the most conservative and uses advanced technologies. Regarding the intertwining of the cultural and the economic, Huggan claims that celebrity involvement in the conservationist agenda constitutes commodifying famous individuals. Indeed, in *Frozen Planet*, though announced as a presenter of the whole series, Attenborough is only seen in the first and last episodes: "That Attenborough has become our principal intermediary with the natural world is shown not as a process aided by his outlasting others, nor as only part of his career, but as an inevitability (Bonner, 2020, p. 17). In a similar vein, Van Krieken (2012) claims celebrities engage in "bundles of attention-capital". This chapter discusses how nature documentaries help raise environmental awareness and disseminate important knowledge about nature's current state. Yet Van Krieken reminds us there is "a surplus of information, data and knowledge" and that viewers' attention becomes a more important resource (Van Krieken, 2012, p. 11). Attenborough's media prominence is truly remarkable. His inclusion into the communicative circuits of BBC production and broadcasting contributes to a "harmonious" growth of ratings that benefit the corporation as well as those who run the corporation.

Continuing with the issue of commodifying nature documentary production and distribution, I turn to the phenomena of celebrity endorsement of progressive attitudes towards nature conservation and, more generally, environmental awareness. Celebrities' ability to provide documentaries with scientific authority, and the added thrill of their fame, has made it the favourite approach among producers and distributors. Thus, celebrity endorsement-based tactics and strategies have expanded significantly. Companies also capitalize on the celebrity's fan base, which social media grows. Such publicity can make celebrities more powerful and popular than ever. In the marketing literature, celebrity trustworthiness has been discussed as a strong component of celebrity effectiveness (Pornpitakpan, 2004; Tzoumaka et al., 2016). Though audiences listen to celebrities, celebrity endorsement of salient societal and political issues is ironic, as one can never be sure whose interests are at stake: "The more celebrities invest their energies in addressing salient issues, the more we can infer that their motives are not to impact society but to use public affairs in the development of their personal brands" (Atkinson & DeWitt, 2016, pp. 2–3). However, charisma, old age, and reputation in the wildlife documentary field make Attenborough's impact and contribution quite persuasive. However, completely identifying with this narrator and presenter – a living legend – might be difficult for audiences. Yet, the personification of the story *Frozen Planet* tells will likely transport audiences, producing awe and admiration for those who spent months filming in these rugged regions.

Scholars justifiably introduce, albeit in the post-humanitarianism context, the concept of the performativity of celebrity. This concept is understood as a distinct form of textuality – marked by ambivalence and combining 'impersonation' (the celebrity's personal testimony) with 'personification' (the celebrity's charisma) (Chouliaraki, 2011, p. 366). Attenborough is often seen and read as justifying corporate decisions (such as the footage of bear cubs born in artificial settings). Celebrity conservationists may not only harness the company's brand visibility but, as a frontline communicator, defend it against public criticism. For Attenborough, these decisions make him the target of criticism, including comments about having long ignored the gravity of environmental issues and preferring awe to concern (Barkham, 2021; Hook, 2018).

One of the series' most remarkable features is the trope of "team" actively engaged in its narrative. The simplest explanation is that while Attenborough does all talking, the stunning footage was obtained by an impressive crew, including alpinists-cameramen and experts in glaciers. Not only do we hear that "the greatest challenge for the team making *Frozen Planet* was the extreme remoteness of their locations" ("To the Ends of the Earth", *Frozen Planet*, BBC One, 2011a), but we are informed this team is astonishingly well-equipped: "The team had to use a ship capable of breaking through the pack ice around the 2000-mile-long coastline of Svalbard ("Summer", *Frozen Planet*, BBC One, 2011a). The team members' skills and daringness are acknowledged: "The team is used to handling multiple cameras, but they don't usually have to dodge ten-tonne ice floes at the same time" ("On Thin Ice", *Frozen Planet*, BBC One, 2011a). Diversification of the team's tasks is also part of the series' narrative: There is a cave team, underwater team, aerial team, and so forth. Most importantly, the contributions of the presenter and the rest of the team are seamlessly combined:

> The increasing unpredictability of the ice was a big issue for the *Frozen Planet* team, who spent three years working on top of it. Whether on sea, land, lake or river, the state of the ice was the first concern for most filming crews. Unexpected break-ups left many a cameraman in need of a swift rescue. Sometimes, help came by boat, and sometimes by air. I had a chance to see the changing ice conditions for myself, when I visited the North Pole. I flew with the team to a temporary camp that is set up every year in the centre of the frozen Arctic Ocean to support expeditions to the Pole. I had never visited the North Pole before, so this was a great highlight for me. But it was hard going in temperatures of minus 40, so as soon as filming finished, we flew south. Little did we know that we had made it out just in time" ("On Thin Ice," *Frozen Planet*, BBC One, 2011a).

The traces of the accounts about the first polar expeditions can be felt in these adventurous discourses and tropes that continue positioning the BBC's exploits in the world's remote corners. The BBC films this frontier that must be protected from climate change and documented, so the footage can be shown to global audiences.

4.5 Conclusion

I have analyzed the aesthetic-political and political-economic implications of the well-known BBC series to demonstrate the mechanisms governing the well-honed tropes of the corporation's leadership in producing top content in the wildlife documentary field. Identifiable actors within these tropes (the legendary front men, polar regions, ice, snow and polar bears, the locals, and the violent clashes of large animals over scarce resources) are depicted with more detail, vividness, and scrutiny than the processes and practices contributing to the immense ice loss. This clash between artistic and commercial needs and scientific, educational, and environmental priorities characterizes the history of wildlife documentaries. These documentaries are meant to represent nature's reality but can only do so by relying on sophisticated artistry and technology. The fact that climate change results in the region rapid change prove critical for the planet, engaging viewers in the dangerous magic of these impossible-to-reach places. However, the issues of the balance among art, technology and reality aside, current viewers need not only be provided with a stunning visual escape but prompted to reflect on the serious and even catastrophic state of the environment. This awareness has been accomplished, I argue, by demonstrating that broadcasters and scientists are keen to examine the degree to which dramatic changes are happening in the Arctic and Antarctic. The popular ways of doing wildlife cinematography – emphasizing the sensational and spectacular coexisting, with novel ways of depicting the environment – are designed to create a sense of responsibility and concern about the planet's future.

I have outlined how Sir David Attenborough and the creators of *Frozen Planet* use well-honed tropes and narrative conventions to sustain their leadership in the wildlife documentary field. Engaging the topic of the Arctic and the global significance of its climate changes create the impression that the BBC is attuned to the global climate agenda. I argue that such may be connected to the corporation's communication strategies. I have traced the links among *Frozen Planet* and earlier programs to show that their production and dissemination continues a well-established tradition of taking viewers to far-flung corners of the world and their nonhuman inhabitants. Attenborough's scientific and TV authority is shown as embedded into the economy and interests of broadcasters, explaining the series' optimistic outlook.

The series, capitalizing on the popularity of the theme of environmental crisis, continued with the aesthetic and broadcasting "feel-good" practices of giving viewers a chance to hope things will be fine while none of the production (including TV series production) and consumption patterns underwent a substantive change. The viewers remain the *subjects* to whom the wonders of the natural world are conveniently delivered as stunning imagery and mostly comforting comments. Mastery over the natural world is reproduced as even the most remote Northern corners are artistically appropriated for corporate gain. The alarming statements about melting Arctic ice and its threat to the planet coexist with the structures and processes generating climate change.

References

Documentaries

BBC One. (2011a). *Frozen planet.* [Viewed 25 January 2022]. Available from: https://www.bbc.co.uk/programmes/b00mfl7n

BBC One. (2011b). *On location.* [Viewed 16 April 2021]. *BBC.* Available from: https://www.bbc.co.uk/programmes/profiles/5j6MfkptvwZ2HpSF676NLJS/on-location

Frozen planet. (2011–2012). *TV-PG|5h 33 min|documentary|TV mini-series (2011–2012) BBC one television*, October 26.

Literature

Atkinson, M., & DeWitt, D. (2016). Celebrity political endorsements matter. *Celebrity Studies, 7*, 119–121. https://doi.org/10.1080/19392397.2016.1131014

Bagust, P. (2008). 'Screen natures': Special effects and edutainment in 'new' hybrid wildlife documentary. *Continuum: Journal of Media & Cultural Studies, 22*(2), 213–226. https://doi.org/10.1080/10304310701861564

Barbas, T. A., Paraskevopoulos, S., & Stamou, A. G. (2009). The effect of nature documentaries on students' environmental sensitivity: A case study. *Learning, Media and Technology, 34*(1), 61–69. https://doi.org/10.1080/17439880902759943

Barkham, D. (2021). The real David Attenborough. *Guardian*, October 22. [viewed 16 April 2021]. Available from: https://www.theguardian.com/tv-and-radio/2019/oct/22/david-attenborough-climate-change-bbc

Baumeister, R. F., Bratslavsky, E., Finkenauer, C., & Vohs, K. D. (2001). Bad is stronger than good. *Review of General Psychology, 5*, 323–370. https://doi.org/10.1037/1089-2680.5.4.323

Bonner, F. (2020). A circumscribed (natural) history: British television celebrates David Attenborough. *Historical Journal of Film, Radio and Television, 40*(4), 875–894. https://doi.org/10.1080/01439685.2020.1730561

Bourgeon, S. (2018). The invisible threat of Arctic pollution. In P. Wassmann (Ed.), *At the edge* (pp. 277–284). Orkana forlag.

Bouse, D. (1995). True life fantasies: Storytelling traditions in animated features and wildlife films. *Animation Journal, 3*(2), 19–39.

Bousé, D. (2000). *Wildlife films*. University of Pennsylvania Press.

Brady-Brown, A. (2021). *David Attenborough documentary the year earth changed suggests silver lining ahead of earth day 2021.* A ABC arts/BC news politics. [viewed 16 April 2021]. Available from: https://www.abc.net.au/news/2021-04-16/the-year-earth-changed-review-david-attenborough-documentary/100062704

Chakrabarty, D. (2009). The climate of history: Four theses. *Critical Inquiry, 35*, 197–222. https://doi.org/10.1086/596640

Chapman, A., Corner, A., Webster, R., & Markowitz, E. (2016). Climate visuals: A mixed methods investigation of public perceptions of climate images in three countries. *Global Environmental Change, 41*, 172–182. https://doi.org/10.1016/j.gloenvcha.2016.10.003

Chris, C. (2006). *Watching wildlife*. University of Minnesota Press.

Christophers, B. (2006). Visions of nature, spaces of empire: Framing natural history programmes within geometries of power. *Geoforum, 37*, 973–985. https://doi.org/10.1016/j.geoforum.2006.03.005

Chouliaraki, L. (2011). The theatricality of humanitarianism: A critique of celebrity advocacy. *Communication and Critical/Cultural Studies, 9*, 1–21.

Dahl, J. M. I. (2015). Assessments, models and international politics of the Arctic: Why the "new north" narrative includes only bomber, polar bear, oil, and gas deposit models, and no original parts or an assembly manual. *The Polar Journal, 5*(1), 35–58. https://doi.org/10.1080/2154896X.2015.1025491

Derksen, C., & LeDrew, E. (2000). Variability and change in terrestrial snow cover: Data acquisition and links to the atmosphere. *Progress in Physical Geography, 24*(4), 469–498. https://doi.org/10.1177/030913330002400401

Estes, Z., & Adelman, J. S. (2008). Automatic vigilance for negative words in lexical decision and naming: Comment on Larsen, Mercer, and Balota (2006). *Emotion, 8*, 441–444. disc. 445–457. https://doi.org/10.1037/1528-3542.8.4.441

Fortner, R. (1985). Relative effectiveness of classroom and documentary film presentations on marine mammals. *Journal of Research in Science Teaching, 21*(2), 115–125.

Hartje, L. (2020). OU academic named in the Queen's new Year's Honours. *OU news* [online]. December 31 [Viewed 5 April 2021]. Available from: https://ounews.co/science-mct/ou-academic-named-in-the-queens-news-years-honours/

Hook, L. (2018). David Attenborough, nature's defender comes under attack. *Financial times*, November 9. [viewed 5 April 2021]. Available from: https://www.ft.com/content/1e11b36c-e37e-11e8-8e70-5e22a430c1ad

Huggan, G. (2013). *Nature's saviours: Celebrity conservationists in the television age*. Routledge.

Jeffries, M. (2003). BBC natural history versus science paradigms. *Science as Culture, 12*(4), 527–545. https://doi.org/10.1080/0950543032000150346

Jouzel, J., Lorius, C., & Raynaud, D. (2013). *The white planet: The evolution and future of our frozen world*. Princeton University Press.

Kaganovsky, L., MacKenzie, S., & Westerstahl Stenport, A. (2019). Introduction: The documentary ethos and the Arctic. In L. Kaganovsky, S. MacKenzie, & A. Westerstahl Stenport (Eds.), *Arctic cinemas and the documentary ethos* (pp. 1–28). Indiana University Press.

Kantor, B. J. (2020). Performing 'the authoritative account': How the BBC's *horizon* produces epistemic authority. *Social Studies of Science, 1–21*. https://doi.org/10.1177/0306312720947181

Levenson, T., Grant, W. R., & Apsell, P. (1988). Nova, television, and informal science education. In M. Druger (Ed.), *Science for the fun of it: A guide to informal science education* (pp. 40–44). National Science Teachers Association.

Mitman, G. (1999). *Reel nature America's romance with wildlife on film*. Harvard University Press.

NECR. (2021). Turning trailblazing science into agenda-setting TV. *Natural Environment Research Council*. [viewed 5 April 2021]. Available from: https://nerc.ukri.org/research/impact/casestudies/society/tv/

O'Neill, S., & Nicholson-Cole, S. (2009). "Fear won't do it" promoting positive engagement with climate change through visual and iconic representations. *Science Communication, 30*(3), 355–379. https://doi.org/10.1177/1075547008329201

Perrott, L. (2002). Rethinking the documentary audience: Reimagining the New Zealand wars. *Media International Australia, 104*(1), 67–79. https://doi.org/10.1177/1329878X0210400109

Pornpitakpan, C. (2004). The effect of celebrity endorsers' perceived credibility on product purchase intention. *Journal of International Consumer Marketing, 16*(2), 55–74. https://doi.org/10.1300/J046v16n02_04

Scambos, T. (2011). Earth's ice: Sea level, climate, and our future commitment. *Bulletin of the Atomic Scientists, 67*(1), 28–40. https://doi.org/10.1177/0096340210392965

Silverstone, R. (1999). *Why study the media?* SAGE.

Stephen, L. (2010). 'At last, the family is together': Reproductive futurism in *march of the penguins*. *Social Identities, 16*(1), 103–118. https://doi.org/10.1080/13504630903465944

The Open University. (2016). *The frozen planet*. Kindle Edition. Available at http://www.amazon.com/kindlestore (Downloaded: June 17, 2020). [Viewed 5 April 2021]. Available from: https://www.amazon.com/frozen-planet-Open-University-ebook/dp/B01D8X6ON2/ref=sr_1_2?dchild=1&keywords=Frozen+Planet&qid=1617778706&s=digital-text&sr=1-2

Thomson, S. (2019). The BBC and discovery: Getting back together. *Digital TV Europe*, 5 April. [Viewed 5 April 2021]. Available from: https://www.digitaltveurope.com/comment/the-bbc-and-discovery-getting-back-together/

Tzoumaka, E., Tsiotsou, R. H., & Siomkos, G. (2016). Delineating the role of endorser's perceived qualities and consumer characteristics on celebrity endorsement effectiveness. *Journal of Marketing Communications, 22*(3), 307–326. https://doi.org/10.1080/13527266.2014.894931

Van Krieken, R. (2012). *Celebrity society*. Routledge.

Walker, P. (2011). *Frozen planet*'s polar bear footage was standard practice, claims BBC. *The Guardian*, December 12. [Viewed 5 April 2021]. Available from: https://www.theguardian.com/media/2011/dec/12/frozen-planet-polar-bear-bbc

Wilson, A. (1992). The culture of nature: North American landscape from Disney to the Exxon Valdez. .

Part II
Hydrological Space and Politics

Chapter 5
The Voice of Ice in the Turku Archipelago: Narrating Icegraphy with Environmental Ethnography

Jaana Kouri and Kirsi Sonck-Rautio

Abstract Winters in the Turku archipelago have become milder. During the last few decades, inhabitants have witnessed a relatively sudden loss of permanent ice cover during the winter months. In our chapter, we use ethnographic tools to conceptualise local skills, experience-based knowledge and tacit knowledge concerning ice as an environmental heritage – and explore the role of nostalgic narration in the process of environmental heritage, including an historical perspective, in times of change. We examine nostalgic narration as future-oriented action in the production of environmental heritage. This chapter builds on prior ethnographic fieldwork done by the authors in the Turku archipelago. Our methodological starting point is to assess how environmental ethnography is produced in intra-actions between human and non-human actors, such as ice. We examine the myriad ways in which archipelago dwellers have defined ice and the presence of ice in different local places. We also describe how people in the archipelago have encountered ice over numerous decades and how they have learned to utilise and co-exist with ice – how they, for example, found paths when walking on the ice, how they allegorised drowning in their nostalgic narration practices or how ice conditions have impacted fishing practices in the archipelago.

Keywords Ice · Icegraphy · Aquagraphy · Environment · Ethnography · Nostalgia · Narration · Environmental heritage · Archipelago · The Baltic Sea

5.1 Introduction

The Turku archipelago, consisting of approximately 20,000 islands, is part of the Baltic Sea, more specifically the Archipelago Sea area. The Baltic Sea contains brackish water and is connected to the Atlantic Ocean through the Danish straits. As

J. Kouri (✉) · K. Sonck-Rautio
Åbo Akademi University, Turku, Finland
e-mail: jaana.kouri@abo.fi; kirsi.sonck-rautio@abo.fi

M. Lehtimäki et al. (eds.), *Cold Waters*, Springer Polar Sciences,
https://doi.org/10.1007/978-3-031-10149-6_5

the Archipelago Sea is in the Northern Hemisphere, much of it tends to freeze in the winter. However, the effects of global warming can be detected in the Baltic Sea area, and the duration and the thickness of the ice cover, as well as its extent, have notably decreased during the last few decades (e.g. Lépy, 2012; Korpinen et al., 2019). The inhabitants of the Archipelago area have both witnessed and experienced the diminishing ice and how the environment and landscape in which they are living has changed, perhaps even drastically.

In this article, we discuss the ways in which archipelago dwellers have intra-acted with ice and how this intra-action has formed over time. We ask what kinds of specific environmental knowledge these intra-actions have produced and how this knowledge has been transformed into *icegraphy* through local narratives. In addition, we explore how local knowledge can become environmental heritage, which, in turn, could be applied when adapting to changing environmental conditions. From a methodological standpoint, we apply environmental ethnography, focusing especially on the intra-actions between human and icy landscapes. We build on our studies based on ethnographic fieldwork in the Turku archipelago in southwestern Finland (see Fig. 5.1; Kouri, 2017; Sonck-Rautio, 2019). Our data consist of ethnographic materials gathered primarily from the communities of Lypyrtti and Rymättylä, both of which have strong traditions and livelihoods long connected to the sea, such as fishing and piloting. Cold winters and the freezing Archipelago Sea

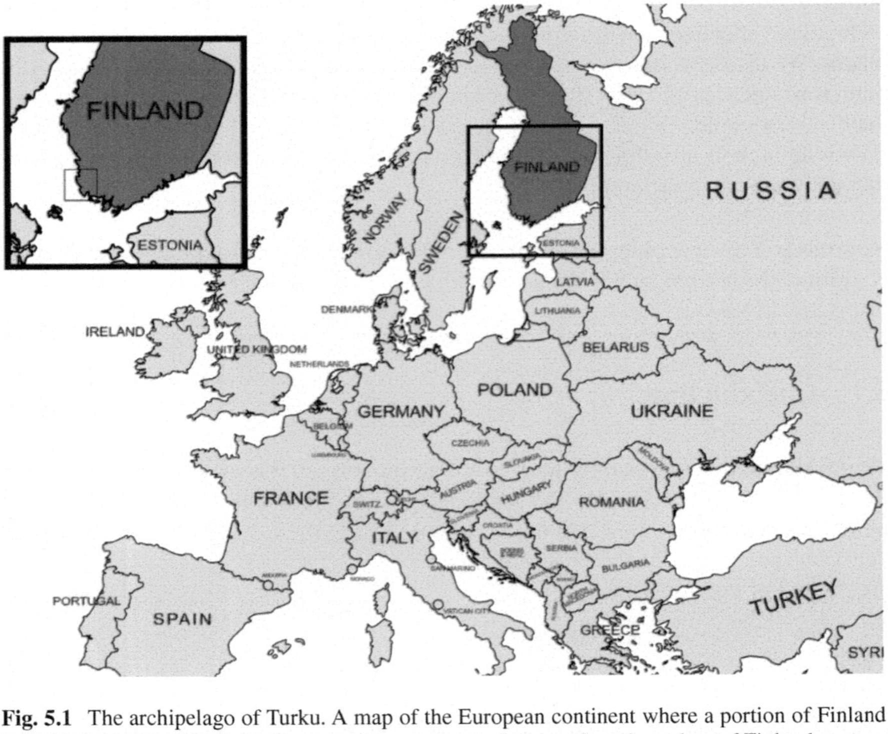

Fig. 5.1 The archipelago of Turku. A map of the European continent where a portion of Finland is highlighted. The highlighted part focuses on a part of the southern boundary of Finland

created a landscape where several species, including humans, have interacted and resided. Ice proved to be an essential element because it enabled, e.g. seal hunting, but also winter fishing. During the winter months, when the archipelago was covered with a thick and vast ice cover, archipelago dwellers practiced winter-seining – a form of fishing where people harvested schools of Baltic herring under the ice cover through holes in the ice with enormous seines. Winter-seining required a large work force, as the catch could often weigh many tonnes. Winter -seining was a communal form of fishing, and it was practiced at least from the fourteenth century onwards. Winter-seining had, however, lost its importance by the 1990s due to such global trends as urbanisation, globalisation and increasing competition (e.g. Sonck-Rautio, 2018, 2020).

Besides fishing, the sea provided its coastal dwellers with other livelihoods as well, such as seafaring and piloting. In the archipelago, people often made an adequate living by combining several types of work, such as piloting, small-scale farming and husbandry, and hunting and fishing, according to the rhythms of the seasons. In previous centuries, local peasants with detailed knowledge of the waterways served as pilots. In Lypyrtti village, piloting took on more importance in the beginning of the nineteenth century (Kouri, 2017). Closer to the present, foreign ships are been required to employ pilots in Finnish territorial waters, as the Archipelago with its narrows and rocks is a quite challenging area to navigate. Nevertheless, in 1961 the Finnish National Board of Navigation abolished the pilotage system in Ströömi, a narrow straight that had traditionally been an important passage through the islands, and it also closed the Lypyrtti pilot station, thus requiring ships carrying larger cargoes to seek deeper waterways for safety reasons. Consequently, the village lost work and became almost solely a summer resort area for the families that had to leave to find new employment and for other people who came between the 1920s and 1970s (Kouri, 2017). The local Ströömi waterway has had a fundamental impact on village life. Ströömi contains narrow passages that affect the formation of strong currents, which affect when, how and where the water freezes and when it is possible to cross Ströömi during the winter period (Fig. 5.2).

The ethnographic material utilised here consists of interviews, participant observation, field journals and auto-ethnography. The material was gathered in the years 2006–2008 and between 2015 and 2016. Altogether, we conducted 69 semi-structured in-depth interviews, where the participants were given an opportunity to freely address the issues they felt meaningful for them. The producing of ethnographic material was a reciprocal signification process in many ways; it was an act of interaction and dialogue between the locals, the environment and the researchers (Kouri, 2016). The interviews revealed that the participants most often wanted to reminisce about the changing circumstances involving ice and winters. During the ten-year span of our fieldwork, we observed clear changes in the water environment. Local and global problems intersect and become intertwined in such water bodies. Changing conditions – the eutrophication of water, the milder winters and the concomitant loss of some primary local livelihoods, such as piloting in Lypyrtti village or winter-seining in Rymättylä – caused the locals to seek continuity with their past way of life. Piloting and winter-seining are both livelihoods entailing a special

Fig. 5.2 Lypertö island, in Lypyrtti village. Thin ice that does not break apart due to the powerful waves is called "calf skin". (Photo: Jaana Kouri)

relationship with the archipelago's water environment. People recollected how the preceding generations had managed under different circumstances, i.e. "practiced the place" (de Certeau, 1984). They recalled, for example, how past generations had observed nature, used rowboats or walked on the ice during the wintertime. Even though the current summer residents of Lypyrtti have roots in the archipelago, they rarely live in the village year round, and thus, they do not experience the cycle of the seasons in the archipelago. Similarly, the old winter seiners in Rymättylä who used to have skills and knowledge regarding this traditional livelihood are ageing, with many of them already deceased. People are afraid of not only losing the ice or the clear water, but also of losing their knowledge of how to adapt to the environment. Such concerns also had an impact on what and how the participants in the research study narrated the environment and their ways of acting in and with nature.

In this chapter, we first introduce the concept of environmental ethnography, especially how we view ecology, intra-action, environmental heritage and local ecological knowledge in relation to this method. Next, we discuss nostalgic narration and its contribution to the environmental heritage process. The following section is dedicated to icegraphy – namely, how the locals narrated ice and how it was present in their everyday lives and their landscapes. Finally, we conclude by focusing on how icegraphy, when contextualised with environmental ethnography, can contribute to understandings of how people adapt to and how scholars study future changes in local landscapes.

5.2 Environmental Ethnography

Our methodological starting point involved exploring how environmental ethnography is produced in intra-action and encounters between human and non-human actors, such as ice. Intra-action as a term is used to replace interaction and signifies that the entangled agencies are mutually constituted and that such agencies do in fact emerge through intra-action rather than preceding it (Barad, 2007, p. 33). For us, these intra-actions and encounters occur within landscapes (see Haraway, 2008; Aisher & Damaradan, 2016), which in our context is the wintry coastal areas and islands of the Archipelago Sea. We employed environmental ethnography to understand the ways in which archipelago communities can adapt to changing environments. In this section, we focus on the part played by water, especially ice, in the making and unmaking of fluid worlds, or waterworlds (see Hastrup & Hastrup, 2017). We examine the human environmental relationship, the *connection* to one's landscape, the main environmental actors, their voice in ethnographic studies and the material co-produced by them. We apply the idea of 'polyphony' in ethnographic fieldwork – the many different 'voices' present in the actual discussions and dialogues through which ethnographic understandings are constructed (Gupta & Ferguson, 1977, pp. 2–3). We are especially interested in the voice of ice in narrations. Ice as a non-human actor is in many ways entangled in the lives of human beings – and all beings – and in the nostalgic narration about disappearing wintry ice landscapes and living with them. We examine the myriad ways in which archipelago dwellers have experienced ice and how ice has impacted and shaped different local places. We also describe how people in the archipelago have encountered ice over many decades and how they have learned to utilise and co-exist with ice.

The longing for a happy place outlined in another time has been considered characteristic of modern times, when it has been possible to think that things could be different and that the current situation is not the only possible one. Nostalgia is, above all, a modern concept (Grönholm & Paalumäki, 2015; Vihma, 2021). Hence, we examine nostalgic narration as a future-oriented act of environmental heritage production, one that advocates for adapting to environmental and climatic changes both now and in the future. The concept of heritage refers to forms of action, modes of doing and ways of being engaged in the world (Guttormsen & Swensen, 2016, p. 5). Heritage is dynamic: it is constantly evolving, based on what people perceive to be meaningful in the past and present and worth preserving, or nurturing, for the future. At the same time, it is a performance in which certain cultural and social meanings and values are identified, reaffirmed or rejected (Smith, 2011). The concept of environmental heritage refers to the process of acquiring the experience, local ecological knowledge, experience-based knowledge, tacit knowledge and skills constructed through living and acting in a particular environment in a particular point of time.

We are interested in the meanings that emerge through narration, in what sociologist Bruno Latour calls "matters of concern". Latour emphasises that this allows us to renew from top to bottom the very scene of empiricism – and hence the division between "natural" and "social" (Latour, 2005, p. 114). Similarly, we do not draw a line between ecological relationships and networks of meaning-making – or nature and culture – but instead see that meanings arise in practical actions, which is essential to experience-based and tacit knowledge. We explore Ingold's idea of a "dwelling perspective": humans are brought into existence as organism-persons within a world, an environment or a lifeworld that is inhabited by manifold kinds of beings, both human and non-human. Therefore, relations among humans, which we are accustomed to call social, are but a subset of ecological relations (Ingold, 2000, pp. 5 & 153).

By paying attention to cultural dimensions of the social and claiming agency for human and nonhuman agents, we want to stress that ecology as a term refers to the totality of relations among human agents, nonhuman agents and their environment, as was formulated by ethnographer Phillip Vannini (2009b, p. 73). In a similar vein, multispecies ethnographers have used the Deleuzian term "assemblage" (Deleuze & Guattari, 1980), or "entanglements" (Kirksey & Helmreich, 2010), to highlight that ecology is not a structure but an ongoing process. Perspectives need to be sensitive to change, adaptation, integration, reintegration and disintegration, and rather than focusing on agency alone, it is worth turning to creative ways in which humans and non-humans relate to one another. Understood in this way, agency is not limited just to something that a human being has, but rather should be viewed as the diffused potential for action present in a particular social and material setting. In this sense, to speak of diffused agency is also to invoke an ecology of interaction from the level of practice to the level of conceptualization (Vannini, 2009b). What is central to such a view is an understanding of sociality and culture as a form of making, doing and acting as well as an understanding of sociality through pragmatic, sensuous intentionality and interactions between human actors and non-human actors (Vannini 2009a, p. 4). Nature and culture should not be thought of as distinct from each other. We see them, like the philosopher Karen Barad (2003), as a single holistic world in which nature and humans act together in the same process of meaning-making.

5.2.1 Producing Local Ecological Knowledge

Making sense of meaning-making in socio-ecological processes implies paying attention to knowledge as communication – including with non-human actors. It emphasises the importance of "knowing with" and "knowing how", complementary to "knowing that", which involves skills that are difficult to put into words. Such cultural knowledge connects global environmental changes, such as climate change, with local changes, since this type of knowledge is constructed in connection with tangible changes in nature, including the prevalence of extreme climatic phenomena, changes in flora and fauna, and seasonal changes. The construction of heritage is thus a material-discursive process. In environmental heritage, the different ways of knowing – the "know what", "know why", "know how", "know who" and the "know with" – are all interconnected with each other (see, e.g. Polanyi, 1966;

Polanyi & Prosch, 1975). "Knowing with" the environment draws on the effort to overcome the conventional dichotomy between naturally given and culturally constructed worlds. For example, a particular landscape becomes part of us through living in it. It is through this relational and compositional context of people's engagement with the world that each place draws its unique significance (Phillips, 2005).

Ethnography especially throws into relief intangible cultural heritage, which represents the living actions, skills and thoughts present in and between individuals. It is a form of activity that is both purposeful and unconscious of its end goal and that contains the power to change the world (Siivonen, 2019). In ethnography, textualisation, particularly a researcher's writing about and in the field, has been at the centre of research for decades (see, e.g. Kouri, 2017). A good example here is anthropologist Clifford Geertz's (1973/1993) idea of culture as a kind of text, with the ethnographer's aim being to write a "thick description" of it. Ethnographer Karen McCarthy Brown (1991/2001, p. 14) notes that a researcher should strike a balance between the tradition of interpreting one – the researcher's – meaning-making network and another, aesthetically different meaning-making network among interviewees in the field. With an emphasis on co-producing meanings, we want to highlight the voice of non-human actors through use of the concept environmental ethnography. We use the concept here to describe that kind of ethnography that specifically examines the ways socio-ecological processes become intertwined, informing the environments and experiences of those environments in the past, present and in the future.

Environmental ethnography is both a method and a methodical attitude that makes it possible to examine the ethnographic heritage process of meaning-making networks. We take seriously in this account, first, different kinds of knowledges, and second, all the actors that participated in the study, whether through ethnographical interviews or researcher observation. We examine how various actors mediated information in the ethnographical process and influenced one another, thereby extending conceptualisations of agency from humans to nonhuman actors. By doing that, we examine in detail their ecological relations, e.g. their actions and co-actors in the environment (see Ingold, 2000).

Local ecological knowledge is often defined as tacit and experience-based knowledge that is acquired by spending a significant amount of time in a specific environment, and thus, observing the surroundings in detail. This knowledge is predominantly transmitted orally and through everyday practices (e.g. Cruikshank, 2005, p. 9). Local ecological knowledge is often also referred to as traditional ecological knowledge, and although the knowledge we describe here could be regarded as traditional, the holders of this knowledge can very well have knowledge that is mixed with scientific knowledge they acquired through the education system. This means the knowledge may not only be traditional knowledge. Therefore, we find local ecological knowledge to be the most precise term to apply here. Since local ecological knowledge is built upon local observations of the environment – even when the changes are global – it also offers a vast contextual way of comprehending the landscape and interconnections and relationships within the environment.

5.3 Nostalgic Narration

When people talk about dwelling and living in the landscape, it is a narration of their "spatial practice", as Michel De Certeau (1984) has termed it. Humanistic researchers have viewed landscape as a fundamental reference system in which individual consciousness of the world and social identities are anchored (Tilley, 1994, p. 40; see also Tilley & Cameron-Daum, 2017). We depart here from Ingold's (2000) idea that meaning is intrinsic to the contexts in which people pragmatically engage with the world. He basically rejects the divisions between inner and outer worlds – respectively, of mind and matter, meaning and substance. For him, the focus is the familiar domain of our dwelling and how, through living in it, the landscape becomes a part of us, just as we are a part of it. It is through this relational and compositional context of people's engagement with the world that each place draws its unique significance.

In this section, we explore the voice of ice in narration. Ice is an actor in the oral history of those dwelling in the archipelago; it helps highlight the functional aspect of environmental relations as epitomised in stories and performances. Although ice is one of the states of water, in the archipelago the surface water areas are generally not in a total liquid state or in the form of solid ice; rather, surface water constantly varies between these states. Ice as a non-human actor participates and has participated in the lives of those dwelling in the archipelago in both the present and past, and it features in the contemporary narration of disappearing wintry ice landscapes and living with such changes by villagers.

The latest studies in oral history have pointed to the importance of expressions of emotion during the narration (e.g. Tuomaala, 2006; see also Boym, 2001; Grönholm & Paalumäki, 2015). Emotions can be interpreted as signalling what the narrators consider meaningful. Emotions manifested themselves in many of our interviews included anxiety or nostalgia over losing one's livelihood and especially the way of life connected to the watery environment. This has occurred due to various global and local trends, or – as was the case in Lypyrtti – due to the cessation of piloting, which meant most of the locals could no longer live year round in the village. As Svetlana Boym (2001, p. xv) writes, at first glance it looks like nostalgia is a longing for a place, but actually it is a yearning for a different time. Fred Davis has concluded that nostalgia is both an emotion and a form of consciousness, basically a cognitive process. When people understand that they are engaging in nostalgia, they have moved from the level of emotion to the level of verbalisation, to recognising the relevance of their moods in relation to a changed reality (Davis, 1979, pp. 16–29, 122–124). Without taking a position on the nature or levels of the process, we emphasise the role of nostalgic narrative in ethnographic research. Winters characterised by ice became emblematic of a time no longer the present, but a time reminisced about, a nostalgic time of the past. The nostalgic narration of the interviewees offers ways to participate in negotiating possible wintry futures.

Nostalgic narration about the past takes place when a lifestyle or practice important to people's identity is threatened. It occurs when the lifestyle or practice is on the threshold of being lost but is still barely here (Korkiakangas, 2006, p. 140). Ethnographic texts of the twentieth century have been criticised for constructing an eternal ethnographic present, for creating an unchanging image of culture that became solidified in the past, when cultures are in reality in a constant state of change. Such texts also promoted the concept of linear time, viewing those cultures as separate, unchanged islets to which there ultimately was no hope for return. The loss of traditional cultures is rooted as allegory in the ethnographic way of writing, but also in the way of reading it. While cultures do disappear, the researchers should be careful about promoting the idea of disappearance in their texts. Similarly, scientists are warned of the idea of saving culture contained in the pastoral, a narrative form that emphasises the rural idyll built into the process of textualisation. However, it becomes a school of thought when such writing or textualisation is understood to depict a collective culture rather than a produced dialogue, as is the case with oral history (Clifford, 1986, pp. 111–113). Our ethnographic study is an example of oral history.

Nostalgia is not always about the past. It can be retrospective but also prospective. Fantasies of the past determined by needs of the present have a direct impact on the realities of the future (Boym, 2001, xvi). Boym talks about restorative nostalgia and reflective nostalgia. In our research, the nostalgia expressed by our interviewees was of the second type. According to Boym, reflective nostalgia cherishes shattered fragments of memory and temporalises space. It can be ironic and humorous and reveal that longing and critical thinking are not opposed to one another, as affective memories do not absolve one from compassion, judgment or critical reflection (2001, pp. 49–50).

5.4 Icegraphy – Living with Ice

Icegraphy is a subcategory of aquagraphy. Each state of water introduces its own *graphy* to the discussion. The voice(s) of ice can be heard in many ways. According to the Oxford English Dictionary (OED), the English suffix *-graphy* comes from the Greek *graphia*, writing. Some words with this ending denote processes or styles of writing, drawing or graphic representation, but more commonly they are names of descriptive sciences. Icegraphy incorporates all aspects. It is a type of art and science of ice as well as a process or method of writing or narrating about/with ice. More specifically, introduce here the term *icegraphy* to depict how ice writes its myriad stories into the landscape, and reciprocally, how humans read ice through interacting with it. Icegraphy has a dual purpose in this study; it also describes the way we apply environmental ethnography to re-read and re-interpret our ethnographic materials, i.e. interviews, observations and research diaries from the viewpoint of ice. The icegraphy of nostalgic narration included different themes based on the types of action they described: sensing and reading ice when walking on ice, finding different routes over the ice or falling through the ice, crossing molten or open waters between ice zones, or drilling through the ice for winter fishing. It is good to remember that one's relationship to the environment is a historical phenomenon: it changes according to the environment, the place and the time. At the same time, different people may have a different relationship with the same environment and its specific actors, such as ice.

5.4.1 Sensing and Reading the Ice

Many participants talked about "reading the ice" (*jään lukeminen* in Finnish) and "sensing the ice" (*jään kokeminen* in Finnish). The reading of ice not only involves watching; it is a multisensory act in the present moment. The interviewees talked about sensing and experiencing the ice through, for example, testing the hardness of the ice with a special kind of stick with a sharp metal end. At the same time, they spoke of the need to be conscious of the water level during the day and of being concerned about the shape of the bottom of the water place. They stressed how the environment changes every day, and that the seasons are never the same. The strength of the ice may change throughout the day: ice is usually stronger in the morning, after the freezing night, and weakest in the afternoon. Ice might, for example, be quite durable at sunrise in springtime, but in the afternoon only be deceptively hard under the influence of heat. The stream "eats" ice from underneath. There is no single or regular circumstance in which, by just following the ice, one could find the most durable route each time or always at the same point. It is important to always search for the appropriate path by sensing and testing the ice (Kouri, 2017). Snowfall has an impact as well – the ice might not thicken under a thick layer of snow. To find the safest and thickest ice, the new snow should be cleared away with wooden sticks, and only the next morning – given that the temperature dropped below freezing at night – the ice will be safer to walk on (TYKL/aud/1267). Here, local experience-based knowledge, knowing how and knowing with are indispensable. Intangible environmental heritage changes when respective environmental conditions change. One must be able to forecast changes in the weather and know their effects. Observing changes in the environment and adapting to them is an important aspect of the environmental heritage of those living in the archipelago.

During the winter months, ice cover can be more than 50 cm thick. Schools of Baltic herring, one of the key species in the Baltic Sea, swim under this ice cover. Harvesting the herring with nets for domestic use was a common practice for centuries, with those living in Rymättylä having practiced winter-seining commercially from the 1880s onwards. Winter-seining requires strong knowledge of the environment, knowledge acquired since childhood. 'Sensing the ice' became the lifeblood of many living in the archipelago not only because they could use the ice to travel from island to island during the wintertime, but also because winter-seining constituted an important source of income during the rough freezing winter months from at least the fifteenth century until the end of the twentieth century.

Ice dictated the rhythm of the winter. Not only did ice play a major role in subsistence, but it also influenced people's leisure-time activities and celebrations. In this way, the ice, the herring and the water all had significant agency in shaping the social and cultural surroundings for the fishing community, and the people's lives were very much entangled with these non-human elements and species. For example, every year the ice disappeared, the people then gathered and stored the seines and organised a big end of season feast (*peijaiset*). They also organised a similar type of feast before the winter-seining season began, when the fishers gathered to repair and assemble the seines after summer storage.

General knowledge of the characteristics of ice is not enough to cope with the changing conditions of winter, though. This point particularly came to light in narrations about non-local jiggers, as one participant demonstrated by quoting a local phrase: "the ice is strong enough when three weeks have elapsed since the first (non-local) jiggers" (TKU/A/10/18/H3). Previous year-round locals also expressed concerned about the occasional visits by summer residents to the village of Lypyrtti in winter. The locals narrated how they watched from the shore as summer residents attempted to walk on the ice and how it looked like "they thought the ice is as thick as it is spread out". This comes to show that knowing the ice includes a strong element of knowing with – people need to be in close connection with the ice to be able to listen to and see with their own eyes how it has developed in order to know where it is safe to walk.

5.4.2 *Making and Breaking Routes with Ice*

During wintertime, when the ice was thick enough to walk on, it connected people. Walking, riding or even driving from island to island was easy via ice roads. On the other hand, ice created a barrier to mobility for ships and boats. Various objects, means and vehicles were used as an aid to cross the ice and the open water fairways between the ice areas depending on the ice situation. Current fibreglass boats are capable of breaking through three-cent ice, but already ice that is one night or a few hours old may be too hard to break. Various customs and aids, such as a boat hook, anchor or ice spike, are still used to traverse the broken ice or ice floes (Kouri, 2017). Icebreakers and fairways created different situations and difficulties for many, and they found ways to adapt to these obstacles. For example, during World War II icebreakers began to open the icy Ströömi which runs through the village of Lypyrtti. One participant recalled piloting operations in the village during the 1950s and 1960s:

> […] they complained about the heavy piloting [operations] in the winter period along the [open water] fairway. You would think it would be easy to go down the open fairway, but it is difficult after the icebreaker. (TKU/A/09/56/H1)

Another interviewee, writing about piloting in the same years, had the following to say in a survey conducted by the Finnish Maritime Museum:

> There was sometimes a disagreement between those who lived on the 'wrong', other side, of the [open water] fairway. Their access to the store and bus became more difficult. When the ice became so thick that it closed off access to the fairway again, the disagreements were forgotten and life returned to the former model. That was the case in the 50s. (Heino, 1996)

Fig. 5.3 The communal wharf of Kekoniity, in Lypyrtti. The time before the ice is permanent is referred to as the "quarrel of the weathers" (*säittenriita*), indicating a clash between different weather systems. (Photo: Jaana Kouri)

Another interviewee recalled how she, along with her grandparents, had crossed the open water fairway between the ice (Fig. 5.3):

> We went on a large sledge. And the fairway was always kind of open because the ships went along there. We also took ice awls with us, and if there was a watery place, then we had to poke [through the ice] and look. [...] we had to go carefully and he [grandpa] tried to tell us where the ice was strong. [...] And the fairway was bad because there was a stream [running through it]. But the fairway was open, and then we had a punt [a small wooden boat] on this side. The punt was pulled over the fairway, rowed, and secured to the ice with a boat hook to get to the other side. And then we proceeded on foot. (TKU/A/09/64)

For winter-seiners, being aware of the fairways was crucial for safe seining. The fairways prevented a thick ice cover from forming and oftentimes cut off the route to their seining posts. One local woman reminisced about her teenage year back in 1950s as follows:

> When I was attending confirmation school,[1] the ice was so weak that we did not catch any fish; the boats just kept breaking the ice, but there were no fish on this side of the fairway. In Nauvo [another island in the archipelago, on the other side of the fairway], there were a lot of fish, but we could not go there with horses over such thin ice. So we went to Hanka

[1] Confirmation class lasts approximately 6 months, during which time the candidates participate in parish activities and reflect on questions of faith and life, both alone and together with their peers. Finnish youth typically attend confirmation class at 15 years of age. Though voluntary, well over 80% of 15-year-old Finns take part in the confirmation class.

> [a small harbour in Rymättylä] and ordered an icebreaker. We took all our horses and seines and people onboard and went to Nauvo, where the ice was very thick. (TYKL/aud/1268)

In the narrations, the narrator discusses human actors, even multiple generations of human actors, whereas the ice, water and stream represent non-human actors, described as various objects. Together, they comprise the actor networks at the moment of ice and water crossings. Still, to paraphrase Latour (2005), the process of textualising environmental heritage includes ambient intangible actants, but during the course of narration and when textualised by an ethnographer they become written and visible actors.[2]

One also had to be careful when skiing on the ice. One participant recalled how she as a child had skied across Strööm. Her father had said, "you should travel on the skis quickly, and you must not stop" (TKU/A/09/194/H1). The interviewees recounted specific characteristics of the ice and water. For example, Strööm, which has a strong current, flows through the village of Lypyrtti. In the wintertime, the current erodes the ice from below, making Strööm dangerous to cross. Therefore, observations about local changes in the ice and currents are crucial. The villagers reminisced about the exact routes over the ice and the points of navigation along the coast that helped them find safe tracks along which to walk. When one of the interviewees was asked about the safest route across Strööm, she said:

> […] if there is a chance that you see how the water freezes gradually [day by day], then you can see where most likely there will be a strong place. Even if you now know that you should start from Punshuvila [a house] towards Eersund rock, it's not exactly on that direct route. It might be a little to one direction and a little to another direction, where the real strongest place is. (TKU/A/20)

Locals also helped summer residents find the correct routes across the ice. They passed ahead of them, marking a safe route with juniper branches. Certain locals were also revered as particularly good connoisseurs of ice. One woman was said to be able to walk on the ice "where not even a cat would go" (TKU/A/09/75), and Rymättylä residents were said to have made a pact with the devil to be able to walk on such thin ice (Field journal, 2007; TYKL/aud/1267). The interviewees highlighted the importance of local ecological knowledge. It is necessary to know how ice behaves precisely in certain local water areas due to the fact that such areas can change based on the presence of water streams and the shapes and patterns underneath the ice. It is an example of local ecological knowledge in practice. Even with extensive local knowledge, accidents still happened and locals went through ice. It was a cold and sometimes fatal reminder that water in its liquid state still exists underneath the ice. In Lypyrtti, participants could name only a few people who had drowned by falling through the ice over long periods time (Kouri, 2017). People always warned each other about the dangers of falling into the water; it was a place

[2] If the actor has not yet "figured", it is an actant. Latour (2005) borrowed the concept of actant from literary studies and incorporated it into his domain theory (ANT). It can refer to a certain structural component, body, individual(s) or a loose grouping of morphemes (pp. 39, 54, 71, 128). See also Kouri, 2017.

"you must not get into" (*ei pitänyt joutua*) (TKU/A/10/19; see also TKU/A/10/9/H2). Some interviewees reported that in the past the people in the archipelago did not know how to swim, in which case they should not even fall into the water. One of the villagers said that as a preschool child, the main thing was to understand that sea ice is always treacherous, especially in the Lypyrtti and Ströömi area (Kairavuo, 2006). Children were warned not to go on the ice, and they were punished, for example beaten, for falling on the ice when they came home from school on skis (TKU/A/10/12). While winter-seiners cannot recall any casualties as a result of their fishing trips, they could recount many near-death experiences. One of the participants recalled how he was yanked under ice when the ice broke under his tractor. He narrowly escaped death that time. Experiences of crossing the ice are also experiences of survival. Moving around the landscape forms spatial stories, narrative forms of comprehension (de Certeau, 1984). Navigating on the ice is the medium and outcome of one spatial practice, a mode of existence in the world (see Tilley, 1994, p. 29). The path can also be a symbol of movement through life. For example, walking in the landscape is comparable to the speaking act in language and is, at the same time, a medium and the result of a spatial practice; it is a way of existing in the world (Tilley, 1994, p. 28). For Christopher Tilley, central to de Certeau's idea is the use of the grammar of space, not as predicted, but as a step-by-step co-composing of the landscape and its actors (p. 86).

5.4.3 Growing with Ice

Ice has a great significance in the lives of all beings within the landscape. It has an effect on the behaviour of fish and seals, for example. Interviewees observed that the occurrence of ice impacts the way fish move, and consequently, the way fishers harvest them. Large numbers of fish can be harvested both when the ice cover first forms and when it melts (TYKL/aud/1274). The interviews with fishers revealed that fish usually follow the ice because they "want to be under the ice cover" (TYKL/aud/1277). The ice, then, has had a major impact on the local community but also on non-human species as well and the intra-actions between human and non-human species. For some children, using ice picks to catch fish was a normal activity done with friends and siblings, and one of the archipelago dwellers noted that during the summer, when it is not a good season to fish, she actually misses the ice and ice fishing. For many, ice was an element of their childhood, and often the talk of childhood and ice unlocked many nostalgic memories. Children are used to playing on the ice, even when it is very thin, so the environmental heritage regarding sensing of ice starts to form at a very young age.

A local woman from Rymättylä recalled how she and her brother would spend the whole day playing on the ice and using an ice pick to catch fish for the house cat:

> Ever since I was little, I've liked it [fishing]. I remember that back then, my brother and I – parents weren't really looking after their children like nowadays – we were on the thin ice trying to catch fish. The ice was all black and the water burst out onto the ice from the

> hole. No one knew we were there. And we were always fishing. Fish for the cats. (TYKL/aud/1268)

One summer resident who had lived as a child in the archipelago reminisced about a time when an older pilot gave him a piece of advice on how to fish for pikes in the wintertime:

> […] you should try to fish in Kieskerrauma because it has a strong current. So the ice could possibly be thinner and the pike can see the bait better. And kick a lot of snow into your new hole in the ice so it doesn't immediately freeze; it helps a lot in your work. But the pike doesn't seem to move so much around in wintertime and during thick ice; it likes more to lie still – but when the spring sun starts to shine around Easter, that's when the pikes eagerly ascend to the surface. (Kairavuo, 2006, pp. 42–43)

In our material, the icegraphy of the archipelago provides a very detailed description of the activities and associated actors. Details are important, not only because of narrative richness, but because they convey the subtle, local-specific qualities of living with ice to the listener. The stories told to others, even before being ethnographically textualised, have served as environmental heritage, sharing experiences and environmental knowhow both with concurrent villagers and the next generations. Children have also grown up with ice themselves – interacting with it has taught them to adapt to changing situations.

5.5 Conclusion

For us, aquagraphy involves different narratives of waterbodies, while icegraphy is still even more focused on winter conditions. Winters are becoming milder, and the loss of ice has been evident and present in the lives of the archipelago dwellers for decades already. Icegraphy then, as it is presented in our environmental ethnography, speaks to the changing landscape and all the human and non-human elements within it, a narrative with mixed emotions of grief, nostalgia and some quite happy memories. It also brings to the forefront local, often tacit, ecological knowledge. This knowledge was born out of daily observations and multisensory experiences regarding ice, which consisted of sensing the ice, watching the ice move, strengthen and develop, and listening to its voice. It is knowing what and knowing how, but most importantly, it is knowing with. With the permanent loss of a regularly freezing Archipelago Sea, this knowledge might not be applicable for future generations, but as climate change increasingly affects local landscapes and ecological conditions, environmental heritage offers tools to learn from, proact and react to, and interact and intra-act with those changes and potentially advocate for adaptations that support sustainable development for the future.

Through intra-actions with the environment, the changing landscape reminds people of their past. They react by feeling sorrow, loss and longing and by actively bringing up nostalgic fragments or memories and observations about the icy landscape. This is the precondition for a local-specific process of creating environmental heritage, and environmental ethnography is a tool to textualise and contextualise such nostalgic memories in a form that could and should complement and add to scientific knowledge about environmental change. Environmental ethnography is not only the weaving together of networks of meaning-making between the researcher and subjects, but also between scientists from different disciplines. The prerequisite for discussions in different disciplines and the weaving together of networks of meaning-making is to have applicable common concepts. In this chapter, we have offered the concept of environmental ethnography for multidisciplinary fieldwork with local actors, both human and non-humans. Ethnography is an important addition to other "graphies" because it provides narratives that emerge from local experiences, offering a wide range of perspectives and perceptions, and more importantly, viewpoints that are quite local-specific and have emerged from both the practicalities of everyday lives and the affective relationship towards a landscape consisting of all non-human actors as well. Moreover, we as researchers are also actors ourselves in these networks when the intangible heritage of the field becomes visible through textualisation. The obligation of researchers is to put the local understanding of actor networks into textual form. Our material verifies that living with ice from childhood to adulthood, from season to season, from day to day, is crucial in understanding and adapting to changes in different water conditions. However, the environmental heritage process does not require the next generations or new inhabitants in the archipelago to absorb all the local ecological knowledge from the previous generations, but rather the important underlying principle of knowing nature. Environmental heritage is a process wherein nature is an inherent component of a local cultural practice and intangible heritage. It should also be noted that not all traditions should be transferred, only elements of tradition that are in a sustainable relationship with the environment and its actors (see Siivonen & Kouri, 2020). Looking at such a process of environmental heritage would be a good starting point for assessing, for example, social decision-making, livelihoods and education and the holistic way of life of people adapting to changing situations.

The environmental heritage process begins in the intra-actions between local individuals and non-human agents in the environment. It becomes a performed social heritage of the community in the present. As our examples demonstrate, the process of heritage creation is closely tied to the surrounding environment and changes in it. The efforts at conserving past experiences, knowledge and ways of meaning-making involve intentional reminiscing about the past. People want to intentionally commemorate and transmit meaningful knowledge to the future in times of change (Kouri, 2017; Siivonen & Kouri, 2020). Through nostalgic narration, our participants recalled meaningful events and practices by verbalising them, a form of oral history. In the narration, they informed us about sustainable ways to adapt to climate change. The nostalgic process also transmitted to the co-actors involved the signification process and tacit knowledge constructed by the villagers through empirical practice. This transmitting of information was not random but intentional. Interviewees wanted to convey the joint activities of different actors in adapting to changes. One of the actors is ice itself. The narrated experience-based everyday knowledge comprises the intangible heritage of a local community, with particular attention directed to the ice and lived environment, to a way of life and meaning-making community that includes both human and non-human actors. As ethnographers, we continue the environmental heritage process by textualising local environmental knowledge for future generations.

References

Research Material

All transcripts of the interviews (in Finnish) are archived in the archives of the School of History, Culture and Arts Studies at the University of Turku (TKU/TYKL). Each transcript has an individual archive code (in brackets).

Field Journal (2007). Kirsi Sonck-Rautio's field journal, in possession of the author.

Heino (1996). Tuula Heino's inscription for the 'Luotsi- ja majakkaperinne talteen' survey by the Finnish maritime museum. Tuula Heino's personal archives.

Kairavuo (2006). Volter Kilven jalanjäljissä / Minun Lypyrttini 1949–2006. Script. Jaana Kouri's personal archives.

Literature

Aisher, A., & Damodaran, V. (2016). Introduction. *Conservation and Society, 14*(4), 293–304.

Barad, K. (2003). Posthumanist performativity: Toward an understanding of how matter comes to matter. *Signs: Journal of Women in Culture and Society, 28*(3), 801–831. http://www.jstor.org/stable/10.1086/345321

Barad, K. (2007). Meeting the universe halfway: Quantum physics and the entanglement of matter and meaning. *Duke University Press*. https://doi.org/10.1515/9780822388128

Boym, S. (2001). *The future of nostalgia*. Basic Books.

Clifford, J. (1986). On ethnographic allegory. In J. Clifford & G. E. Marcus (Eds.), *Writing culture: The poetics and politics of ethnography* (pp. 99–121). University of California Press.

Cruikshank, J. (2005). *Do glaciers listen?: Local knowledge, colonial encounters, and social imagination*. UBC press; University of Washington Press.

Davis, F. (1979). *Yarning for yesterday: A sociology of nostalgia*. Free Press.

De Certeau, M. (1984). *The practice of everyday life*. University of California Press.

Deleuze, G., & Guattari, F. (1980). A thousand plateaus: Capitalism and schizophrenia. Athlone.

Geertz, G. (1993). *The interpretation of culture: Selected essays* (revised reprint). Harper Collins. (Original work published 1973).

Grönholm, P., & Paalumäki, H. (2015). Nostalgian ja utopian risteyksessä: Keskusteluja modernin kaipuun merkityksestä ja aikaulottuvuuksista. In P. Grönholm & H. Paalumäki (Eds.), *Kaipaava moderni: Nostalgian ja utopian kohtaamisia Euroopassa 1600-luvulta 2000-luvulle* (pp. 9–38) Turun Historiallinen yhdistys.

Gupta, A., & Ferguson, J. (1977). Culture, power, place: Ethnography at the end of an era. In A. Gupta & J. Ferguson (Eds.), *Explorations in critical anthropology* (pp. 1–30). Duke University Press.

Guttormsen, T. S., & Swensen, G. (2016). Introduction: Heritage, democracy and the public: Nordic approaches. In T. S. Guttormsen & G. Swensen (Eds.), *Heritage, democracy, and the public* (pp. 1–10). Ashgate.

Haraway, D. J. (2008). *When species meet*. University of Minnesota Press.

Hastup, K., & Hastrup, F. (2017). Introduction: Waterworlds at large. In K. Hastrup & F. Hastrup (Eds.), *Anthropology in fluid environments* (pp. 1–22). Berghahn Books.

OED. *Oxford English Dictionary. The definitive record of the English language*. Oxford University Press. https://www.oed.com

Ingold, T. (2000). *The perception of the environment: Essays on livelihood, dwelling and skill*. Routledge.

Kirksey, S. E., & Helmreich, S. (2010). The emergence of multispecies ethnography. *Cultural Anthropology, 25*(4), 545–576. https://doi.org/10.1111/j.1548-1360.2010.01069.x

Korkiakangas, P. (2006). Etnologisia näkökulmia muistiin ja muisteluun. In O. Fingerroos, R. Haanpää, A. Heimo, & U.-M. Peltonen (Eds.), *Muistitietotutkimus: Metodologisia kysymyksiä* (pp. 120–144) Suomalaisen Kirjallisuuden Seura.

Korpinen, S., Laamanen, M., Suomela, J., Paavilainen, P., Lahtinen, T., & Ekebom, J. (2019). *Suomen meriympäristön tila 2018*. Finnish Environment Institute SYKE.

Kouri, J. (2016). Enchanted environment: Co-composing a village history in the archipelago of southwestern Finland. In P. Ingman, T. Utriainen, T. Hovi, & M. Broo (Eds.), *The relational dynamics of enchantment and sacralization: Changing the terms of the religion versus secularity debate* (pp. 231–249) Equinox.

Kouri, J. (2017). *Vesi kuljettaa ääntä: Autoetnografinen tutkimus Lypyrtin kylän historiantuottamisesta*. Doctoral dissertation. University of Turku.

Latour, B. (2005). *Reassembling the social: An introduction to actor-network-theory*. Oxford University Press.

Lépy, É. (2012). Baltic Sea ice and environmental and societal implications from the comparative analysis of the Bay of Bothnia and the Gulf of Riga. *Fennia: International Journal of Geography, 190*(2), 90–101. https://doi.org/10.11143/4403

McCarthy Brown, K. (2001). *Mama Lola: A vodou priestess in Brooklyn* (3rd ed.). University of California Press. (Original work published 1991).

Phillips, A. (2005). Landscape as a meeting ground: Category V protected landscapes/seascapes and world heritage cultural landscapes. In J. Brown, N. Mitchell, & M. Beresford (Eds.), *The protected landscape approach: Linking nature, culture and community* (pp. 19–35). IUNC – The World Conservation Union. https://portals.iucn.org/library/sites/library/files/documents/2005–006.pdf

Polanyi, M. (1966). *The tacit dimension*. University of Chicago Press.

Polanyi, M., & Prosch, H. (1975). *Meaning*. The University of Chicago Press.

Siivonen, K. (2019). Intangible cultural heritage will revolutionise the future. In S. Eggmann, S. Kolbe, & J. Winkler (Eds.), *Wohin geht die Reise? Where does the journey go to? Eine Geburtstagsgabe für Johanna Rolshoven 2019, a birthday present for Johanna Rolshoven* (pp. 93–97). Akroama Editions. https://www.geruchderzeit.org/siivonen/

Siivonen, K., & Kouri, J. (2020). Tulevaisuusperintö ihmisen kestävän merellisen luontosuhteen perustana. In J. Kouri, T. Räsänen, & N. Tynkkynen (Eds.), *Muutoksen tyrskyt ja kotirannan mainingit: Kulttuurisia näkökulmia merentutkimukseen* (pp. 76–100) Suomalaisen Kirjallisuuden Seura.

Smith, L. (2011). The 'doing' of heritage: Heritage as performance. In A. Jackson & J. Kidd (Eds.), *Performing heritage: Research, practice and innovation in museum theatre and live interpretation* (pp. 69–81). Manchester University Press.

Sonck-Rautio, K. (2018). Adaptation and cultural sustainability of the winter-seining community in the archipelago of southwestern Finland. In I. Birkeland, R. Burton, C. Parra, & K. Siivonen (Eds.), *Cultural sustainability and the nature-culture interface: Livelihoods, policies, and methodologies* (pp. 125–136). Routledge.

Sonck-Rautio, K. (2019). *The fishers of the Archipelago Sea: Resilience, sustainability, knowledge, and agency*. Doctoral dissertation. University of Turku. https://www.utupub.fi/handle/10024/148446

Sonck-Rautio, K. (2020). Saaristomeren kalastajien muuttuva toimintaympäristö ja kulttuurinen kestävyys. In J. Kouri, T. Räsänen, & N. Tynkkynen (Eds.), *Muutoksen tyrskyt ja kotirannan mainingit: Kulttuurisia näkökulmia merentutkimukseen* (pp. 157–181) Suomalaisen Kirjallisuuden Seura.

Tilley, C. (1994). *A phenomenology of landscape. Places, paths and monuments*. Berg.

Tilley, C., & Cameron-Daum, K. (2017). *An anthropology of landscape*. UCL Press. https://doi.org/10.14324/111.9781911307433

Tuomaala, S. (2006). Sukupuolen kokemuksista muistitietohistoriaan. In O. Fingerroos, R. Haanpää, A. Heimo, & U.-M. Peltonen (Eds.), *Muistitietotutkimus. Metodologisia kysymyksiä* (pp. 271–291) Suomalaisen Kirjallisuuden Seura.

Vannini, P. (2009a). Introduction. In P. Vannini (Ed.), *Material culture and technology in everyday life. Ethnographic approaches* (pp. 3–12). Peter Lang.

Vannini, P. (2009b). Material culture studies and the sociology and anthropology of technology. In P. Vannini (Ed.), *Material culture and technology in everyday life: Ehnographic approaches* (pp. 15–26). Peter Lang.

Vihma, A. (2021). Nostalgia: Teoria ja käytäntö.. Teos.

Chapter 6
The "International" in Water–Society Relations: A Case Study of an Arctic Urban Watershed

Monica Tennberg

Abstract Water inspires innovative ways of thinking about circulation, power, and knowledge. Power circulates in water-society relations in different ways: within and between discourses about what the problem with water is, within constituting networks of agencies and alliances to solve problematic water-society relations, and defining different positions in regard to the future of water-society relations. The dimension of "the international" in water-society relations remains understudied, due to the territorially, legally, and institutionally bound approaches in studies of water governance. The watershed in the city of Rovaniemi in northern Finland, a meeting point of two major rivers, provides a case study for discussing the elusive "international" in complex, entangled water-society relations. Located near the Arctic Circle, the urban watershed is a regulated, disciplined, and governmentalized – yet contested – space in numerous ways as a site involving nature protection, energy and climate politics, flood protection, and water management by multilevel governance incorporating local, regional, national, and European dimensions. The current complex and entangled water-society relations in the watershed are analyzed via multisite data to map multiple issues, relations of power, and positions towards the watershed and its future, exemplifying the relationally diverse nature of water-society relations beyond territorial boundaries.

Keywords Arctic · Water governance · Governmentality · Situational analysis

Water resources are a good example of entanglement in human-nature relations. An 'entanglement' refers to how people situate or order themselves in relation to nature through a myriad of world-making relationships. These relationships – material, practical, and ideational –constitute environments for human and non-human agencies and interactions. For example, the watershed in Rovaniemi city, in northern Finland, constitutes "a hydro-social cycle" (Swyngedouw, 2009, 2014) wherein an

M. Tennberg (✉)
Arctic Centre, University of Lapland, Rovaniemi, Finland
e-mail: monica.tennberg@ulapland.fi

M. Lehtimäki et al. (eds.), *Cold Waters*, Springer Polar Sciences,
https://doi.org/10.1007/978-3-031-10149-6_6

entanglement of historical, economic, and demographic developments of water-society relations leads to a specific distribution of people and their activities in the watershed. The watershed is also "a hydro-social territory" (Boelens et al., 2016; Hommes et al., 2019) involving an entanglement of water-related infrastructure, human activities, and technology in the river valley, especially in connection to hydropower development, related technology, and adjacent infrastructure. Moreover, the watershed can be interpreted also as a site of "hydromentality" (Hommes et al., 2020; Usher, 2014), consisting of an entanglement of diverse perceptions about water-society relations and claims of knowledge, expertise, and authority, which constitute the central aspects of governance of the watershed.

This chapter is inspired by the critical water studies that have developed in recent years to "resocialize" water (Mollinga, 2019; Mustafa & Halvorson, 2020; Rusca & Di Baldassarre, 2019; Rattu & Véron, 2016; Wilson et al., 2019). According to Jamie Linton (2014), in the twentieth century water planning and management were characterized by a particular way of understanding and relating to water: "Modern water is a way of knowing, accounting for, and representing water apart from its social context" (p. 111). While critical water studies have shed light on the social complexity of water-society relations, one aspect is little studied: the "international" in water-society relations. The "international" is usually defined as state-centric, focused on legally- and institutionally-separated but interacting levels of rights, obligations, and responsibilities around water governance as an issue of effective governance. The popular, modern understanding of the "international" results in an understanding of water as a resource, a territorialized and unevenly-shared asset, limiting the notion of the "international" in water-society relations to cross-border situations. If one understands the "international" as an aspect in all social relations, then water-society relations can be understood as governmentality (Usher, 2014) – that is, as arranging multiple, entangled human-nature relations according to different societal aims across different governmental boundaries and levels. From this perspective, water defies the two modern understandings – non-social water and non-social international – with its relationally diverse nature.

The Arctic is a good example of the ways the "international" is elusive in water–society relations. The Arctic is mostly water, as most of the region is the Arctic Ocean and major rivers ending in the Arctic Ocean, along with concerns over the fate of ocean currents, icebergs, and melting sea ice due to climate change in the region. Waters in the Arctic are undersood as an issue for marine governance (European Commission, 2021); however, river basin management is also a common, pan-Arctic concern, as more major rivers flow to the Arcic Ocean than to other oceans. Five major rivers flow into the Arctic: the Mackenzie and the Yukon in North America; and the Ob, Yenisey, and Lena Rivers—the three largest in Asia. In the northeast of Finland, Lake Inari discharges via Paatsjoki River into the Barents Sea in the Arctic. Finland is characterized by its position bordering the Baltic Sea, the Gulf of Bothnia, and the Gulf of Finland; it is known as the country for "thousands of lakes", but also thousands of islands and rivers – 647 rivers, to be exact.

In the case of the watershed in Rovaniemi, since the mid-1990s the European Union (EU) has been the main force of govermentalization of the area.

'Governmentalization' in the European context can be understood as an integration of individuals, organizations, authorities, and nature into a collective that shares thoughts, images, rationalities, and belief in a certain apparatus for governing (Walters & Haahr, 2005; see also Agrawal, 2005). In the European Arctic, the EU plays a central environmental role through its environmental policies on various matters, including nature protection, flood protection management, climate politics, and fisheries, all implemented and applied in different national, regional, and local contexts – ideally in an ordered way, following the principles of the EU Water Framework Directive (2000). However, the case of Rovaniemi shows that these different governmental rationalities of protecting nature, fisheries, and renewable energy production in line with the implementation of EU directives and strategies result in a complex, highly politicized situation.

6.1 The Problem of the "International" in Water–Society Relations

Water is a problem of practical problem-solving; for example, securing clean drinking water and treating waste water is not only an issue for common rule-making for the management of border rivers, but also an issue of normative consensus for the sustainable use of water from the perspective of governance. Water defies our understanding of state-centric governance of human–nature relations in numerous ways – as a cross-border issue of inter-state cooperation and conflict, a regional concern for management of water resources, and a referent object for global security debates. Governance is typically understood as collaboration between states and other societal actors to manage common concerns and create opportunities for further governance. Analysis of governance focuses on such questions as: Who governs? What is governance in practice? And what is the problem to be governed? (Colebatch, 2002).

While governance approaches generally acknowledge the role of societal actors in addition to the states, the focus nevertheless remains very state-centric. The spatially-bound political imaginary locates the "international" in the world of sovereign nation-states, with increasing recognition of its shortcomings. The state-centric "international", as we know it, is deeply rooted in the modern imaginary of social relations (Walker, 1992). The *raison d'etre* of modern states is to secure their territories and the well-being of their population. This state-centric focus results in our understanding of the "international" in governance as a specific regime of spatio-temporal relations. Those relations are firmly rooted in fixed state-centric borders, territories, and identities, both individual and collective (Bonditti, 2017, p. 6).

This state-centric, territorially based approach is also part of EU governance. The EU is a unique, complex, and changeable entity of regional governance with a mix of power relations as regards environmental issues, leading to a peculiar role (Knill & Liefferink, 2013; Zito et al., 2019). In our case, the requirements of Finnish legislation on nature protection (Law on nature protection, 1996), flood protection management (Law on flood risk management, 2010), and water management (Law on water, 2011) are based on the EU Water Framework Directive (2000) and related directives, the EU Habitats Directive (EU, 1992), the Flood Directive (EU, 2007), and EU climate policy (European Commission, 2020), since Finland became an EU member in 1995. Finnish environmental management is not only an issue of national legislation: EU legislation has an impact on private subjects without general or specific permission given by the state (Kokko, 2011). Regionally, centers for economic development, transport, and environment are responsible for planning, monitoring, and supervising water management. These centers produce expertise-based risk assessments concerning the state of the river basins, and these reports often refer to various EU environmental legislation as the basis for their assessment of the state of the water environment and measures (For Lapland, see Lapland centre for Economic development, transport and environment, 2015, 2021).

Despite our complex, multilevel systems of governance, water does not respect fixed borders, defined territories, or pre-assigned subjectivities. Power, like water, moves throughout all sectors of society, across scales and across different governing bodies and governed agencies. Here, the analysis – inspired by Foucault's (1991) concept of Governmentality and Agrawal's (2005) application of it to environmental issues as Environmentality – starts with a question: What is the issue (the problem) to be governed? Rather than assuming that water is a problematic object of governance, a governmentality perspective calls into question *how* water and its governance become objects of concern through power/knowledge. The matter to be controlled is not the circulation of individuals but of things and elements, mainly water and air (Foucault, 2000a, p. 148). A second question deals with the network of actors created around the problematization of the issue to be governed. Who governs is not the most important question here; a governing body, like the state or some other body, manages a complex web of people and objects with the intent to strengthen the governing body – the state for example – and to improve the welfare of the population. The third question asks what kind of subjectivities such a network develops. The act of governing takes place through interventions that change the relationship between the governing body and those it governs, mediated through objects of concern such as flood water, endangered fish species, and energy. These interventions aim to securitize the objects to be governed.

6.2 Eventalization as a Method

In our case, governmentality produces different problems, different networks, and different alliances for governing as well as different rationalities. In any event, many power relations are at play: state sovereignty-based legal relations, knowledge- and expertise-based disciplinary relations of power, and governmental and administrative power-based relations to support the well-being of population – here understood also to include the non-human population (Lemke, 2015). A central idea of studying governmentality is eventalization. Foucault defines eventalization as a moment when the self-evident becomes strange and one considers how else things might have been. Eventalization focuses on the event in its historical and situated entirety, including the mundane "little" things and details within the context of wider struggles of power and knowledge. It entails analyzing an event according to the multiple processes that constitute it (Foucault, 2000b). Power relations and their simultaneous interplay(s) in an event make for messy social situations. Situational analysis (SA) is a way to deal with complex social realities and to make sense of them without simplifying social reality (Clarke, 2005).

Here, the most relevant aspect of *SA* is an understanding of the "situation" to mean that all the relevant aspects of the situation exist *in* the situation, not outside nor as a context. The situation to be analyzed here is the complex, messy social world of the watershed in the city of Rovaniemi. Instead of focusing on action- and agent-centered social processes, SA makes the situation the main unit of its analysis, and the understanding of its elements and their relations its analytic focus. "The situation" is constructed empirically by creating three relational maps: situational maps, social worlds/arenas maps, and positional maps. Situational maps articulate all the elements in the situation and examine the relations among them. These maps include all the major human, nonhuman, discursive, historical, symbolic, cultural, political, and other elements in the research situation of concern, showing the messiness of the social world. Social worlds/arenas maps lay out all of the collective actors and the arena(s) of action and negotiations, including the broader situation; these map organizational, institutional, and discursive dimensions. Positional maps depict the major positions taken, and not taken, vis-à-vis particular axes of variation and difference, focus, and controversy found in the situation of concern (Clarke, 2005, pp. 83–108).

The mapping exercises aim to capture the complexities and multiplicities of social life and to produce thick analyses that present all human and non-human elements in the situation. The aim of the maps is to find and analyze the heterogenous, contradictory complexities of social life (Clarke, 2005, pp. 125–136). In SA, the mapping exercies are based on multisite data. In this case study, the data is based on publicly available textual data, which includes EU and government documents, reports in local and national media, statements by various stakeholders – such as associations for nature protection, water management, and environmental organizations, along with energy companies and social media discussions – and social science- based research about the rivers and their development (Suopajärvi, 2001; Krause, 2011, 2012; Alaniska, 2013; Autti, 2013; Vola, 2020). It is notable that the case of the rivers is of national interest that is discussed in the Finnish Parliament in various ministries, including the ministries of agriculture and forestry and of the environment, as well as by national media. The different, official steps in river basin management in northern Finland are well documented on specific websites by the ministry of the environment; these include reports, list of participants and experts, and reports from stakeholder events held in connection to the two rivers and their management (Ymparisto.fi).

The debates over the watershed in Rovaniemi have a long history, with current developments having their roots in the industrial era of development after the Second World War. In this case study, the focus of the analysis is the situation of the watershed in Rovaniemi, Finland between 2019 and 2021. There are two reasons for this focus. First, the year 2018 marked an end to debates in Rovaniemi about the most effective flood protection management measures that dominated the discussion between 2012 and 2018, during the planning period led by the regional center. A government decision in 2018 (State Council, 2018; Supreme Administrative Court, 2019) prevented the building of a new water reservoir, a power station, and a dam located in another municipality in order to prevent floods in Rovaniemi. These decisions turned Rovaniemi's attention back to the city, to develop its urban water management plans.

A second focus of this analysis is the protest by the local politician and activist Vesa Puuronen, a professor at the University of Oulu, who used graffiti on the walls of the nearby power station and dam in the summer 2019 to demand the freeing of the Kemijoki River and the return of salmon to the rivers. Puuronen, who lives by the Kemijoki River downstream of the city, painted graffiti of salmon and slogans such as "Free Kemijoki", "Free Salmon Rivers", and "Save Mother Earth" on the walls of the power plants. The salmon, a culturally and economically important fish, were kept out when the Kemijoki River was dammed for hydropower after the Second World War. The return of salmon and other migratory fish species to the Kemijoki and Ounasjoki rivers is planned, with the use of fish passes to be built by the hydropower company. The hydropower company sued Puuronen, and he was ordered to pay 20,000 euros in fines and compensations for damaging the property in early 2021 (YLE, 2020).

6.3 The Watershed in Rovaniemi

The city of Rovaniemi, with approximately 60,000 people, is located in a river valley where two rivers meet – the Kemijoki and Ounasjoki Rivers. The city is large in geographical size (7582 km^2), almost three times the territory of Luxemburg (2586 km^2). The city is the center of administration, education, and health care in the county of Lapland. Historically, the rivers have been an integral part of urban development: they bring nutrients to fields along the rivers, making the river valley good for farming; they serve as a route of transport for goods, people, and especially timber until 1991; they are a site for fishing salmon and other valuable fish species as well as a place for industrial development, especially for forestry and hydropower development. In addition, the waterfronts are part of the everyday life, recreation, and festivities of the city residents in all seasons.

The 550 km-long Kemijoki River runs through Lapland from the northern fjells to the Gulf of Bothnia. The Ounasjoki River is its largest tributary. The watershed is composed of where the two major rivers meet, together with the nearby areas in

Rovaniemi; it includes the Kemijoki river valley and the Ounasjoki river delta near the city center and several adjacent water bodies, including ponds and canals. Administratively, the rivers belong to one of the seven river basin districts in Finland, designated for the purposes of river basin planning and management as required under the EU Water Framework Directive (EU 2000). The administrative region includes several rivers running to the Bothnia Gulf and sea, the lakes and water reservoirs, and several groundwater areas. The Lapland center for economic development, transportation, and environment administers the region, together together with neigboring countries (Sweden and Norway) in the case of border-crossing rivers.

6.4 The Complex Situation in the Watershed

The different EU policies create different social realities of the river basin and naturalize the river basin as a site of governing through separate, naturalized realities. The protected Ounasjoki river delta and the Kemijoki river valley and adjacent water bodies are important for energy production. However, the situational map of the watershed in Rovaniemi is much more complex, demonstrating the ways power circulates in social relations, the definitions of the issues considered as problems, the different agencies and alliances emerging around the problem, and the diverse positions concerning the watershed – simultaneously in need of nature and flood protection and a potential for production of renewable energy. Figure 6.1. presents the main themes, actors, and administrative processes taking place.

The idea of water management is to maximize circulation, support good circulation, and limit bad circulation. The question in this analysis is: What is the problem of circulation, what circulation needs to be supported and what needs to be limited, and why? These seemingly separate yet connected discourses focus on flood protection, nature protection, and and hydropower development, and they relate to the flow of water, its circulation, and its seasonal changes in the rivers in one way or another. The "bad" circulation is the spring floods in both rivers, typically caused by ice dams. The legendary flood of 1859 is often mentioned in these discourses, as well as the most recent major flood of 1993, both of which flooded the city badly. However, floods are usually mostly harmless in Rovaniemi: they are a popular spectacle gathering city people to the waterfronts, but floods may create challenges for rescue services and need for evacuation, generate problems for property owners, and cause pollution. The development of regional flood protection plans in 2012–2015 led to many concerns, controversies, and divisions among the residents, decision-makers, authorities, and stakeholders in terms of the most effective and appropriate ways to protect the city from major floods. Most importantly, after legal reforms in the early 2010s, local property owners became responsible for buying private flood insurance and for preparing for floods with protective measures, such as building flood embankments on their property, instead of the state compensating for flood damages (Tennberg et al., 2018).

EU Natura 2000 Ounasjoki River	Regional flood protection management plan	EU Flood directive Building flood embankments	Return of salmon to Ounasjoki River	Loss of farming land Local farmers and fishermen	Sand dunes Flood islands
Monitoring EU fish pass strategy	Flood warnings	Flood prone areas in the city	Endangered species	Harmful to fishing	Cultural landscapes
Property owners	Urban residents' associations	10,000 people threatened by floods	EU Water Framework directive	New power plant and dam	Electrification Energy dependencies
Grudge toward the company Kemijoki River	Hydropower company State council Employment and tax income	50-, 100-, and 250-year floods Temporary flood protection measures	Permission to build the adjacent infrastructure Compensation for the company for the loss of energy production	Legal obligations by the company Ministry of Agriculture and Forestry	Electricity security Ministry of the Environment The obligation to consult Sweden
Water level and quality	City authorities Statements and complaints	Cultural sites threatened by floods	Environmental organizations EU Commission	Rowing Demonstrations	Historical trauma Court proceedings
Ice dams	Urban planning	Pollution	Recreation	Climate policies	Water management -related associations
Temporary flood protection measures	Private flood insurances for property owners	New water reservoir in another municipality	Nature tourism Vesa Puuronen Jasper Pääkkönen	Renewable energy Compensation for the loss of salmon	Historical floods Raising the water level of Kemijärvi Lake

Fig. 6.1 A situational map: Kemijoki- and Ounasjoki-related issues, actors, and processes. In the map, in addition to **main actors**, issues and processes are marked with different fonts in relation to **energy production**, **flood protection**, and *nature protection*
A collective image of the primary topics, key players, and administrative procedures that are being carried out in 6 columns

In addition, water flow and its control are important to the energy production along the Kemijoki River. The Kemijoki company, in this chapter called the hydropower company, owns most of the power stations and dams along the Kemijoki River. Since the 1940s, hydropower development has been closely tied to the political and economic development of the Finnish state, as Finland lost both territory and energy-producing facilities in eastern Finland to the Soviet Union after the Second World War (Alaniska, 2013). One of the early power stations and dams, Valajaskoski, built in 1960, is located 10 kilometers downstream from Rovaniemi.

Its building has a major impact on the urban landscape in the city center. Currently, the plans for a new power station and dam on the Kemijoki River upstream of the city, in Sierilä, is the source of heated local, regional, and national debate. (See the website of the Finnish Association for Nature Protection (NA) dedicated to the Sierilä plans.) From the perspective of hydropower, the circulation of water in the rivers is both "good" and "bad". On the one hand, there are concerns about the impact of the new power station and dam to the cultural landscape, endangered species, and local livelihoods, even if the building project might bring employment and income to the region. On the other hand, hydropower is considered climate-friendly renewable energy, important as a means to balance fluctuations in national electricity production and also important to diminishing energy imports, as Finland is poor in terms of domestic sources of energy.

Floods are "good" for the river's natural and cultural landscape, for the flood islands, forests, and fields in the Ounasjoki River. The river's large flood fields and islands in the river delta are examples of Scandinavian river landspaces with unique natural features that are in need of protection. Migratory fish need enough water in the rivers as well as access through the power stations and dams to return to the Kemijoki and Ounasjoki Rivers. Since the 1970s, the environmental movement advocating for the Ounasjoki River has led to the protection of the river and river delta near Rovaniemi, through national legislation in 1983 and later through inclusion in the EU Natura 2000 program (Suopajärvi, 2001). The movement resisted the hydropower company's plans for a similar kind of development on the Ounasjoki River that occurred on the Kemijoki River, which has seen the building of power plants, dams, and water reservoirs since the 1960s. The tradition of protests started with the Ounasjoki movement, with people rowing on the river with small boats, continues today. The latest protest with rowboats was organized in August 2021 against the plans for the new power station and dam along the Kemijoki River upstream of Rovaniemi (Lapin Kansa, 2021).

In these debates about the flows of water in these rivers and the discourses about the rivers and their problems circulate a mix of scientific knowledge held by environmental authorities, academic knowledge about local people's perceptions and experiences of the rivers, local and regional politicians' and administrators' practical knowledge of governance, and the hydropower company's own research on the matter. Deep disagreement persists between regional environmental authorities, experts, and the hydropower company on whether and how salmon and other migratory fish can be returned to the rivers. Another point of disagreement between the environmental authorities and the hydropower company is the quality of water in the rivers. The environmental authorities assess the water quality to be lower than the assessement by the hydropower company. Overall, in the company's view, the requests by the environmental authorities to support the return of migratory fish species with fish passes and to improve the quality of the river's water are "unrealistic and contradictory": it is "impossible" to create a natural, maintained stock of migratory fish in the rivers, and this idea is based on "wrong information" about the reproduction rate of salmon and their survival in such waters (Kemijoki Oy, 2020).

6.5 Mapping of Social Arenas and Worlds in the Watershed

The "international" is not a question of agency of any actor as such, but of a network of actors, alliances, and agencies governing the watershed. In the mapping of the social arenas and social worlds, the main question is: what are the networks and agencies governing the watershed in Rovaniemi? The networks and alliances extend beyound the administrative boundaries of the local, regional, national, and even international. Franz Krause (2012) points out that river management in Rovaniemi illustrates how a riverine conflict can create allies and adversaries – in this particular case, between the local residents, the city, and the Kemijoki company, with its plans for a new water reservoir and a power station. The local property owners, their associations, and the city authorities have been on the same side, supported by the hydropower company, while the regional environmental authorities, environmentalists, and neighboring municipalities have been on the opposing side. Another set of alliances cetners around the nature protection of the river: an alliance between farmers, local people, and environmentalists on one side, and the city and the hydropower company on the other side. Additionally, an alliance has emerged between the local protesters and artists who agreed to sell their work to support Puuronen to pay his fines in spring 2021 (Kansan uutiset, 2021).

Water management issues are controversial in Rovaniemi. First, the city needs to strike a compromise beween the regularly flooding rivers and the needs for urban development in housing and tourism. While the city estimates a high flood risk in the city and requires the highest possible level of flood protection, at the same time it continues urban planning in waterfronts which are known to be flood-prone areas (Tennberg et al., 2018). Second, the city balances the needs of nature protection as part of being an "Arctic city" and supporting more hydropower development as a source of income and employment in the city. The city supports the building of the new power station and dam with permits to build adjacent infrastructure (roads and service buildings), but it also supports the return of salmon to the rivers so as to support regionally important tourism, which can potentially be developed around salmon-fishing. The city supports the idea of reviewing the hydropower company's water management obligations (City of Rovaniemi, 2020).

Nature is "something that runs under, through, and within the exercise of governmentality" (Massumi, 2009, p. 165). Nature – conceptually, materially, and practically – enters the social domain and becomes the essential medium through which power is exercised. This opens up the possibility of a non-human understanding of agency and power (Lemke, 2015). Salmon can be seen as a non-human agent in international water-society relations. Salmon is a referent object of economic, cultural, social, and political significance beyond the city, the region, and the whole of the Arctic, as one pan-Arctic key species. The salmon are part of the problem of flowing waters in the rivers. Salmon are known to move in the spring, in the flooding icy-cold waters near the waterfronts.

The curious aspect here is that salmon have not been in these rivers near Rovaniemi since the late 1940s, but the yearning for salmon and its return is strong as ever. The salmon is a symbol of life in the river, has economic importance, and is viewed as a remedy to a historical injustice. The salmon and their possible return have a long history, going back to the 1940s. The salmon were lost because the building of a hydropower dam prevented them from entering the Kemijoki River from the Gulf of Bothnia. At that time, any notion of an obligation of hydropower companies to build fish passes was struck down. The Water Court (and the Supreme Court) concluded in the 1960s that the production of salmon in the river area can not be compensated by any measure (i.e., fish pass, stocking other species), and therefore power companies were ordered to pay monetary compensation for the loss of salmon. However, the building of fish passes has been passionately discussed since then in several contexts along the Kemijoki River, and the first fish pass to the Kemijoki River was built in 1993, the building of which was primarily funded by the state. (See, e.g., material produced by Lohijokitiimi, an association of municipalities and fishing areas along the two rivers, called the "salmon river team" and established in 2001; they aim for the return of migratory fish to the rivers.)

The city of Rovaniemi can be described as a modern, rational, and development-oriented place – the "Arctic capital city of expertise" – but the emotions of loss and yearning, both individually and collectively, that are expressed in the discussions about the return of salmon are also integral to the urban affective milieu. An example of the emotional power of the salmon is the plea by politicians from all political parties in Lapland to return salmon to the rivers in Rovaniemi in 2021. The plea states that Finland is a country of rule of law and that actions should be taken to follow the law. Finland has twice been notified by the EU about its lack of implementation of European water management policies. The plea states that any permits for hydropower stations need to respect the law and its obligations to support the return of the salmon. Actions are needed to support "the return of life to the rivers" (Matkailutoimittajien kilta, 2021).

6.6 The Future of the Watershed

Governmentality consists of more-or-less calculated governmental interventions for definite but shifting societal ends, with relatively unpredictable consequences, effects, and outcomes; it is undertaken by a multiplicity of authorities and agencies employing different practices of power and forms of knowledge (Dean, 1999, p. 11). In principle, the adoption of the EU Water Framework Directive (EU, 2000) has introduced a legal mechanism for ensuring a strategic approach to river basin management across the different rivers of the EU, regardless of sectoral, administrative, or national constraints and borders. The case of the Rovaniemi watershed shows the difficulties in bringing this idea of coordination into practice. The modern governmental logic is to securitize, but in this case, the question arises: what does it mean, in practice, to securitize the watershed? In short, the different positions in the local, regional, and national debates circulate around the question: "Who needs to be protected from what/whom?" The options are: protecting the hydropower company from the environmental protestors; protecting the river's nature from hydropower development; protecting the city from floods; and protecting hydropower development from the fish pass strategy.

In a video made of the rowing demonstration in summer 2020, a local inhabitant describes the position of the locals as "ants against the hydropower company" (Sieriniemen siunaussoutu 2020). This setting of power relations between the hydropower company, the river, and the local people is captured in a cartoon. Securing the activities of the hydropower company comes first, since domestic law protects private property and company activities need to be protected. Public figures like Jasper Pääkkönen, a Finnish actor and environmentalist, is amazed that a company can resist obligations put in place by the state (YLE, 2019). The environmentalists challenge the idea that building fish passes would entail significant costs for the hydropower company, which has benefited from production for decades. Also, they point out the company's lack of full compensation for the loss of fisheries (WWF, 2020).

Finland has received two notifications from the European Commission (the most recent European Commission, 2019a, b) about improving its implementation of the Waterwork Framework Directive, especially from the ecological perspective, and about coordination between different management plans, such as flood protection management plans and water management plans (State council, 2019). Without EU intervention, it would be unlikely that the debate about the salmon and its return would have become the focus of debate, as it has been in 2019–2021. The environmental authorities have started a legal process requiring power stations to build fish passages across all dams and to restore breeding areas. The company has reservations about following these obligations: "Even if it would be possible to make the salmon to return to the rivers with well planned cooperation, it is another question [as to] can it be caught" (Kemijoki Oy, 2020). The company view is that "we do not resist the obligations to build fish passes, but the regional authorities' expectations are unreasonable" (Kemijoki Oy, 2020). Fish passess (both up- and downstream), additional implantation of fish stocks, and the use of the river by the fish should be priorititzed. The company has no opportunity to make such large investments and bear the losses in production with such measures (Kemijoki Oy, 2020).

6.7 Final Remarks

The "international" in water-society relations is, in the case of Rovaniemi watershed, "European". This chapter sought to illuminate the "international" aspect of water-society relations as relational, not only spatial (Massey, 2007), and as diverse, overlapping, and contradictory rationalities of European governmentality. Governmentality is, in principle, strategic problem-solving with localized interventions. The watershed in Rovaniemi is a "milieu" – that is, a concrete, materially-based environment with multiple social and spatial relations. The flow of river waters is both good and bad, and the ways to maximize the circulation are contested.

In this milieu, the circulatory processes of river waters consist of "things, events and elements" occurring and forming the object of governance of water-society relations (Foucault, 2007, p. 19). While the EU's various directives and strategies contribute to the complex social situation in the watershed, the EU's notifications to Finland about restoring the ecological quality of river waters and supporting the return of salmon with fish passes give much-needed leverage to environmental authorities in legally demanding actions by the hydropower company.

The EU and its manifold environmental policies are part of the local, regional, and national discussions in many ways, and they construct the watershed as a regulated, disciplined, and governmentalized site of power relations. One physical spot, the watershed in the city of Rovaniemi, provides an angle from which to see multi-spatial governmental logic and rationalities. Multiple power relations are at play: sovereignty as legal rights, responsibilities, and obligations, and as knowledge, expertise, and authority in water-related matters and administrative interventions. Power circulates in water-society relations in the urban watershed as definitional power about the problems, as agential power to define the responsible agents, as normative power about the "right" solutions for the future, and as securitizing power—although disagreement prevails as to what most needs to be securitized. This results in a complex mix of governmental rationalities, interventions, and resistance to them. The analysis of the data from 2019 to 2021 show how the past is part of the current debates, which are dominated by different interpretations of past developments in the watershed, the role of different actors in its development, a sense of historical injustices, and different ways to remedy them according to the authorities, company representatives, environmentalists, and local people. The salmon, although not present in these rivers, are still a force to be reckoned with, due to both a yearning for their return to the rivers and a disagreement of the appropriate ways for securing this return.

References

Agrawal, A. (2005). *Environmentality: Technologies of government and the making of subjects*. Duke University Press.

Alaniska, K. (2013). *Kalojen kuninkaan tie sukupuuttoon: Kemijoen voimalaitosrakentaminen ja vaelluskalakysymys* 1943–1964. Universitatis Ouluensis B117. Oulun yliopisto.

Autti, O. (2013). *Valtavirta muutoksessa: Vesivoima ja paikalliset asukkaat Kemijoella*. Universitatis Ouluensis E136. Oulun yliopisto.

Boelens, R., Hoogesteger, J., Swyngedouw, E., Vos, J., & Wester, P. (2016). Hydrosocial territories: A political ecology perspective. *Water International, 41*(1), 1–14.

Bonditti, P. (2017). Introduction: The international as an object for thought. In P. Bonditti, D. Bigo, & F. Gros (Eds.), *Foucault and the modern international: Silences and legacies for the study of world politics* (pp. 1–12). Palgrave Macmillan.

Clarke, A. (2005). *Situational analysis: Grounded theory after the postmodern turn*. Sage.

Colebatch, H. K. (2002). Government and governmentality: Using multiple approaches to the analysis of government. *Australian Journal of Political Science, 37*(3), 417–435.

Dean, M. (1999). *Governmentality: Power and rule in modern society*. Sage.

EU Floods Directive. (2007). Directive 2007/60/EC of the European Parliament and of the Council of 23 October 2007 on the assessment and management of flood risks. https://ec.europa.eu/environment/water/flood_risk/.

EU Habitats Directive. (1992). *Council Directive 92/43/EEC of 21 May 1992 on the conservation of natural habitats and of wild fauna and flora*. https://eur-lex.europa.eu/legal-content/EN/TXT/PDF/?uri=CELEX:31992L0043&from=EN.

EU Water Framework Directive, WFD. (2000). Directive 2000/60/EC of the European Parliament and of the Council establishing a framework for the Community action in the field of water policy. https://ec.europa.eu/environment/water/water-framework/index_en.html.

European Commission. (2019a). *COM (2019) 95 final*. Komission kertomus Euroopan parlamentille ja neuvostolle vesipolitiikan puitedirektiivin (2000/60/EY) ja tulvadirektiivin (2007/60/EY) täytäntöönpanosta. https://eur-lex.europa.eu/legal-content/FI/TXT/HTML/?uri=COM:2019:95:FIN&from=EN.

European Commission. (2019b). *COM (2019) 95 final*. Liite komission kertomukseen Euroopan parlamentille ja neuvostolle vesipolitiikan puitedirektiivin (2000/60/EY) ja tulvadirektiivin (2007/60/EY) täytäntöönpanosta. https://eur-lex.europa.eu/resource.html?uri=cellar:bee2c9d9-39d2-11e9-8d04-01aa75ed71a1.0016.02/DOC_2&format=PDF.

European Commission. (2020). *Stepping up Europe's 2030 climate ambition investing in a climate neutral future for the benefit of our people*. Communication from the commission to the European Parliament, the council, the European economic and social committee and the committee of the regions. COM/2020/562 final. https://eur-lex.europa.eu/legal-content/EN/TXT/PDF/?uri=CELEX:52020DC0562&from=en.

European Commission. (2021). *A stronger EU engagement for a peaceful, sustainable and prosperous Arctic*. Joint communication to the European Parliament, the Council and the European Economic and Social committee and the Committee of the Regions. Brussels, 13.10.2021 JOIN (2021) 27 final. https://eeas.europa.eu/sites/default/files/2_en_act_part1_v7.pdf.

Foucault, M. (1991). Chapter 4: Governmentality. In G. Burchell, C. Gordon, & P. Miller (Eds.), *The Foucault effect: Studies in governmentality*. The University of Chicago Press.

Foucault, M. (2000a). The birth of social medicine. In J. D. Faubion (Ed.), *Power: Essential works of Foucault 1954–1984, volume 3* (pp. 134–156) Allen Lane.

Foucault, M. (2000b). Questions of method. In J. D. Faubion (Ed.), *Power: Essential works of Foucault, 1954–1984, volume 3* (pp. 223–238) Allen Lane.

Foucault, M. (2007). *Security, territory, population*. Palgrave Macmillan.

Hommes, L., Boelens, R., Harris, L. M., & Veldwisch, G. J. (2019). Rural–urban water struggles: Urbanizing hydrosocial territories and evolving connections, discourses and identities. *Water International, 44*(2), 81–94.

Hommes, L., Boelens, R., Bleeker, S., Duarte-Abadia, B., Stoltenborg, D., & Vos, J. (2020). Water governmentalities: The shaping of hydrosocial territories, water transfers and rural–urban subjects in Latin America. *Environment and Planning E: Nature and Space, 3*(2), 399–422.

Finnish association for nature conservation, Suomen luonnonsuojeluliitto (NA) Vapaa Kemijoki. https://www.sll.fi/mita-me-teemme/vedet/nain-toimimme/vapaa-kemijoki/.

Sieriniemen siunaussoutu 17.–18. August, 2020. Youtube video https://www.youtube.com/watch?v=HBbvygdd75c.

Kansan uutiset. (2021). Pohjoisen taiteilijat ryhtyivät isolla joukolla puolustamaan vaelluskaloja – Vesa Puurosen sakkotuomio lohigraffiteista käynnisti taidehuutokaupan. *Kansan uutiset*, 1 June. https://www.kansanuutiset.fi/artikkeli/4519899-pohjoisen-taiteilijat-ryhtyivat-isolla-joukolla-puolustamaan-vaelluskaloja-vesa-puurosen-sakkotuomio-lohigraffiteista-kaynnisti-taidehuutokaupan.

Kemijoki Oy. (2020). Kemijoki Oy on jättänyt muistutuksen koskien Lapin ELY-keskuksen hakemusta Kemijoen voimalaitosten kalatalousvelvoitteiden muuttamiseksi. 10.8.2020. https://www.kemijoki.fi/viestinta/tiedotteet-ja-uutiset/kemijoki-oy-on-jattanyt-muistutuksen-koskien-lapin-ely-keskuksen-hakemusta-kemijoen-voimalaitosten-kalatalousvelvoitteiden-muuttamiseksi.html?opened=988.

Knill, C., & Liefferink, D. (2013). *Environmental politics in the European Union*. Manchester University Press.

Kokko, K. (2011). Eurooppalainen ympäristöoikeus korkeimman hallinto-oikeuden lainkäytössä. *Lakimies, 3*, 475–503.

Krause, F. (2011). *River management: Technological challenge or conceptual illusion?* Salmon weirs and hydroelectric dams on the Kemi River in northern Finland. https://doi.org/10.1007/978-3-540-77568-3_19.

Krause, F. (2012). Managing floods, managing people: A political ecology of watercourse regulation on the Kemijoki. *Nordia Geographical Publications, 41*(5), 57–68.

Lapin Kansa. (2021). Voimalaitoksen vastustajat soutivat jälleen viikonloppuna vapaan Sieriniemen puolesta. Heille ympäristöuhkana näyttäytyvä voimalaitos olisi vesivoimayhtiön mukaan osa ilmastokriisin ratkaisua. *Lapin Kansa*, 11 August. https://www.lapinkansa.fi/taistelu-sierilasta-jatkuu-kemijoki-oy-hakee-jatko/3871702?fbclid=IwAR1_Cy4MwwMatUaCnNyYlYsZe4pmnOusy-9gs3R9Q8_-FW3Q_Cc0ERmLEkA.

Lapland centre of economic development, transportation and environment, Lapin ELY. (2015). Vesien tila hyväksi yhdessä. Kemijoen vesienhoitoalueen vesienhoitosuunnitelma vuosiksi 2016–2021. Raportteja 89. https://www.doria.fi/bitstream/handle/10024/124007/Raportteja%2089%202015.pdf?sequence=2&isAllowed=y.

Lapland centre of economic development, transportation and environment, Lapin ELY. (2021). Ehdotus Kemijoen vesistöalueen tulvariskien hallintasuunnitelmaksi. LAPPI_TRHS_Kemijoki_2022_2027.pdf.

Law on flood risk management, Laki tulvariskien hallinnasta. (2010). https://www.finlex.fi/fi/laki/ajantasa/2010/20100620

Law on nature protection, Luonnonsuojelulaki. (1996). https://www.finlex.fi/fi/laki/ajantasa/1996/19961096#L10P65.

Law on water, Vesilaki. (2011). https://www.finlex.fi/fi/laki/ajantasa/2011/20110587?search%5Btype%5D=pika&search%5Bpika%5D=vesivoima.

Lemke, T. (2015). New materialisms: Foucault and the 'government of things'. *Theory, Culture & Society, 32*(4), 3–25.

Linton, J. (2014). Modern water and its discontents: A history of hydro-social renewal. *WIREs Water, 1*, 111–120.

Lohijokitiimi, Salmon river team. (2021). Historia. https://lohijokitiimi.fi/.

Massey, D. (2007). *For space*. Sage.

Massumi, B. (2009). National enterprise emergency: Steps toward an ecology of powers. *Theory, Culture & Society, 26*(6), 153–185.

Matkailutoimittajien kilta. (2021). Lapin puolueet vetoavat lohen puolesta. 21 June. file:///C:/Data/monica/AINEISTOT/Vapaa%20lohijoki/Lapin%20puolueet%20vetoavat%20lohen%20puolesta.mhtml.

Mollinga, P. P. (2019). Cultural political economy and critical water studies: An introduction to the special themed section. *Water Alternatives, 12*(3), 788–801.

Mustafa, D., & Halvorson, S. J. (2020). Critical water geographies: From histories to affect. *Water*. https://doi.org/10.3390/w12072001

Rattu, P., & Véron, R. (2016). Towards a Foucauldian urban political ecology of water: Rethinking the hydro-social cycle and scholars' critical engagement. *Foucault Studies, 21*, 138–158.

Rovaniemi city, Rovaniemen kaupunki. (2020). Ympäristölautakunta, 26 August.

Rusca, M., & Di Baldassarre, G. (2019). Interdisciplinary critical geographies of water: Capturing the mutual shaping of society and hydrological flows. *Water*. https://doi.org/10.3390/w11101973

State Council, Valtioneuvosto. (2018). Valtioneuvoston päätös. DN:o YM4/577/2016. Annettu Helsingissä 1 päivänä helmikuuta 2018. Asia: Luonnonsuojelulain 66 § 2 momentin mukainen valtioneuvoston päätös koskien Rovaniemen ja Itä-Lapin maakuntakaavan hyväksymistä. https://valtioneuvosto.fi/paatokset/paatos?decisionId=0900908f805730e9.

State Council, Valtioneuvosto. (2019). Komission suositukset Suomelle vesienhoidon ja tulvadirektiivin täytäntöönpanosta. 14 March. https://valtioneuvosto.fi/-/1410837/komission-suositukset-suomelle-vesienhoidon-ja-tulvadirektiivin-taytantoonpanosta.

Suopajärvi, L. (2001). Vuotos- ja Ounasjoki-kamppailujen kentät ja merkitykset Lapissa. Lapin yliopisto.

Supreme administrative court, Korkein hallinto-oikeus. (2019). Valtioneuvoston päätöstä Natura-poikkeuksesta koskeva valitus (Rovaniemen ja Itä-Lapin maakuntakaava, Kemihaaran suot), https://www.kho.fi/fi/index/paatoksia/muitapaatoksia/muupaatos/1554870994926.html.

Swyngedouw, E. (2009). Political economy and political ecology of the hydro-social cycle. *Journal of Contempory Water Research & Education, 142*, 46–69.

Swyngedouw, E. (2014). 'Not a drop of water…': State, modernity and the production of nature in Spain, 1898–2010. *Environment and History, 20*, 67–92.

Tennberg, M., Vuojala-Magga, T., Vola, J., Sinevaara-Niskanen, H., & Turunen, M. (2018). Negotiating risk and responsibility: Political economy of flood protection management in Northern Finland. In T. Hiyama & H. Takakura (Eds.), *Global warming and human–nature dimension in northern Eurasia* (pp. 207–221). Springer.

Usher, M. (2014). Veins of concrete, cities of flow: Reasserting the centrality of circulation in Foucault's analytics of government. *Mobilities, 9*(4), 550–569.

Vogler, J. (1999). The European Union as an actor in international environmental politics. *Environmental Politics, 8*(3), 24–48.

Vola, J. (2020). Untied resource as a threa/–/t/–d for social fabric(ation). In M. Tennberg, H. Lempinen, & S. Pirnes (Eds.), *Resources, social and cultural sustainabilities in the Arctic* (pp. 41–55). Routledge.

Walker, R. B. J. (1992). *Inside/outside: International relations as political theory*. Cambridge University Press.

Walters, W., & Haahr, J. (2005). *Governing Europe: Discourse, governmentality and European integration*. Routledge.

Wilson, N. J., Harris, L. M., Nelson, J., & Shah, S. H. (2019). Re-theorizing politics in water governance. *Water*. https://doi.org/10.3390/w11071470

WWF. (2020). Pohjois-Suomen aluehallintovirastolle. Asia: Lapin elinkeino-, liikenne- ja ympäristökeskuksen hakemus, Kemijoen Isohaaran, Taivalkosken, Ossauskosken, Petäjäskosken, Valajaskosken, Vanttauskosken, Pirttikosken ja Seitakorvan sekä Raudanjoen Permantokosken voimalaitosten kalatalousvelvoitteiden muuttaminen. Dnro PSAVI/932/2017. 10 August, 2021. https://wwf.fi/app/uploads/8/4/h/6zs8d9bkep67jjm6m6ql0f/wwf_lausunto_kemijoen_kalatalousvelvoitteiden_muuttaminen_psavi9322017.pdf.

YLE. (2019). Jasper Pääkkönen Kemijoen kalateistä: "Käsittämätön tilanne, että valtionyhtiö taistelee valtion asettamia tavoitteita vastaan." 6 November. https://yle.fi/uutiset/3-11053917.

YLE. (2020). Kemijoki Oy:n rakennuksia spraymaalattiin vastalauseena vesivoimalle – teosta epäilty Oulun yliopiston professori kiistää rikoksen. 3 December. https://yle.fi/uutiset/3-11680307.

Ymparisto.fi. (n.d.). Tulviin varautuminen. https://www.ymparisto.fi/fi-FI/Vesi/Tulviin_varautuminen/Tulvariskien_hallinta/Tulvariskien_hallinnan_suunnittelu/Tulvariskien_hallintasuunnitelmat/Kemijoen_vesistoalueen_tulvariskien_hall%2829318%29.

Zito, A. R., Burns, C., & Lenschow, A. (2019). Is the trajectory of European Union environmental policy less certain? *Environmental Politics, 28*(2), 187–207.

Chapter 7
Living by the River: Means, Meanings and Sense of Place

Anna Stammler-Gossmann

Abstract The chapter discusses the river's central role for the residents of the small northern settlement in Sakha Yakutia (Northeastern Russia) and how living-by-the-river gives people more conscious thoughts on the distinctiveness of their "river place". The Amma River has an iconic image as the region's cleanest and most beautiful river, by which people proudly situate themselves. However, highly aestheticized narratives of the river through which this place is actively sensed do not entirely concur with the practical ways residents experience "fluid" qualities of the riverine environment. Local waterscape also demonstrates its "troubled" properties, which villagers are painfully aware of. By some standards, the community is deprived of resources because the riverine environment seems to provide little material value regarding the main economic activity – animal husbandry – and is exposed to increased flooding. This chapter analyzes how a place may anchor human lives despite the tension between "means": "insufficient" physical settings of waterscape scenery and its rendered meanings. The chapter examines the complexity of how a place is sensed and makes sense through various forces, forms of agency, and encounters between people and the river.

Keywords Human-water relationship · Republic of Sakha Yakutia · Sense of place · Northern livelihood · Flooding · Spirituality · Human–more-than-human

---

A. Stammler-Gossmann (✉)
Arctic Centre, University of Lapland, Rovaniemi, Finland
e-mail: anna.stammler-gossmann@ulapland.fi

M. Lehtimäki et al. (eds.), *Cold Waters*, Springer Polar Sciences,
https://doi.org/10.1007/978-3-031-10149-6_7

7.1 Introduction

The visible particularity of local watery surroundings and the beauty of this location on the Amma River (Amga–Russian) cannot be overlooked. This place in Northeastern Russia (Republic of Sakha Yakutia, Churapcha District)[1] has an exceptionally distinct character that holds itself firmly – a character its residents explicitly shelter: during several conversations with locals I heard the expression "we are river people". This area's distinct atmosphere, its *genius loci*, is "sensed together" by locals and outsiders. The Amma River has an iconic image as the region's cleanest and most beautiful river and is widely celebrated in Sakha Yakutia.

Nevertheless, during my prolonged stay in the settlement, I realized the aesthetics of the watery environment broadly defining this place does not entirely concur with the practical ways residents experience its fluid qualities. In daily life, where the river provides a means to a different end, water also demonstrates its "troubled" properties, which villagers are painfully aware of.

The riverine environment's physicality is considered insufficient for the profitability of their single economic activity – animal husbandry. Riverine grass's poor natural mineral content negatively affects the meat's quality and, accordingly, people's income.[2] Furthermore, the Amma River is liable to dangerous floods. From a dreamlike state, its water has turned with increasing frequency and ferocity in the last decade, resulting in fearsome and destructive floods. Moreover, disastrous occurrences (structure fires, suicide, deadly diseases) happened in a chain the year before my fieldwork, stressing anew the community's human–water relations in an unexpectedly spiritual way.

This settlement is endangered and deprived of resources and services by some standards. A paradox of these sites is that their tangible "means" – viewed through the lens of productivity, economic logic, or risk exposure – seemingly oppose highly aestheticized narratives through which these areas are actively sensed. However, these places continue holding their meaningful significance. What is their power?

There is still much to learn about how human and more-than-human world make themselves at home in contested settings. This research field is still in its infancy. Understanding the sense and resilience of these locations is important regarding the growing number of environmentally exposed, disaster-prone, or declining and abandoned sites.

Based on the Sakha case study, this chapter explores how a place may anchor human lives despite the tense relationship between the material-physical settings and their rendered meanings. It analyzes why, over time, the river's meanings form

[1] The Republic of Sakha Yakutia is the largest administrative entity within the Russian Federation and the world's largest country subdivision.

[2] The information on meat quality and other events described in this chapter was trustfully shared with me. Due to the matter's sensitive nature, the settlement's name is purposely unmentioned, following the principles of confidentiality and ethical guidelines for anthropological research by AAA and IASSA.

the symbiotic character of the human-water relationship, what the Sakha people[3] claim to their dwelling site, and how they facilitate cohabiting practices and continuity of place.

Considering this area's strong distinctiveness, I frame my study under an umbrella concept – "sense of place" – which can be approached from many different angles. In this account, "sense of place" refers to the experiential and expressive ways that place is voiced, held, contested, and struggled over (Feld & Basso, 1996, p. ii). This attitude allows me to settle the binding and not concurring tangible values and nonmaterial features of the place.

My assignment on this site is to view it through two environmental aspects that, as Marsh highlighted, must be addressed by any broad concept of place: "means" (the physical support it provides) and "meanings" (the intangible rewards it offers) (Marsh, 1987). Utilizing this approach, I examine how village residents try balancing the deficiency of "means" (tangible) and "meanings" (intangible) to make the seeming tension as slight as possible. This study is based on materials collected in Sakha Yakutia between 2013–2019 and 2021, including intensive fieldwork on location and trips in the Churapcha District.

7.2 Place and Water

Addressing people, water, and place from a "synthesized" perspective and linking it to positive and negative settings is a large and challenging task. Different empirical studies in diverse cultural contexts deal with this perspective's particular components, demonstrating the complexity of how places naturalize other worlds of sense – from areas of contentment and proudness to ones of tropes of marginality and anxieties (Feld & Basso, 1996; Twigger-Ross & Uzzell, 1996; Marsh, 1987; Montefrio et al., 2016). A multiplicity of perspectives of sensing places resonates in greater diversity when we address the matter of meanings through its watery relations. Challenging the modernist nature-culture dualism has opened new perspectives on humans' relationship with water. This de-centring turn, away from humans, forces researchers to rethink the categories of analysis. An increasing number of theoretical concepts refers to the complex interdependencies between "countless beings that share terrestrial dwelling" (Abram, 1996). Empirical research on seeing water as "other than human" focuses on analyzing conflicting frameworks in valuing water. These frameworks are at play when concerning water governance and managing different types of knowledge (see Anderson et al., 2019; Strang, 2016).

Russian regional studies particularly emphasize the role of Siberian rivers as a dominant category in the cosmological worldview, the centrality of the "river factor" in the forming regional material cultures, and as an element of the ideological

[3] The Sakha (Yakut) people comprise approximately half the population, roughly one million. Sakha is a self-designation of the northernmost Turkic-speaking ethnic group in northeastern Russia; "Yakut" is the ethnonym in Russian.

construct (see Pavlinskaia, 2007; Pudov, 2020). Increasing flood events in Sakha Yakutia bring on the research agenda a growing interest in the issues related to water as a "troubled" element in the external world in which an individual culture resides (Hiyama & Takakura, 2018; Stammler-Gossmann, 2012; Vinokurova et al., 2016).

This chapter is less about such a place as the site of a power struggle than a focus on understanding the human–river relationship – universally valid for contemporary residents of the settlement under consideration. The cross-cutting perspective of the cultural, social and more-than-human dimensions of nature offers the possibility to sense a place by encountering those cohabiting there.

7.3 "Means" of Living-by-the-River

Affinity with the river is of an individual and collective nature, coupled with the pride in belonging to the "most beautiful place" (personal communication). When people introduce themselves to those "out-of-place," they start by pointing out their belongingness, using the fixed expression: "I am from the village on the beautiful Amma River." In online forums, on the Churapcha District website, the location's scenery is the most highly elaborated and acknowledged point regarding the friendly rivalry between communities. However, this designated "positive distinctiveness" of place might not always appear convincing as a suitable site for living. Rather, this distinctiveness could be seen as "discrepant" from the perspective of its material settings and physical details.

Seventy kilometers by car from the district's administrative center to this place took us – the passengers – unexpectedly around 12 hours. Spotty summer rain turned the only road connecting the village with the outside world into a driving adventure. Some damaged cars cast to the roadside pointed out that the geographical proximity measures here hold alternative representations of distance. The village totally depends on this single transport line. Access to essential goods and services becomes quite challenging for the community in spring and autumn when the road is unfit for traffic. Inadequate transport connectivity in the northern low-density remote areas significantly constrains rural residents.

Normative standards of life in the village differ considerably from the local center and regional capital: low salaries, limited cash, no food or other supplies for extended periods, high production costs due to transportation expenses, and no running water or sewage facilities. Built on the Amma River in 1932, this village of approximately 842 people (Report, 2021) shares, with many other places in Sakha Yakutia, a history of Soviet settlement. Vinokurova points out that many of these places were established near a river via the mindset of Soviet planners and their experiences from Central Russia, who did not consider that the northern river regimes differ strongly from the upland rivers of the southern regions in seasonality and ice thickness (Vinokurova, 2018).

The concept of closeness to a river as an economic resource fulfilling various functions – production (fishery), carrier (waterway), and regulation (irrigation) – was also not always in correspondence with Sakha pastoralists' engagement with nature. The Sakha people's traditional dwelling was clustered around the lakes rich with carp, surrounded by fruitful pastures for animals and forests for hunting (personal communication with Vinokurova Uliana). The dispersed dwelling places in these traditional sites were entirely abandoned by the end of the 1950s with the emergence of centralized supra-local communities (see Filippova, 2007).

The actual biophysical living settings by the Amma River offer less, as might be expected, to the villagers for whom cattle breeding is the base of the local economy. Riverine grasslands as the food base for grazing animals and hay grounds are concerning. Poor-quality grass causes severe problems: emerging diseases and the birth of weak calves. Cows and horses are in nearly every household and are entirely grass-fed. Supplementary feeding is hardly practiced because of the supplements' high costs and transportation problems. The place seems deprived of the primary resource crucial for successfully raising livestock, considering its almost absolute dependency on natural grazing.

Moreover, the Amma River provides a home for fish. Although it might sound unusual, fishing in the river is uncommon, as the fish are rarely consumed: Sakha culinary traditions value other fish that is locally unavailable, such as whitefish species or carp from a lake. Freshwater fish from local lakes or the lower reaches of the large rivers are considered a delicacy in central parts of Sakha Yakutia, not an everyday food, unlike meat and dairy products. "Of course, you can eat fish from the river, or my father can catch it for our cat," a young woman told me, "but we prefer carp from a lake and good fish from the Far North. As we do not have relatives in the northern districts who could supply us with the white fish, we only buy it for some special occasions." Thus, the availability of river fish does not compensate for the "low" quality of the local meat.

The place's "fluid physicality" is a growing concern among locals from another perspective: The Amma River has increasingly become a volatile neighbor. Every spring, the villagers are alarmed by the ice jam that could result in water levels rapidly rising. Sakha Yakutia has been one of Russia's regions most exposed to flooding in the last two decades. According to the regional reports, Amma River floods have become more frequent over the last ten years (see Mozolevskaia, 2019). Also, an annual spring flood with a naturally high water level in the Amma River, so-called "black water" (*khara uu*), regularly splits the settlement into parts, causing connectivity problems.

From the rationalist perspective, the location seems poor in compelling terms. However, people live in this lean environment, feeling it is congruent with their values and aspirations.

7.4 Living by the River and Drinking Water

Living by the river is one way that makes human-river interaction here particular. Residing close to the river belongs to a special category of river experience, as the river's energy "flows before us, around us, or over us" (McMillin, 2011, p. 6). This proximity, according to McMillin, is the simplest, most basic, and most common way of "making contact with a river." Similarly, people who live alongside the Amma River have the same relationship with water: "work and rest," "drink and bath," "flirt, honeymoon and fool" around the river, as McMillin (2011, p. 27). However, the most obvious and specific reality about living by the river here is the practice of "harvesting" drinking water – the substance most vital to everyone's continued existence. The village has no water supply system, treatment facility, or water wells.

In many of the region's small remote settlements, clean drinking water comes not from the tap via a piped network but in ice blocks taken primarily from the river or lake. Immobilized in ice for most of the year, the Amma River is a water source for the year. Hard physical work requires a multi-layered collective effort, good knowledge about the river's ice conditions – "right" ice thickness and "right" time for harvesting – and preparing the ice ground before harvesting.

Preparing the ice blocks also helps the community's members in need. Relatives living elsewhere might join this activity in November, when the ice harvesting usually starts. Later, when the weather warms up, the ice blocks stored outside must be stockpiled in the permafrost underground cellars. Cellars might be shared among households without cellars or with those whose cellars were damaged by flooding. "Ice water" (*muus uuta*), can be shared among relatives when ice water runs out. Drinking water, extracted in its solid form, becomes a series of events, granting continuity of life, linking community, and strengthening family ties. Some emphasize the repeating feature of ice harvesting, saying, "and so is every year" (personal communication).

Thus, the river's nourishing materiality shapes the village's social life. As a consumable and biological unit for locals, drinking water in its solid form turns into a substance imbued with meanings, values, and ideas. The quality of "ice water" is a matter of pride in the community, especially compared to other places where drinking water might have a taste of industrial contaminants. Water quality monitoring by the regional Ministry of Ecology showed that the Amma River water is far cleaner than other central water sources in the republic (Ministry of Ecology, 2020). The term "clean" accompanies the epithet "beautiful" awarded the river.

When initially visiting the place, a woman from the neighboring district conveyed her impression when she saw an ice cube from the river: "I was surprised by the crystal-clear bluish tint to the ice cube and the very good taste of the water. The water is so clean that the locals get passionate when they talk about their water. They cannot drink any other water and even find it disgusting to swim somewhere else than in the Amma River." In another story I heard about a couple who greatly missed "their" water after moving from the village to the administrative center. Whenever they visited relatives in the village, they took ice blocks back with them.

The possessive pronoun was quite common among the people I met when they talked about their place – "our" or "my" river, "our" or "my" water. These reflections have a particular reason, considering the quality of drinking water from other water sources in the district that have been rated nonstandard and of poor quality – naturally and due to contamination (Report, 2021). Many in the administrative center depend on bottled water from the shop.

In the Republic of Sakha Yakutia, where over 60 mining companies operate, the pollution of the large and small rivers has become a particularly acute problem in the last few years (see Burtseva et al., 2022). This issue is the focus of public attention and the main force of increased ecological activism in Sakha Yakutia (see REGNUM, 2018). For locals who relate to the river through habitual everyday practices and the social life of the river, these relationships raise their awareness of their personal and collective responsibility to care for "their" water.

Although the Amma River, unlike other rivers and lakes in the district and the whole region, belongs to the cleanest and "undomesticated" waters, local worries about potential harm from engineering interventions, deforestation plans, and tourism activities are extensive. In many other cases, realizing damages to water might be delayed compared to the destruction of other natural components, such as forests. Such damages might be less visible and unnoticeable. However, water's essentiality for community existence and proximity to the river has made locals more receptive to environmental changes, primarily helping reduce the gap between the not-too-late versus too-late dichotomy.

This sensitivity also exists because the locals see the Amma River in regular public discussions concerning the catastrophic failures of hydroelectric facilities, industrial contamination of waters, or water's unpredictability in the area, fluctuating between drought and deluge. Experienced or potential critical situations with water amplify local fears related to the ambitious Soviet idea of turning northern rivers southwards, which materialized in the Churapcha District (Stammler-Gossmann, 2012). This project transporting water from the Lena River, the largest in the region, to the small Tatta River in the neighboring district through Churapcha, occurred between 1992 and 2004 (REGNUM, 2004), resulting in establishing a heavily developed net of water reservoirs, locks, dams, pipelines, and bridges.

Today, the idea of hydropower – the symbol of human supremacy over nature and progress in political reference to the Soviet time – has been largely replaced by a growing awareness of the impact of human interventions. Thus, just as humans "make things of water" (Linton, 2010), water in the village makes things of humans, too. Linton's thesis that "neither can water exist without us" (ibid., p. 4) is reflected in the symbiotic character of the community's relationship with the river determining the local environmental agenda. People's concern can take on a very emotional expression as the title of a paper by a well-known scientist and environmentalist in the region indicates: "I will fight for the Amma River until my last breath" (Maksimov, 2020).

Accompanying environmental activism metaphors attributed to the Amma River – like "masterpiece of nature," "pearl of Sakha Yakutia," "sacred charm of our nature," "one of the last cleanest rivers of the world" – reinforce local concerns. Treating a river's meanings as "having both matter and energy," as McMillin suggests, sets things in motion (McMillin, 2011). Protective thoughts are driving forces for environmental movements like grassroots associations "Youth of Amma River" and "Protect Amma River," or activities established in the village's ecological summer school.

Living by the river with the "cleanest" water provides an icon largely contributing to maintaining "in-place distinctiveness," while practical aspects around drinking water issues raise community awareness of the fragility of relations between people and material settings.

7.5 Between "Mundane" and "Aesthetic"

The river's ubiquity and visible appearance here are more straightforward than in another Amma River settlement I visited – where I knew the river was out there somewhere. Maybe that is why I heard more often in this place how people had initially approached the river, which is always feminine in the Sakha language. Usually, rural Sakha respectfully call the river *ebe*, "grandmother", in their daily or symbolic practices: "Are you planning to visit the grandmother? How is the grandmother today? How is the water situation by the grandmother this spring?"

The Amma River, in its aestheticized form, voiced by poets, musicians, artists, and writers, is represented as a *kyys* – a beautiful, innocent, and gentle "young woman" – while the river itself is used as a metaphor for time: the "river of life" (Strang, 2005). The river's feminine nature can symbolize a woman's life cycle. The "Girl Amma" sculpture in the neighboring Amma District was deemed "an expression of our respect and worship of the Mother River, our Grandmother" (Monument, 2011).

A distinguished image of the river within the republic was acknowledged on the National Day of the Amma River, introduced in 2018. Water's sensual primacy emerges in the regional naming practices of administrative districts and settlements, most of which derive their names from rivers or lakes. The district's name Churapcha comes from Lake Churapcha. However, villagers tend to associate themselves more with the Amma District by sharing the river with their river neighbors, as the settlement was once administrative unit of this district.

I felt the local renderings of the place based on aestheticized narratives and poetic forms strongly connected to the river's dramatic visual presence. McMillin writes that closeness to water flow does not involve going anywhere and brings different ways of thinking about and experiencing rivers (McMillin, 2011, p. 27). Seeing the physical stream just before us includes following McMillin's idea of "using it as a tool for thinking" (ibid., p. 32). Remarkably, the small remote village gives the impression that the river, with its picturesque appearance, forcefully links water and everyday experiences to aesthetic forms.

The Amma River's uniqueness and beauty might be recognized in the special attention villagers pay the built environment's aesthetics. The river's aesthetic power resonates with the emphasis on nature as a place's foundation and the enjoyment of a quality scenic environment. Although the main economic activity, animal husbandry, viewed through the normative lens of profitability, is unsuccessful, by no means does the village leave a mark of being depressive. Rather, the housing patterns and beautifully decorated houses competing with another might impress visitors. Nevertheless, personalized dwellings and the homes' individual features reveal community-based values and the settlement's structure.

Locals are pleased about the place planning where the river is central, with all the essential social institutions clustered around it: kindergarten, school, administration offices, post office, shops, and other public units. This place at the river is a fixed point where things happen, and every human-river movement and dynamic is easy to follow. The river's nearness carries a kind of subconscious "internalization" of its aestheticized representative forms into mundane life. This internalisation could be considered an active process facilitating human movement through its environment, which may go, as Ingold states, "along multiple sensory pathways" (Ingold, 2000, p. 18). One "sensory pathway" might be observed in the impressive creativity and intensity of this rural place's social life. Various formal and informal groups include the Neighborhoods' Council, Elders' Council, handicraft club, folk theater, Nordic walking group, singing and dancing groups, and an environmental initiative group. Small ateliers produce handmade furniture, jewelry, carpets, and hats.

We might see the connection between symbolic representations of geographical distinctiveness by which residents identify themselves and specific activity patterns for the community's well-being. The Neighborhoods' Council forced a local municipality to a "full sobriety" rule in 2016, with a complete ban on selling alcohol. In 2018, this village's initiative, cooperating with seven other municipal units from other districts, was presented at the State Assembly of Sakha Yakutia session with a request to illegalize alcohol sales. A Russian central newspaper published a report about another initiative of a farmer from the village who recently started growing vegetables. He had supported people in the community and all single fathers in the district with his potatoes and carrots (Serebrovskaia, 2020).

However, not all initiatives contributing to creating a sense of place are shared among the villagers, especially concerning the growing number of tourists. Recently, this location's picturesque natural surroundings have attracted increasingly more non-organized tourists, and some ideas about developing more controlled and reliable tourism activities are in the works. People's reaction in the community and full-blown essays in regional online forums have mainly focused on the need to protect the place from visitors for the river's sake. Nevertheless, the growth of tourism has been in the municipal and regional development plans for years. Some younger people consider wildlife tourists' spots elsewhere inspiring. Others may see shielding the place's charming particularity as a more appropriate strategy than satisfying the temptation of tourists.

As the scale of "outsiders' invasion" strongly depends on the summer season and road conditions, this issue is framed until now as a question of potential threats and not (yet) as a matter of specific power relations. While some are unhappy about the pollution left on sandy beaches by the "party-visitors," others appeal to the need to increase income and introduce organized tourism instead of having irresponsible campers. The reluctance regarding this issue could be seen in the "frozen" tourism development plans, such as no promotional activities, and the appeals to keep tourists away or, in the sentence I heard, "…better we would not have the road; our nature would be safer."

7.6 River and Grass

The human and more-than-human bond includes other central figures in the settlement: cows and horses. As noted, the quality of the riverine grass as a primary resource for cattle breeding is insufficient to be profitable. Living-by-the-river also means the river determines soil fertility and terrestrial production. I was told that regularly overflooding the grasslands "takes minerals away from the grass and grass cover is not good either." As revealed by the laboratory investigation of grass and water quality in 2012 (see Socio-economic characteristics, n.d.), these deficiencies cause the ongoing problem of maintaining a healthy brood cow. Moreover, Sakha people, especially rural residents, believe the riverine grass has dramatic consequences for meat taste and fat content.

Many assert that *yriakh oto*, "river grass," cannot compare with highly valued "juicy" grass from *alaas*, an open space formed due to thermokarst depression with fertile meadowland and lake surrounded by forests. According to widespread opinion, "You can immediately taste the difference between the meat from *alaas* and the meat from the not *alaas* area" (personal communication). Therefore, it was unsurprising to hear people say that "nobody wants to buy our meat" in a region where the commodified meat products are tightly bound with their genesis. In Yakutsk, Sakha Yakutia's capital, regional meat and dairy products for sale have a label with the place of production: the district's name.

The meat from Churapcha *alasses* has a good reputation, but not from the river place deemed inferior. When I asked a woman from another village if she bought meat from this place, she replied, "No, absolutely not. I bought it just once and used all possible spices and herbs during cooking, but it did not improve the taste at all." The "bad" quality of the grass and, accordingly, the meat's quality is a burden of the place, which even the locals acknowledge. Thus, even those who produce meat just might prefer to buy non-local meat for their consumption.

The unfavorable reputation of the local products has another implication. The matter of grasslands goes beyond laboratory measurements and has a particular cultural meaning for the Sakha people. "Riverine grass" cannot compete with valuable grassland from *alaas*, which symbolizes pastoralist culture. Images of *alaas* are visualized in regional museums exhibitions, theater performances, songs, and art (see Stammler-Gossmann, 2010). The representational power of *alaas* has more pronounced ethnic and nostalgic components of rural Sakha. Sometimes, people respectfully approach *alaas* like the river as *ebe*, a "grandmother" (see Prokopeva, 2015, p. 138).

Grass quality is essential for humans and cattle, but its availability is vital to continue animal husbandry, too. The local "grass stories" have always emphasized this aspect as complementary to the animal and the environment. Living by the river can give the image of an inferior grass another meaning: lending a more distinctive feature of the place, justifying its benefits, and even making people proud.

Unlike other grassland areas in the Churapcha District often suffering from drought, no shortage of productive grassland in the community exists. The availability of pastures turns the grass issue into an advantage. The village residents emphasize this advantage, for example, by drawing the line between those born there and those considered newcomers: "They say they are local, but in reality, they moved here with their cows because of the dry summers and because there was no grass in their places" (personal communication). Local interpretations may redirect the grassland quality issue from the river's closeness to the consequences of "bad management." I heard people say that "in the Soviet time, a good work was done for pastures: leveling the floodplain soil and removing willow bushes."

In this case, quantity compensates for "low" quality. A municipal report stated that "although animal husbandry is not highly profitable in the village, people consider it satisfactory to meet local needs" (Programme, 2007). An official from the administration mentioned that the local population is reluctant to governmental attempts to establish extensive grain cultivation as a supporting cattle breeding sector and a possibility of diversifying economic mono-dimensionality in the community (personal communication).

In the "grass stories," the farmers have repeatedly emphasized their ability to cope with the challenges. One practice is keeping grass physicality and animal needs in balance by carefully monitoring the grass where the river is an integral part. Knowledge of grass reflects in the grass terminology the village uses that is not poorer than animal vocabulary. The Sakha people commonly use a few terms for grass. However, the local lexicon extends this terminology by applying phrases connected to the river: "grass after the first flooding when the ice starts to move," "grass after the black water of the second flooding is gone," or "grass after the water went back quickly" (personal communication).

Grass monitoring is an integral part of the community's social life, including different ways of sharing information. An elderly man told me, "In the early days, I went very often to the river. I still go there from time to time, but nowadays, you can also get information from the internet, which is shared by talking around." Compensative arrangements in the sensitive matter of "river grass" embrace another strategy: One family member often has a job in the budgetary system or runs a small business.

A place at the river providing little economic value still has the powerful meaning as a place where only cattle breeding as the leading economic activity and as a base of the community's continuity is possible.

7.7 River and Floods

Every spring, for the last two decades, many in Sakha Yakutia worried that vast blocks of "jammed" ice breaking up on the rivers would unleash an excess of water. Over 90 settlements in Sakha Yakutia were in the government's relocation plan of 2010 due to the threat of overflooding (Decree, 2010). The last severe flooding in the region was in 2018 when 15 districts and 63 settlements (including the Amma River village) were affected (see Tananaev et al., 2021, p. 126). Ice jams are highly unpredictable and difficult to anticipate; during the spring freshet, the water in the Amma River can rise as fast as several meters per day (Tananaev et al., 2021, p. 130). Because the Amma River is free from hydroengineering structures, floods occur naturally, primarily following changes in the ice regime and ice-jam formation. However, *khalaan uuta*, "flood water," could be a blessing or disaster.

A plate on the tree of the riverbank in the village reminds of the most severe flood in 1998. Two months before my first visit in 2013, there was less dramatic but still massive flooding. A farmer, whose house had been the closest to the river, reported losing eight of his ten cows, which drowned in the river that spring. Others remembered how vast piles of ice damaged the houses and the fences around them.

Remarkably, the flooding issue is not reflected anywhere in the publicly accessible annual local and district administrative reports. Drought, not excess water, is considered a contemporary challenge of the Churapcha District. The district's main territory has low rainfall and hydrology dominated by lakes and water reservoirs. Only a few remote settlements out of 30 districts have experienced water's destructive power.

In 2018, after extensive flooding in this village and another area close to a water reservoir came the extreme arid summer of 2021. Like many other places in the Sakha Yakutia region, some of the Churapcha District's areas experienced record-breaking, lasting forest fires. A woman from the site near Churapcha, whom I met in the capital city, described the severe problem of 2021concerning hay and having to purchase it elsewhere: "The family could not make hay for winter; everybody was fighting fires, day and night. Thus, the summer of 2022 should also be arid."

"River people" often refer to their "water place" in the context of drought – a primary hazard for most of the district population. This perspective contributes to generating a specific mindset that considers the balance between the "good" and the "bad" of liminal waters. People try different ways to make sense of the fear, uncertainty, and blessings connected to the river. Though locals are very aware of the river's violent destroying power, they recognize that the other villages in the district suffer mainly from drought, not excess. Under these circumstances, people give even more thought to their sense of place and water's practical realities as a life-giving rather than a life-threatening element.

As well as reporting the damages from the massive flood in 2013, people described the year as a "wet" year, blessing the community with excellent grass growth (personal communication). Only a few houses on the riverbank were moved back to a safer place from then until now. In 2021, the owners of the houses near the river stated that they "somehow still stay; flooding is an episode, and after a while, the water is gone, and life continues" (personal communication). Those who decided to move back and left their houses on the riverbank continue benefitting from the river's nearness by using their exposed houses as summer cottages. Their yards after flooding are covered with gorgeous flowers, symbolizing the meaning that people give the space.

The disruptions the river cause may be framed by the regularity of flooding and the exceptions in the rules of nature: "Floods are often here, but a big water is only once in your life." The farmer who lost his stock during the flood did not turn this sad fact against the river. He explained this loss by the animals' behavior: "The cows did not understand the situation and went to their grazing ground that was covered by icy water. We could not get them on land." Increased irregularities of the previous years are often attributed not to nature but human actions like deforestation or mining, turning the river into a dangerous substance.

Noticeably, the idea of water as a liminal space of loss and life may appear in local narratives as a period of "reworking," where destruction is paired with redemption. One example of this thinking may involve taking the river, though sometimes fearsome, as a "prompt for transcendent vision" (McMillin, 2011, p. 32). Although few in the community declare a particular interest in religious matters, the river, as another invisible world, belongs to the register of thinking "by-the-river" thoughts, as I will show in another section.

Acknowledging the river's "otherness" begets respect concerning its destructive nature. People also approach the river as *ebe*, or grandmother, in hard times. "You should not talk negatively about the river," as parents may warn children. During the devastating flooding in 2018, the sculpture of the Amma River in the neighboring district was undamaged. A local journalist subscribed preservation of the monument from damage to "river intelligence" (Bulchukei, 2018).

The fear of violent water is intense, but the river's sublime nature dominates people's minds, which could lead to discounting the risks of living in a threatened area. Several studies on environmental risk and place attachment reveal that people tend to underestimate the probability of disastrous events, considering big floods as a once-in-a-lifetime event (see Caretta et al., 2021; Dominicis et al., 2015).

People in the village pointed out that "nobody believed that the flood of 1998 would be so devastating, even though we got warning" (personal communication). People appear to have a short memory, as the same behavior occurred before the massive flood of 2013: "We also got a warning; even an insurance representative was here." However, no one took an insurance offer. What's more, "all men left for hunting, and only women were in the village when water came up."

Nevertheless, the river can teach about changes, too. The knowledge people obtain using their closeness to the river encourages them to avoid the risk. Especially after the massive flood in 2018, the community has transitioned to more actively preparing for potential flooding. Villagers monitor the water level directly and via social networks, intensified with the availability of the internet. Before the spring flood, residents bring their cattle to safer pastures and move household articles to the homes of relatives or friends. Locals say the river might be flooding, but they can cope with it.

Beyond the physicality of flooding experiences and preparatory and redemption practices, local narratives reflect the sense of commonality during these events. People remember these times in terms of stress and great solidarity – a renewal of their community's emotional strength and resilience. In this way, thoughts are given to the place when threatened, while these same thoughts enhance the meaning of place experienced daily.

7.8 People and River Biographies

Discussing the human-river relationship without encompassing biography is impossible. People connect water and place to make other connections: between place and the outside world, self and place, or between community or family members. These connections resound for villagers with memories of collective history and individual biographies. Memories of the place add another meaningful dimension to the same body of water as a nurturing, central, or dangerous force. However, these memories are not just afterthoughts but a forceful element that pulls a sense of place along the dynamic sensory path through fascination and fear, as well as physicality and non-tangible values.

The river's crucial role in the place's history becomes its distinct point in connection to a historical event during World War II known as the "tragedy of Churapcha." In 1942, nearly 5000 people, almost every third person in the area, were forcefully relocated to the northernmost areas, mainly on the coast of the Arctic Ocean. Pastoralists from the region's central part were involved in a large-scale fishery to supply the Soviet army. Around 2000 people died within two years after relocating (see Mestnikova et al., 2019; Vinokurov, 2017). As one whose father was involved in mobilizing people as a local collective farm leader recalled, "People were just thrown on the cold Arctic coast without any experience of commercial fishing – without anything."

Battlefront needs for fish and the Churapcha region's extreme drought were justifications for the mass relocation. The drought resulted in devastating famine throughout the area. A woman from Churapcha, whose parents-in-law were relocated, described the extreme scarcity of food: "Without water, cattle perished, *alaas* lakes dried out completely, no fish was available, and people were starving. Only those on the Amma River still had their pastures and fish" (personal communication).

Such was the most extensive relocation in Sakha Yakutia's history. Losing so many people from the sparsely populated district was a tragedy that remains a collective trauma in people's memory. This historical fact first became public in the 1990s as one of the Soviet past's hidden histories. Memorials, museums of the relocation, theatre plays, schoolbooks, and annual Memorial Day remembrances serve as current reminders for everyone in the district. Processing this historical shock continues, questioning its legacy and consequences for the younger generation in the Churapcha area (Mestnikova et al., 2019).

The village could not escape the relocation. Fortunately, however, only a few families were affected and had to go to the Arctic coast. Memories of an essential connection between the place and the tragic episode of the local history are still present in the community, strengthening the community's passion for the river that saved human lives. Locals are very aware avoiding relocation was possible only thanks to the river: "Our river has rescued us: We had fish, grass, and cows, and our village did not suffer like others" (personal communication).

The life-supporting features of water and time also merge with nostalgic thoughts on other relations – between river and road. These features refer to the river and human biographies in the community's "golden age." The community's entire Soviet generation built their lives by relying on the access to the outside world that the river, as a navigable stream, provided. A small harbor and storage hall for delivering goods on the riverbank were the central points of community life.

The waterway is associated with and remembered as the most vibrant time of settlement life. Locals recount the navigation season's opening as an integral part of the river and the community's yearly life cycle. Residents describe different central events in the former community's life as following the river's flow: ice jam, first flooding, navigation opening, the arrival of the first ship, new goods, new people, and the beginning of hunting season. This time is remembered nostalgically as "a big joy for everybody in the village." Many young men from Churapcha arrived during the navigation period for seasonal work at the harbor and "contributed to the increase of community demographic statistics by falling in love with the local girls."

A storage hall provided many jobs for the villagers. The seasonality of navigation did not mean a break in the community rhythm. Goods delivered in the summer had to be stored, sorted, rearranged, transported to the local shop, and distributed to other places via the winter road, even to the administrative town, enabling more prolonged economic engagement than the navigation period itself. The waterway was associated with the village's regular and stable connection to the "bigger" world.

In the 1990s, the road replaced the waterway, radically changing the community's social life. Many lost their jobs, and the flood of 1998 destroyed the storage hall. One villager described this change as a dramatic chapter of his and the river's biography: "I worked for so many years as a storage hall worker, but everything was in vain, and the river as a waterway degraded completely."

All my conversations within the community about the road related to human mobility's absolute dependency on weather and safety, as well as high car repair costs, dust pollution, and unsuitable travel conditions for more comfortable "foreign brand" cars (*inomarki*). The road's condition is the most discussed topic on the district website's online forums. For years, the road has been a critical point in municipal and regional reports (see Nogovitsyn, 2017; Socio-economic development, 2012).

Replacing the waterway by the road turned the place from a goods distributor to a goods receiver. Despite increased mobility and interconnectedness, the road marginalized the area, taking its economic power and rearranging the community's social relationships – individuals skilled in repairing cars or who have more robust vehicles could enhance their position in the village. Construction of the road also changed the sense of spatiality, as the road ends in the village.

Nevertheless, people still call the empty place on the riverbank the "storage hall" – in remembrance of the glorious time by the river regarding the community collective and individual biography. Moreover, some villagers may say, "If something happens with the road, we still have our river."

7.9 Voice of the River: Human–More-than-Human Relationship

Several theoretical and empirical studies have advanced our knowledge about the different beings coexisting in our world and their habitats (see the Sect. 7.2). These studies more frequently use the definition "more-than-human" instead of "non-human," emphasizing the non-divorcing characteristics of the human-nature relationship. The more-than-human term confronts human exceptionalism more strongly, avoiding the prefix *non* that implies the absence or lack of something. However, immersion in these contexts of other selves remains open questions for further examination: Can more-than-human beings speak for themselves? Can they talk back, and if so, how?

The voice of something "other" is hard to grasp. The "other" does not communicate with us in the human tongue. This communication may include a variety of forces offering certain resistance to an explanation. In the following account, I present a story where the human-river relationship seems to involve a pronounced agency in this two-way communication. The river here appears as a visible source of power and not just as a backdrop for human action.

Values that people subscribed to the Amma River in economic terms and memories of the Soviet past have been more straightforward. More challenging to understand was why misfortunes mentioned in different conversations during my stay pointed again to the river's primacy. These misfortunes occurred randomly and in a shorter time: a suicide, a house fire, and fatal diseases. Important public buildings also burned down.

People did not perceive all these occurrences as separate cases but as a series of extraordinarily conspicuous matters with a common origin and possible cyclic patterns. Death and significant losses, in this case, happened in a row, affecting the whole community. Their scale frightened the people. Accidents and diseases hit particular persons and households. However, these events aroused the feeling that these mishaps were interconnected and could happen to anyone.

No one could find a rational explanation for these events and identify their causes; no one could elaborate on the perceived interrelation between matters. At the community's urging, the head of the municipality invited a prominent woman to identify the misfortunes' causes and find a possible solution as soon as possible.

The person invited was a spiritual practitioner whose powerful ability to communicate with nature is broadly recognized in the republic.

Such should not necessarily ring any religious bells, although religious contexts find their way in the community. I heard expressions emphasizing people's secular views concerning this story: "I am a modern person" and "I grew up in the Soviet Union, so I am atheist." However, even those whose religious bias may be unrecognizable to outsiders did not deny reducing a series of misfortunes to a spiritual interpretation.

Many values and forms of environmental engagement by rural Sakha have not lost their vitality due to their atheistic past. Some traditional beliefs were deliberately revived after the Soviet Union's collapse. Popular Sakha rituals function as a symbolic reenactment and the desire for certainty. Leaving small items in places with sacred meanings while driving long distances is seen as an assurance for safe travel. Putting small pancakes on the river shore during the ice movement might be connected to wishing individual well-being. In this context, Russians and other ethnic groups may also conduct traditional Sakha rituals.

Simultaneously, Sakha beliefs incorporate elements from Russian Orthodox Christianity. Every January, the celebration of Orthodox Epiphany occurs in the village at the river. An elderly man reported that he always takes water from the river during this event, and "water remains in the house for a long time, unspoiled" (personal communication). Hence, in this syncretistic context, different views assert an underlying unity. As this case related to the misfortunes shows, the boundary between secular and spiritually oriented people might be blurred. Belief and disbelief in the abilities of the invited person to solve the problem were ultimately intertwined and corresponded with fear.

I was unaware that the sad events, which I only noticed in passing, were interconnected. This episode in community life is not among the stories local people share with outsiders when discussing their place but starkly contrasts the narratives where they proudly celebrate their location's distinctiveness. Unlike "Western" approaches teaching us not to conceal sorrows, in Sakha cultural codes, sad events might be more purposely silenced for protection from possible adverse developments. First, after listening to the "spiritual expertise" story, I could understand the link between the village's seemingly disjointed occurrences and their connectedness to the river.

The spiritual authority identified a common reason for the misfortunes: The sauna built on the riverbank disturbed the river spirit. In her interpretation, all sad events related to the troubles humans caused the river: the noises, dirt, and smoke coming from the sauna. She pointed out that it was particularly distressing for the river in the wintertime when it sleeps. "It is only our human fault; you cannot blame nature for the occurring disasters like a flood or fire" was the message of the person whose animistic worldview has gained tremendous popularity in the region (personal communication).

In this way, a secular public object unexpectedly turned out to have a life-endangering spiritual essence. The community did not foresee the diagnosis, much less what the famous expert prescribed to cure the place. The quickest removal of the sauna should guarantee a return to order in the community, which greatly confused the residents: "We have been enjoying the sauna for three years, and suddenly we had to get rid of it within three days!"

Locals greatly appreciated the sauna. Years later, they nostalgically recall how much they enjoyed it. The sauna house was built by a well-known skilled worker and even consecrated by an orthodox priest. People enjoyed it year-round and could go to the ice hole on the river after their sauna session in the winter. Local TV programs cheerfully reported about the new public service. Launching the sauna was a community wish and an outcome of the municipal development program "to establish a recreational place for the inhabitants and tourism growth in the area" (Programme, 2007).

The pragmatic "means" of the attractive service came to the discrepancy with the meanings. Making a decision became a big challenge for the villagers and local administration, dividing the community into two parties, with one part urging the other to follow given advice, while the other tried justifying keeping the sauna but using it carefully: "Water from the sauna did not go directly to the river, it was kept aside and cleaned by us."

Finally, although not everybody in the village believed removing the sauna was a suitable solution, the community majority ultimately decided to remove the sauna. Those against the removal have not expressed any doubt concerning the connectedness between the river and human behavior. Considering the scale of dire happenings and the spiritual expert's powerful currency, who transmitted the voice of the river, community acknowledged this choice. The order by the mediating person who detected this strong connection between river disturbances and a series of bad events proved more potent than the blessing by the orthodox priest.

People living by a river, lake, sea, or ocean know how to behave on and around water. My friend in Finland said, "I don't know why, but my father told me when I was a child that we should not be too loud at the water. I suppose because of the echo, so we do not disturb neighbors. However, I taught my children the same way, just saying, 'Do not scream." Such is how knowledge is often transferred, and matters are remembered, often, with no justification.

While in a boat on the Amma River, I learned in the same way from a local person about being more silent when on water – listening to a story about the deadly accidents of the past: "Maybe the motorboat was too loud, maybe people were too loud, maybe some words were not spoken with a certain care – and they drowned. The river took them." His message was strong and powerful, without any single word in a language of technique or calculation.

The sauna episode made the community more conscious of the tourism development plans. The expanding flow of tourists, fascinated by the river's beauty, constrained villagers' welcome. This expansion has made the villagers more aware of possible consequences of the presence of those without a long-term commitment to the place and the people living there: "They came from the other places and behaved wrongly."

Moreover, this awareness fed the impulse to defend the river against other planned economic activities. Being passionate and proud about the river that "gives us everything," locals have become more concerned about the possible industrial use of river resources like sand, limestone, and pea gravel. The economic value of the proposed tourist activities and minerals' processing for the construction industry has not necessarily complied with the meanings subscribed to the place, becoming the community's next challenge. From the Amma River lesson, people learned that a relevant social explanation for certain things does not always exist. This knowledge about river's agency is used to make decisions and plans for the village's future.

The river was not the cause of happenings; the happenings were the outcome of human practices. Their river's clear, direct voice ensures the relationship between human and more-than-human remains balanced and reciprocal. The community removed the sauna charged with the "evil" spirit from the riverbank, but nobody dared use it elsewhere. During my fieldwork in the region in 2021, I discovered that the abandoned but robust sauna house has remained in the same place for years. Like the previous municipality leaders, the current head does not want to take any responsibility for reviving this once-popular community service. Why does the sauna house remain undemolished? Locals do not openly debate much about it.

Maybe the undestroyed sauna is a reminder that the spiritual "curing treatment" with the sauna removal was not the last treatment needed to keep human and more-than-human coexistence in balance. Perhaps any remaining uncertainty mirrors the ambiguity that is intrinsic to this relationship. Or maybe the abandoned sauna is about our incomplete knowledge of the other beings, which should be interpreted as a call to heighten our receptivity to a more-than-human world. However, people told me that afterwards, the bad things stopped, and life in the village returned to normal – that is, until now.

7.10 Conclusion

Northern waters are cold, and summers are short. Living by the river with slow-moving water and sandy shores does not make local people the most enthusiastic swimmers. Locals say, "Swimming is for the tourists, and kids go for splashing, but adults usually do not swim. You do not know what is beneath the water. There might be holes you can get washed into." Such does not sound like a distrust of the water but a reverence. The children are allowed to go to the river, but they should have a respectful fear and learn "not to be loud, especially in the evening."

The untamed Amma River is free-flowing and runs through everything for local people, even over the land. People can "battle" with it or for it, but there is no other way than a symbiotic one of living by the river. The villagers accept the challenges and fluctuations as a natural course of life. The village's distinctiveness rests not just on the presence of a beautiful river or the atmosphere intentionally created by the people but reveals itself through many layers of interaction between the human and more-than-human dwelling there.

A variety of forces, forms of agency, and encounters in this relationship exist, for which people must be constantly aware, as Boomgaard puts it, switching "from an open, welcome approach towards it to a defensive one when necessary" (Boomgaard,

2007, p. 2). Nevertheless, the unfavorable physical settings and challenges of the sometimes unpredictable river can be viewed and used by people in different ways to balance the economic "unproductivity" of place. The case study demonstrates other ways to relate "inadequate means."

Meanings applied by the community to material deficiencies of place have turned "means" into valued and worthwhile cognition. The rational means and intangible meanings of place shape each other and are mutually interdependent. No sharp barrier between their apparent contradictories exists. Such has enabled people to accept "discrepant" settings; nobody in the village has felt they live in an unmanageable environment. Moreover, the local agency has turned "discrepancy" into a rewarding distinctiveness of place, and the village residents have positioned themselves as river people versus not-river people.

To date, community receptivity to the river's rhetoric has supported the balance between materiality and values and the continuity of this water place. By sorting through discrepancies of "means" and "meanings" of place, people have created a space of coexistence of beings – a place of river people.

Acknowledgements With deep respect, I thank the residents of the Churapcha District for sharing their knowledge and generous hospitality. Special thanks to Borisova Maria Afanas'evna, Zakharkina Galina Lukinichna, Bopposova Zoia Petrovna, Vinokurova Uliana Alekseevna and Marjut and Eero Pajula for their expertise and invaluable conversations. I am also grateful for the insightful comments by Prof. Julie Cruikshank and the editors of the book.

References

Abram, D. (1996). *The spell of the sensuous: Perception and language in a more-than-human world.* Vintage Books.

Anderson, E. P., et al. (2019). Understanding rivers and their social relations: A critical step to advance environmental water management. *WIREs Water, 6.* https://doi.org/10.1002/wat2.1381

Boomgaard, P. (2007). In a state of flux: Water as a deadly and life-giving force in Southeast Asia. In P. Boomgard (Ed.), *A world of water: Rain, rivers and seas in Southern Asian history* (pp. 1–26). KITLV Press.

Bulchukei, S. (2018). *Day of the Amma River.* https://sakhalife.ru/uchrezhden-den-reki-amga/

Burtseva, E., Sleptsov, A., Bysyina, A., Fedorova, A., Dyachkovski, G., & Pavlova, A. (2022). Mining industry of the Republic of Sakha (Yakutia) and problems of environmental and social security of indigenous peoples. *Land, 11*(105). https://doi.org/10.3390/land11010105

Caretta, M. A., et al. (2021). Flooding hazard and vulnerability: An interdisciplinary experimental approach for the study of the 2016 West Virginia floods. *Frontiers in Water.* https://doi.org/10.3389/frwa.2021.656417

Decree. (2010). Concept of protection of the settlements and economic units from flooding. No. 253, May 27. https://docs.cntd.ru/document/815003558

Dominicis, D. S., et al. (2015). We are at risk, and so what? Place attachment, environmental risk perceptions and preventive coping behaviours. *Journal of Environmental Psychology, 43*, 66–78.

Feld, S., & Basso, K. H. (1996). Introduction. In S. Feld & K. H. Basso (Eds.), *Senses of place* (pp. 3–12). School of American Research Press.

Filippova, B. B. (2007). Indigenous people of northern Yakutia in the changing livelihood's world. *Nauka*. (In Russ.).

Hiyama, T., & Takakura, H. (Eds.). (2018). *Global warming and human-nature dimension in Northern Eurasia*. Springer.

Ingold, T. (2000). *The perception of the environment: Essays in livelihood, dwelling and skill*. Routledge.

Linton, J. (2010). *What is water?* UBC Press.

Maksimov, J. (2020). *I will fight for Amma River until my last breath*. https://www.sakhaparliament.ru/ru/obshchestvo/1915-iogan-maksimov-za-amgu-budu-borotsya-do-poslednego-vzdokha. (In Russ.).

Marsh, B. (1987). Continuity and decline in the anthracite towns of Pennsylvania. *Annals of the Association of American Geographers, 77*(3), 337–352.

McMillin, T. S. (2011). *The meaning of rivers: Flow and reflection in American literature*. University of Iowa Press.

Mestnikova, A., et al. (2019). Transformation of collective memory of descendants of Churapcha settlers. *Social and Behaviour Sciences*. https://doi.org/10.15405/epsbs.2019.12.04.299

Ministry of Ecology. (2020). Amma *River is under the state protection*. April 20. (In Russ). Retrieved December 19, 2021, from https://minpriroda.sakha.gov.ru/news/front/view/id/3154190

Montefrio, M., Dressler, J. F., & Wolfram, H. (2016). The green economy and constructions of the "idle" and "unproductive" uplands in the Philippines. *World Development 79*, 114–126.

Monument "Kyys Amma". (2011). Retrieved October 7, 2021, from http://amga360.ru/монумент/кыыс/амма/. (In Russ.).

Mozolevskaia, G. (2019). Spring water: What is the reason of the spring floods in Yakutia? (In Russ.). Retrieved December 5, 2021, from https://www.sakha.gov.ru/news/front/view/id/3011455

Nogovitsyn, A. T. (2017). Cooperation in rural development: Report of the head of the Churapcha District. (In Russ.). http://apksakha.ru/wp-content/uploads/2017/07/06-CHurapcha-9-16.pdf

Pavlinskaia, L. P. (2007). Rivers and people of Siberia. *Nauka*. (In Russ.).

Programme of socio-economic development of municipal unit Bolugurskii nasleg (Churapcha District) for 2007–2011. (2007). (In Russ.). https://pandia.ru/text/78/173/72759-2.php

Prokopeva, A. N. (2015). The alaas as an element of sacral landscape in the culture of modern Yakut people. *Teoriia i Praktika Obshestvennogo Razvitiia, 3*, 138–140. (In Russ.).

Pudov, A. G. (2020). Mythological conceptualisation in films. In A. O. Balabanova (Ed.), *Interpretation of the "River" film. Observatoriia Kul'tury* (Vol. 17(3), pp. 262–277). (In Russ.).

Regnum IA. (2004). Water from Lena River arrived to Churapcha and Tatta. September 1. (In Russ.). https://regnum.ru/news/316970.html

Regnum IA. (2018). The pollution of the rivers of Yakutia by ALROSA. (In Russ.). https://regnum.ru/news/accidents/2471387.html

Report on socio-economic development of the Churapcha District in 2020. (2021). (In Russ.). Retrieved January 9, 2021, from https://mo-churapchinsky.sakha.gov.ru/files/front/download/id/1743335

Serebrovskaia, E. (2020). Sakha farmer helps with vegetables the single-fathers. *Komsomolskaia Pravda*. April 20, 2020. https://www.krsk.kp.ru/daily/27120/4201948/

Socio-economic characteristics of the Churapcha District. (n.d.). (In Russ.). Retrieved January 9, 2021, from https://studbooks.net/2440593/tehnika/sotsialno_ekonomicheskaya_harakteristika_churapchinskogo_ulusa_respubliki_saha_yakutiya

Socio-economic development plan of the Churapcha District (2012–2016). (2012). (In Russ.). https://pandia.ru/text/78/635/70870.php

Stammler-Gossmann, A. (2010). Political animals of Sakha Yakutia. In F. Stammler & H. Takakura (Eds.), *Good to eat, good to live with: Nomads and animals in Northern Eurasia and Africa* (pp. 153–178). Center for Northeast Asian Studies.

Stammler-Gossmann, A. (2012). The big water of a small river: Flood experiences and a community agenda for change. Chapter 4. In M. Tennberg (Ed.), *Governing the uncertain in the Arctic: Adaptation and climate change in Russia and Finland* (pp. 55–82). Springer.

Strang, V. (2005). Common senses: Water, sensory experience and the generation of meaning. *Journal of Material Culture, 10*(1), 92–120.

Strang, V. (2016). Re-imagined communities: A new ethical approach to water policy. In K. Conca & E. Weinthal (Eds.), *The Oxford handbook of water politics and policy*. Oxford University Press. https://doi.org/10.1093/oxfordhb/9780199335084.013.4

Tananaev, N. I., et al. (2021). Assessment of the community vulnerability to extreme spring floods: The case of the Amga River, central Yakutia, Siberia. *Hydrology Research, 52*(1), 125–141.

Twigger-Ross, C., & Uzzell, D.L. (1996). Place and identity processes. *Journal of Environmental Psychology 16*, 205–220.

Vinokurov, A. (2017). "Fish – for the front!" the tragedy of Churapcha, 1942–1944. From the history of labour mobilization in the Yakut ASSR. *RSUH/RGGU Bulletin, 1*, 90–97. (In Russ.).

Vinokurova, U. A. (2018). River worldview as the factor of ethno-pedagogic innovations. In A. B. Pankin & V. K. Pichugina (Eds.), *Ethno-pedagogic complementation of the development of creative potential among the students: Methodology, theory and practice* (pp. 55–59). Izd-vo Kalm. Un-ta.

Vinokurova, L. I., et al. (2016). Waiting for the ice movement: Social aspects of the destructive floods in rural Yakutia (late 20th and early 21st centuries). *Gumanitarnye Issledovaniia v Vostochnoi Sibiri i na Dal'nem Vostoke, 1*, 28–40. (In Russ.).

Chapter 8
Emerging Trends in Arctic North America's Maritime Security Agenda: From Ice to Water

Heather N. Nicol and Barry Scott Zellen

Abstract Dramatic transformations in geopolitical thought have accompanied equally dramatic transformations of the physical environment, prompting a significant shift in the nature of geopolitics and the role of the state in defence and security. Many traditional geopolitical assumptions are no longer universally or perpetually applicable – perhaps they never were. But at the same time, the final consequences of the current polar thaw remain conjectural and cannot be predicted with any degree of certainty. This chapter explores the intersection between contemporary geopolitical thought and security narratives regarding Arctic North America's maritime spaces. Its focus is on change in the physical state of water under conditions of climate change and subsequent reassessments of strategic location, and the nature and origin of resulting threats to defence and human security.

Keywords Geopolitics · Climate change · Arctic · Maritime security · Sea power

Remote from global centres and located in a perpetually frozen environment, the North American Arctic has been positioned throughout much of history as peripheral to geopolitical assessments of world order. Today, however, it has become much more central to the global future that those early geopolitical theorists thought to define (Vuković, 2020). Dramatic transformations in geopolitical thought have accompanied dramatic transformations of the physical environment, prompting a significant shift in the nature of geopolitics and the role of the state in defence and security. Many traditional geopolitical assumptions are no longer universally applicable – perhaps they never were. But at the same time, the final consequences of the current polar thaw remain conjectural and cannot be predicted with any degree of certainty. This chapter explores the intersection between contemporary geopolitical thought and security

H. N. Nicol (✉)
Trent University, Peterborough, ON, Canada
e-mail: heathernicol@trentu.ca

B. S. Zellen
University of Connecticut, Mansfield, CT, USA

M. Lehtimäki et al. (eds.), *Cold Waters*, Springer Polar Sciences,
https://doi.org/10.1007/978-3-031-10149-6_8

narratives regarding Arctic North America's maritime spaces. Its focus is on change in the physical state of water under conditions of climate change and subsequent reassessments of the Arctic's strategic location, and the nature and origin of resulting threats to defence and human security. We posit that all of these are inter-related.

As late as World War II and the Cold War that followed, the most strategic sites for defence purposes within the North American North were on *terra firma* (for example, the D.E.W. Line and other military installations for strategic resources such as the Canol Pipeline, the Alaska Highway, and numerous extractive sites of industrial minerals). Today's Arctic security investments, however, are increasingly oriented towards maritime spaces and passages, coastlines and islands, in recognition of the propensity for melting ice to create new opportunities for shipping, tourism, and strategic mineral and energy resources, including hydrocarbons. This has significantly altered maritime security activities and narratives. There is now a greater emphasis on environmental impacts and disasters, pollution, search and rescue activities at sea, and mitigating other non-conventional threats resulting from significant and sudden environmental change. The objectives of security are now more oriented towards communities and safety, rather than an exclusive focus on the state, than any time in the past. They reflect new understandings and imperatives about the meaning of security and the strategic importance of Arctic places (Steinberg & Williams-Reed, 2017). They also reflect new geopolitical assessments.

Our discussion is inspired by the observation that despite significant scholarship about rapid environmental change and resulting shifts in security paradigms (Kee, 2019; Sfraga, 2021; Zellen, 2021), commensurately little attention has been devoted to understanding how environment, security, and geopolitical narratives intersect within maritime regions in the North American Arctic. In this chapter, we argue that shifts within traditional definitions of security (Nicol, 2020), and the changing geostrategic assessments that result, are highly interconnected and evolving. North America's Arctic "water world" is embedded in, and reflective of, geopolitical theory in ways which have not yet been fully addressed, except perhaps through the lens of critical geopolitics (Dittmer et al., 2011; Dodds, 2019; Dodds & Nuttall, 2015), which is arguably more concerned with deconstructing the power relationships that undergird regional security than understanding the intersections between geopolitics, climate change and security assessments. We make no claim that this observation is unique to North America, but rather that the North American experience sheds light on the nature of shifting Arctic geopolitical assessments and security policy and practice in general, in an era of climate-change.

We begin with examination of classical geopolitical theory, before assessing current changes to the strategic role of Arctic maritime spaces. We then question the assumption that water and ice are binary opposites whose ontological state determines the balance of world order within the region (Steinberg et al., 2020), and examine how the intersections between climate change and geopolitical thought have contributed to transformation in definitions of traditional security and have created new strategic spaces and subjects of security within the North. This is a story that leads us to reconsider the geopolitical saliency of Arctic's vast frozen Oceans – to trace their transformation from marginal to strategic places – and to explore how the concept of security has changed in keeping with these new geopolitical (and geo-economic) assessments.

8.1 Setting the Geopolitical Stage: The North in Classical Geopolitics

Early exploration in the North American North resulted from a wave of voyages initiated in European centres or from Russian explorers and fur-traders coming from Kamchatka and eastern Siberia (Nicol & Chater, 2021). Although finding the fabled passage to the orient was the overarching mission, interest in its importance waned in the wake of the 19th century Franklin expedition. Nonetheless, Franklin's search parties were perhaps the most significant 19th century force in mapping and exploring the North American Arctic. By 1880, however, Britain had transferred the Arctic Archipelago (claimed through previous expeditions) to the new Dominion of Canada. It had apparently lost interest in the North American North as the century drew to a close, if faith is put in the work of Sir Halford J. Mackinder (1904, 1943), a pioneering British theorist of geopolitical thought in the late 19th and early 20th centuries, who drew attention to the parts of the world that he deemed strategic. Mackinder's classic 1904 tome, *Geographical Pivot in History*, described the terrestrial Eurasian heartland as history's central pivot area, the inner or marginal crescent abutting the heartland as peripheral to land power, and the more remote northern region (dubbed Lenaland) as even more peripheral from the ebb and flow of world history as great powers battled for control of the heartland (Fig. 8.1).

Fig. 8.1 A view of the World map. Most of the north and western Russia and parts of the Middle East make the pivot area. The curved area from Europe passing through the Middle East, India, and up to eastern Russia is the marginal crescent. The larger curved area from the top of northern America through South America, Africa, and Australia is the insular crescent. (Source: Mackinder, 1904)

Lenaland is, of course, a reference to its location in proximity of the great Northern Russian River and its Siberian site. But it is not too much of a stretch to suggest that Lenaland could conceptually, although not literally, be seen to have also included much of the Russian Far East and the Northeast Passage, as well as all the islands lying offshore. All of these were, before the current era of global climate change, every bit as remote.

Throughout the 20th century, geopolitical theories were much more explicit about the role of the Atlantic, or most specifically the mid-Atlantic as the forum for maintaining global balance, but had little to say about the North itself (Heininen & Nicol, 2008). Indeed, for much of the 20th century there was complete lack of recognition about how the Arctic region and its world of ice fit within the broader geographical or circumpolar zone, or how they, in turn, fit into a broader global order. Heininen and Nicol (2008, p. 6) suggest that while Mackinder eventually added the Russian far north to the pivot area of the Eurasian north in 1919, and in doing so brought this whole region into a strategic zone which required containment (with reference to Fig. 8.1), "the North American North [still] remained in the outer crescent, a virtual Rim around the more strategic areas." Moreover, while "[l]ater theories were to reassess the importance of North America, specifically the U.S. in this equation" they did not do so "in any way which significantly included the North outside of its relationship to the USSR" (ibid., p. 6). Instead, Mackinder's geopolitical theory doomed it to remain of minimal strategic importance – certainly as long as Ice Age conditions persisted in the Arctic. Although subsequent geopoliticians, such as Mahan, reassessed the strategic value of the margins of Eurasia and the North American continent, they continued the practice of marginalization inherent in these cartographic projections of power. Mahan drew the attention of the world to the strategic value of maritime spaces (Pollock, 1982), even as Mackinder continued to refine his original "World Island Theory" to include American power interests – although not the American Arctic (Heininen & Nicol 2008, p. 5). That said, if in the early 20th century, the strategic value of the Arctic and Lenaland was more for seals and whales than nuclear submarines and fighter jets, two great events changed this status in the late 20th and early 21st century. One was the changing role of technology and the redefinition of what constituted strategic spaces. The other was the recognition that the ice-covered polar seas were melting.

In the first instance, changing technology in the early twentieth century played an important role. Air power refined and extended land and sea power. Emerging as the principal rival to Mackinder in the post war years, Spykman (1942a, b, 1944), Spykman & Rollins, 1938a, b, 1939a, b) renamed Mackinder's "marginal crescent" the "rimland", elevating its strategic importance for the ages of sea and air power. It was then that the North American Arctic, remote, frozen and on the margin, gained real strategic value – perhaps in a way not seen since the golden age of polar exploration. In elevating the rimland in strategic recognition and consequence, and by postulating that it was now the age of "rimland supremacy", Spykman inverted Mackinder's classical geopolitical logic that control of the heartland was of paramount importance. And it is here that the Arctic became increasingly important (Heininen & Nicol, 2008). Central to the geopolitical framework developed through

Spykman's assessments were new postulations regarding global struggle. As a result, post-World War II, the location of the rimland shifted from Atlantic and Indo-Pacific to include the North American Arctic. The Cold War's ballistic missile threat recalibrated the significance of the North American Arctic as a buffer zone countering the Soviet threat (Coates et al., 2008; Nicol, 2015; Nicol & Chater, 2021), and placed it on the cartographical rim of the Eurasian continent. This was a circumstance unforeseen by Mackinder's geopolitical lens. But, neither Mackinder, nor Spykman could have foreseen the thawing of the polar ice, nor the integration of the Arctic Ocean into the world ocean economically and strategically that followed (Greaves, 2021; Sfraga, 2021; Heininen and Southcott, 2010).

8.2 Classical Geopolitical Futures in a Melting Arctic?

From the perspective of classical geopolitical theory, the Arctic, or rather the Arctic's water word, has become increasingly strategic. Mackinder's twenty-first century "heartland" is now more likely to be focused on a "rimland", and a maritime domain, rather than a terrestrial domain at the heart of a continent. The work of two naval and maritime-geopolitical theorists – Mahan and Corbett – are useful to understand this cartographic transformation. Mahan has been called one of the foremost thinkers on naval warfare and maritime strategy, providing "the essential starting point for studying the course and conduct of war at sea and for understanding the strategic importance of the maritime commons in determining the rise and fall of great powers" (Maurer, 2016, n.p.). He believed that "war and change in world politics was rooted in competitions among the great powers". Thus, "the great commercial seafaring states, in particular, would play a leading role in world politics because of the wealth they generated from international trade" (ibid., n.p.). He also believed the United States needed to play a bigger role "in upholding the balance of power on a global scale" (ibid., n.p.). Mahan "exhorted naval commanders to seek out and defeat enemy battle fleets, thereby winning command of the sea" (Holmes, 2011, n.p.) and generally assumed that "permanent, absolute command of important expanses was possible" (ibid.).

Contrast this with the thinking of Mahan's principal rival, British historian Corbett, who "agreed that 'permanent general control' was a worthy goal, but he also insisted it might prove unattainable". Instead, the "normal position" was an "uncommanded sea," if only "because no navy was big [...] enough to be at all places at all times" (Holmes, 2011, n.p.). While for both Mahan and Corbett, the tenets of classical geopolitical thought to which they both subscribed would predict that as the Arctic region emerged as a new pivot area for world politics, its maritime zones would become more important to naval competition and strategic rivalry, they could not have foreseen that with the additional complexities of a dynamic geophysical transition from an icy polar sea to a more liquid, open, navigable, and commerce-friendly Blue Arctic a circumpolar "rimland" would become of global geopolitical importance in both eastern and western hemispheres.

Pivotal to our analysis of emerging geopolitical assessments of the North American Arctic is, therefore, the changing strategic importance of coastal and maritime regions as ice melts and climate change proceeds. It requires consideration of the dynamic change in the Arctic Ocean's frozen state. Rather than essentialize geopolitical thought in relation to its predictive ability, however, the point here is that Arctic geography, being largely insular and archipelagic, matters more today in the era of polar thaw than in the prior era of near-permanent polar pack ice. Historically the many islands and archipelagos that defined the region's physical geography remained locked in ice, in some cases all year round – accessible by air and snow mobile, perhaps, but mainly off the beaten path for most of the year. Because of conditions favoring natural inaccessibility, they were also perceived as strategic buffers circumventing regional conflicts during the Cold War, for example, due to their inaccessibility. More suitable for travel by dog sled to hunt on sea ice than the advancement of platoons, the climate of the North American North was seen to provide the best possible means of continental defence from the Soviet threat and the Nazi and Imperial Japanese threats before that.

However, climate change has accelerated a shift in both the strategic saliency of Arctic maritime spaces and the nature of threat itself. The seabed of the Arctic Ocean has been mapped with some precision, while defining the details of its contours is ongoing, with ever-greater degrees of precision possible using satellite imagery and new forms of technology. Of equal concern, this mapping has also begun to reveal the degree to which ice is transforming to water – on a seasonal or even permanent basis. Models and predictions vary, but in general it is clear that already today, significant portions of North America's Arctic coastline and the Northwest Passage will remain ice free over summer months. Predictions are that the Central Arctic Ocean will be ice-free during summer months by approximately 2050, while routes through the Northeast and Northwest Passage may be open even sooner. All of this affects the accessibility of coastal regions and islands, and transforms their accessibility, having consequences for strategic theories concerning both global maritime order and the changing position of the North American Arctic. Associated with this shift, in symbolic if not actual strategic importance, is the changing relevance of coastal and archipelagic regions, passages and maritime zones.

There are essentially two different ways to explore the impacts of this transformation. The first is strategic in nature and in regard to defence activities, and raises a number of questions. Will changing geopolitical assessments be reflected mainly through new assessments driven by the exceptional nature of Arctic defence under conditions of climate change? Will these be strategically focused upon the coastal and maritime regions of the North? If so, how? Although the Arctic is a region currently not stabilized by a single naval hegemon, but by its strong tradition of multilateral, collaborative governance, will this struggle for control continue to allowing for a more Corbettian, multipolar balance of power?

The second way of assessing the future is through examination of traditional defence and security interests as they intersect with shifting geopolitical assessments and environmental change. As ice melts and new maritime spaces emerge, geopolitical certainties unravel. New security threats develop that are quite unlike those identified by classical geopolitical theorists. What are the implications? We discuss the first set of strategic and defence impacts in the section below, returning afterwards to the second idea, that new understandings of security cannot be disassociated from geostrategic analysis and defence praxis in the contemporary North American Arctic.

8.3 Transformations

Transformations to geopolitical thought result from shifts in the strategic value of places – generally for defence or economic reasons. From a North American perspective, coastal and insular regions of the Arctic are becoming increasingly important. They facilitate the "opening of a new ocean" that "joins the Atlantic Ocean, Gulf of Mexico, and the Pacific Ocean" and makes it a coherent geopolitical space. From a U.S. perspective, and arguably a North American one too, the opening of the new coastal, insular and maritime Arctic spaces is "a critical, geographic component of our country's maritime ring of economic, homeland, and national security" (Sfraga, 2021, p. 6). In this sense, the current geopolitical significance of North America's Arctic and its insular and coastal regions under conditions of climate change does not depart significantly from that predicted by modern geopolitical theory. The distinctive and strategic geopolitical importance of islands, archipelagos, and island chains reflects and reinforces much thinking with regard to maritime and naval competition around the world. This is particularly evident in recent years in the Arctic Ocean, where amidst the polar thaw islands have played an outsized role, with increasingly accessible island chains, coastal waters and passages forming natural bridges between the continents.

It is within these insular zones, some of which are under claim as internal coastal waters and passageways, that the potential for conflict is said to exist. It represents a nexus in as yet "uncommanded" space: an "important crossroad where issues of climate change, international trade and global security meet" (Canada, 2019, n.p.). Because of the centrality of the Arctic Ocean to the future world maritime order, in the future we might anticipate a dynamic balance-of-power emerging, as illustrated by the current tensions and evolving relationships among Russia, China, the "West" and the other Arctic states – who are either members of, or partners with, NATO. This might in turn precipitate numerous security dilemmas as Arctic states fortify their remote Arctic insular territories, as Russia is presently doing, while Arctic NATO members (and their non-NATO partners) respond, in turn. Just as China has successfully embraced and deftly implemented its own island-chain strategies to counter those of the United States and its Pacific allies, might we anticipate a similar effort in Arctic waters, as long isolated Arctic islands and archipelagos are "rediscovered" for their strategic value, and fortified in turn? This could stimulate a race to develop Arctic strategic infrastructure reminiscent of World War II and the Cold War, and result in a more active and robust naval presence by the Arctic states, their partners, and rivals.

But whether this would be a Mahanian struggle for domination and mastery, or a more balanced, multilateral and Corbettian absence of control, is still open to debate. Much depends upon our understanding of how polar thaw will affect maritime access and control.

Some analysts suggest that melting sea ice will change geopolitical interest in the region, and create flash points for conflict between great and medium powers alike. Great Powers like China may be drawn to the region in search of resources, while others, like Russia, may continue to militarize the region – ostensibly for similar purposes. The replacement of sea ice by open seas could nullify existing arrangements for managing territorial claims within certain areas of the Arctic Ocean – in many cases those archipelagoes where territorial control of one kind or another is managed by claimant states which rely upon sea ice provisions within the Law of the Sea to assert authority – for example Canada and Russia.

Yet these understandings suggest that ice and water are static, binary opposites and that each is endowed with stable territorial properties. Steinberg et al. (2020) argue, instead, that these binary definitions, especially that of sea ice, are rife with "ontological indeterminancies" (p. 86). These indeterminacies, in turn, raise clear challenges to classical geopolitical theory. They complexify a direct correlation between changing legal status and the corresponding right to control water regardless as to its frozen and liquid properties. The debate around the status of Section 8, Article 234 of the Law of the Sea Convention (UNCLOS) that deals with the special context of ice-covered waters is a case in point. Article 234 states that:

> Coastal States have the right to adopt and enforce non-discriminatory laws and regulations for the prevention, reduction and control of marine pollution from vessels in ice-covered areas within the limits of the exclusive economic zone, where particularly severe climatic conditions and the presence of ice covering such areas for most of the year create obstructions or exceptional hazards to navigation, and pollution of the marine environment could cause major harm to or irreversible disturbance of the ecological balance. Such laws and regulations shall have due regard to navigation and the protection and preservation of the marine environment based on the best available scientific evidence. (UNCLOS).

Yet some analysts assert that Article 234 of UNCLOS is really being used to justify actions by certain Arctic coastal states to adopt and enforce their own national regulations in the Arctic Ocean and its ancillary seas and passages. For example, Canada's claim to the Arctic Archipelago and Northwest Passage and Russia's claim Northern Sea Route. The US, in particular, is opposed to a broad application of the definition of terms of this Article, since its position regarding the status of the Northeast and Northwest Passages is at odds with Russia and Canada, respectively. It is in this sense that the melting status of ice poses new challenges in non-conventional ways. While the regulatory and legal frameworks that undergird the current right of coastal states to impose a protective regime on Arctic waters is consistent with both the Polar Code and the intent of Article 234, such regulatory frameworks are neither static nor uncontestable in a changing Arctic (Bartenstein, 2018). What happens when these waters are no longer perceived by the world community to be ice-covered areas? Will Article 234 still apply (Schreiber, 2019)? Opinions are mixed.

Some experts suggest that it is important to separate the physical state of water (frozen/liquid) from the stable area covered by regulatory frameworks such as

Article 234 because the Article refers to a sensitive condition of place, rather than the condition of ice cover (Dremliuga, 2017). Thus, the application of such regulatory law in the future could allow for more, rather than less oversight, particularly in the area of environmental and safety practice. This because "sea ice decline has many implications," including increased accessibility to coastlines, but also "increased risk to mariners; stronger and more frequent storms; threats to coastal communities due to coastline and permafrost degradation and shifting subsistence patterns" (Sfraga, 2021, p. 7).

For these reasons, the implications of the polar thaw for geopolitical conflict are uncertain. Whether the region its regulatory framework and the territorial control that it supports will remain relatively stable, or witness a classical struggle for domination and mastery, or emerge with a strong tradition of multilateral, collaborative governance under the region's existing institutional embrace of consensus – remains to be seen.

8.4 From Ice to Water: New Security Frameworks

The idea that melting ice will result in conflict builds upon a *Realpolitik* assessment. In doing so it sidelines other institutional and strategic features of the region that could be useful in mitigating this outcome. The first is that North American states are members of the Arctic Council, an intergovernmental forum for regional cooperation that has met since the mid-1990s to discuss issues of mutual concern. Dominating discussions within this regional forum are ways to understand and address regional environmental threats, among them concerns regarding safety and protection of the Arctic's maritime regions. This unity of purpose and commonality of values, reflected in governing institutions and international fora across the Arctic strengthens the Arctic as a whole and the intra-Arctic bonds between the sovereign states of the Arctic region. While recent events, especially the Russian invasion of Urkaine have seen the Arctic Council suspended for the duration of the Russian Chair, peaceful cooperation has held thus far. Prospects for stability are also supplemented by a host of other agreements that create international cooperation on sea as well as land, in recognition of the growing importance of the maritime spaces that surround island chains and coastal regions. This includes a relatively recent search and rescue treaty and binding agreements concerning pollution control and prevention. Moreover, by strengthening ties within the Arctic states to their Indigenous communities, as well as their relationship with fellow Arctic states, the members of the Arctic Council greatly reduce the likelihood of experiencing new rounds of competitive geopolitics. Active participants in regional institutions such as the Arctic Council, the Barents Euro-Arctic Council, and the Arctic Coast Guard Forum, which already have established a solid foundation for enduring intra-Arctic collaboration, are especially well positioned to take the lead on these initiatives, with deep and broad traditions of indigenous engagement to build upon.

At one and the same time that regional cooperation has flourished, the definition of security itself has expanded to account for a world of water, and all of the marine environmental and ecological challenges encountered, if not a world island using Mackinder's terms. This brings us to the question of water itself. We have seen that geopolitical trends within the North American Arctic have oriented concern towards the status of waterways and the strategic location of insular and coastal areas in binary ways. But this is not the entire story. Matthews et al. (2021, p. 56) advocate a broader view of Arctic geopolitics, security and international relations, noting that "a growing body of research urges recognition of a more holistic perspective in which the interacting dimensions of the Arctic's 'soft security' characteristics, such as economic, food, health and environmental security are used to assess the region's overall resilience." In short, yet another part of the geopolitical equation has to do with how definitions of security have evolved since the end of the Cold War. The Arctic has become a theatre for speculation about conflict at the same time that other types of threats have emerged as security concerns (Heininen & Exner-Pirot, 2020). Many of these are environmental, in recognition of rapid changes to climate and its impact upon communities and ecosystems in the North, and of earlier pollution-related concerns that accompanied the dissolution of the USSR. In other words, "environment" itself has become a concern affecting human security, broadly defined, and has been a fixture of defence policy. An important feature of polar thaw thus has been adjustments to traditional definitions of security within the Arctic. There is growing recognition of the constituent and intertwined roles of human security and state security within the Arctic. The concept of human security has for decades privileged the role of state security within security narratives (Gjørv Hoogensen, 2005; Gjørv Hoogensen et al., 2013, 2020) and historically the prevailing language of security ignored the role and importance of non-state actors in the Northern regions. The introduction of concepts of environmental stability, sustainability and resilience has, however, moved the needle significantly.

While today security threats can still be identified in relation to Russia's seeming remilitarization if its Arctic region, or Beijing's apparent interest in this polar region (see Huebert, 2010), there is recognition that non-conventional security threats are growing in importance and indeed the greatest existential threat to human security in the region now exists in the form of climate change. At the root is the way in which climate change has changed conditions for human security. Today's existential threats are more likely to come in the form of rapid shifts in previously stable environments than from military aggression. This realization has accelerated a shifting geopolitical focus from the classical terrestrial heartlands to the coastal areas of the Arctic. Although the traditional understanding has not disappeared entirely, the role of armed forces for civilian support, and the definition of security have clearly expanded (Kee, 2019). Arctic cooperation now focuses as much upon marine pollution prevention and search and rescue capacity, as it does upon maintaining political stability. Moreover, the strategic importance of the Arctic's islands, coasts and maritime spaces identified within classical geopolitical theory, is now mirrored in concern with their heightened vulnerability to climate change and human security.

One key outcome of this change is that the idea of community has become an important component of Arctic security – not just as the object of security response in the face of disaster, but as an interested partner wishing to be engaged in the delivery of security. Gjørv Hoogensen (2005, 2020) has identified this trend as security from the "bottom up." But it also means that communities – and particularly Indigenous communities – have been able to take on a much larger role in the North – especially in regard to ecological knowledge production and environmental observations regarding climate change and environmental security (Krupnik & Jolly, 2002; Kee, 2019). Indigenous voices and communities, organizations and institutions now speak to the comprehensive nature of threats to human well-being in the face of environmental change.

8.5 Environmental Security

Because of the speed at which change is occurring, and the dramatic region-wide consequences of environmental change, the nature and tone of debate about sustainability has shifted to one about resilience and survivability (Arctic Council, 2016). In the latter case, a new relationship between human and ecosystems is postulated. While sustainability speaks to keeping balance between social, economic and environmental spheres of activity, resilience speaks to surviving change which is neither balanced nor predictable. Measuring the 'resilience' of Arctic communities to climatic and environmental change is a significantly different activity from that of advocating for sustainability, although not impossible to reconcile (Vlasova et al., 2021). From both a sociological and environmental perspective, the dynamic nature of change – its suddenness and unpredictability – within the Arctic region refocuses our attention on water, in its liquid and frozen state. More frequent, unpredictable and violent storms are occurring within the region, ice melt is accelerating to the point of changing coastal subsistence patterns, and melting permafrost and coastal erosion and implosion all threaten communities and settlements.

This brings us specifically to the changing nature of environmental threat, and the rapidity at which change is unfolding, especially for coastal regions that are particularly vulnerable: "Two central hazards are generally associated with climate change and coastal areas: coastal flooding and erosion. These threaten settlement infrastructure, such as housing, ports, roads, energy generation and distribution infrastructure, strategic facilities, and more generally reduce the potential uses of coastal areas" (Larsen et al., 2021, p. 2 of 17).

This threat to security is indeed a very different one from Mackinder's concept of heartland/pivot area, Spykman's rimland, or Mahan's strategic maritime space, as previously discussed. Climate change is refocusing the attention of governments and state security agencies to the safety of maritime spaces and coastal communities as well as to the strategic role of such regions in global geopolitics. Security is no longer exclusively concerned with defence, and indeed threats associated with extreme meteorological events are infusing security narratives and security planning in the North American Arctic region (Nicol, 2020). While such factors include coastal erosion and melting permafrost damage to infrastructure, the impact of violent and extreme weather events and rapidly changing ecosystems will create increasing degrees of insecurity for human communities, meaning that it is not just the nature of threat that makes for an unconventional security regime (Kee, 2019; Sfraga, 2021) it is also the capacity for imagining or premediating security challenges where communities and non-state actors also find themselves on the front lines (Heininen & Nicol, 2016). While there still remain opinions suggesting that traditional security is the tantamount challenge, most Arctic scholars are now having an increasingly difficult time discerning the difference in urgency between traditional definitions of security and broader human security agendas (Gjørv Hoogensen et al., 2013; Kee, 2019; Nicol, 2020). State defence activities and organizations are increasingly embracing more holistic understandings – as reflected in new dialogues concerning domain awareness, civilian-military engagement in disaster mitigation, and the phenomenon of "grey ships for civilian society" (see Kee, 2019; Treadwell & Holshouser, 2019). Indeed, the concept of domain awareness which now informs many North American defence activities, is increasingly referencing environmental conditions within the North American Arctic region and its maritime places.

8.6 Intersections

While initially the idea that melting multi-year ice would liberate resources and new spaces for geopolitical contestation shaped dialogue, more recently, the perception of conflict as an immediate outcome has receded. Indeed, the propensity is seen by some as acutely regional rather than circumpolar in nature (Østhagen, 2021). Nonetheless, security itself remains an important concept, but inherently reconceptualized to address both state and non-state agency and actors (Nicol, 2020). The changing conditions of the Arctic Ocean and surrounding coastal regions "advance the need for security and defense professionals to inclusively seek to account for environmental factors, in reducing risk to their own activities and to better accomplish their overall mission to secure and defend" (Kee, 2019, p. 19). At the heart of this is the need to understand where and how traditional and non-traditional security threats can overlap or reinforce each other – and the narratives that derive from their intersection. In other words, security no longer can be viewed as meaning simply the military security of the state. Rather, many new and unconventional referent objects – from the economy, to the environment, to military relations – now represent security challenges. As Kee (2019), Sfraga (2021) and others remind us, this new security relationship includes water and maritime regions. Indeed, in the North American Arctic, the concept of maritime domain awareness has emerged to cover this contingency, and the intersection between human security and state security is most profoundly taking place within its insular and coastal regions.

One particularly important new intersection now occurs at the defence-development interface. A warming Arctic means that the nature of security in the Arctic is rapidly transforming. A region that was on the periphery of political will and long neglected by central authorities, is now front and center to global politics. Previous neglect has exposed gaps in wellbeing that make the region more vulnerable to external actors who might seek to destabilize the diplomatic, strategic and political balance in the North. The historic struggle for the human terrain of the Arctic is thus of great importance to the future stability of the region, and requires forward thinking investment, respectful relationship-building and sustainment, and a continuous process of confidence-building measures to ensure that the legitimacy of the rule of the sovereign states of the Arctic remains intact and uncontested. Because northern Indigenous homelands have been imperfectly integrated with the political economies of the Arctic states, despite much progress and ongoing efforts in the last half century, this remains a near universal fault line across the Arctic, and a challenge faced by the seven Arctic states that have indigenous populations engaged in long-term processes of cultural renewal, economic development, and restoration of land rights. There is a growing consensus that defence activities can be better directed toward acknowledging how environmental threats and climate change pose broad and existentially relevant security challenges in ways consistent with an emerging human security agenda (NATO, 2010). This understanding reflects a convergence of modern security narratives post-9/11, and the particular circumstances of security in Arctic regions.

This defence-development interface is reflected not only in terms of broad regional narratives, but in domestic policies and practices, too. For example, 2010 US Department of Defence's 2010 Quadrennial Defense Review (US Department of Defence, 2010) observes that:

> The effect of changing climate on the Department's operating environment is evident in the maritime commons in the Arctic. The opening of the Arctic waters in the decades ahead which will permit seasonal commerce and transit presents a unique opportunity to work collaboratively in multilateral forums to promote a balanced approach to improving human and environmental security in the region.

Today, it was common to speak of "domain awareness" as an important tool for successful defence activity, and indeed the notion of maritime domain awareness has become binimical with responsibilities of maritime security agencies including navies and coast guards in North America, clearly focusing the defence-development intersection on the regions coast and maritime zones. Policy documents hint that a focus on more explicit consideration of the intersection between maritime space and environmental change has become normative in defence and broader human security considerations across the region (see Heininen et al., 2020). Nicol (2020) argues that this is reflected in the increasing concern with regional security operators regarding community safety and well-being, while. Welsh (2020, p. 480) suggests: "Traditional geopolitics may not be at stake in the Arctic, but nontraditional geopolitics most certainly is, and the lesson would seem to be that we must move forward gingerly to maximize the benefits and minimize the costs."

Inherent in the changing recognition of the importance of human security in the North is both to the awareness of eminent threat to the infrastructure and food security of coastal communities in the wake of erosion, permafrost melt and violent weather, as well as what can only be called the decolonization of Arctic politics and knowledge systems.

8.7 Conclusions

Until recently, the water world of the North American Arctic has always been less understood than its world of snow and ice. The region's islands and coastlines were not completely mapped until the late nineteenth century, but even then, many areas remained poorly defined. Since then there has also been significant change – the creation of new marine straits and the new perception of strategic saliency and geopolitical importance of the region's islands has occurred.

Climate change has greatly accelerated a shift in both the strategic saliency of Arctic maritime spaces, and the existential nature of the threat itself. Models and predictions vary, but in general it is clear that already today, significant portions of North American's Arctic coastline and the Northwest Passage will remain ice free over summer months. It is likely that the Central Arctic Ocean will be ice-free during summer months by approximately 2050, while routes through the Northeast Passage and Northern Sea Route have opened even sooner with assistance from Russian nuclear-powered icebreakers (Stephenson & Pincus, 2017). All of this affects the accessibility of Arctic coastal regions and islands, and transforms their accessibility, and has consequences for strategic theories concerning both global maritime order and the changing position of the North American Arctic.

By understanding the shifting geopolitical significance marine geographical structures, and indeed by understanding the shifting meaning of geopolitics and security, and their enduring importance to a stable world order, we can better contextualize the emerging strategic importance of the Arctic region. Emerging from a framework which positioned the North American Arctic as a periphery, current geopolitical thinking has adjusted to both changing strategic frameworks of place, and changing understandings of security. Their intersection creates new frameworks for Arctic defence – both in realpolitik and human security perspectives, and more importantly, in the space of intersection between the two.

Overall, the security challenges to the top of the world today, combined with the new emphasis on defence interventions at the community level (for individuals rather than state survival), while not determined by melting ice alone, is heavily shaped by it. There are indeed three broad dimensions of these processes – which include a knowledge transformation in understanding Northern environments as dynamic and changing terrestrial environments; a related conceptual transformation, in understanding the corresponding relationship between human security and environment in relation to climate change and the transformation in the object of security from state to community; and finally transformation in the strategic valuing of maritime spaces through geopolitical theory. These transformations occur in tandem with a melting Arctic and its emerging maritime environments. They are reflected in the development of geopolitical and security paradigms with an increasing focus upon a "liquid" North as well as the classical "terrestrial" North and its coastal communities. Indeed, in the North American Arctic, with the polar thaw, maritime geopolitical structures are re-emerging from the ice in their primordial insular form, transforming the Arctic region and potentially fostering its reconnection to the world ocean in classical geopolitical terms. And here, the reality of the regional stability and cooperation endures.

References

Arctic Council. (2016). Arctic resilience report. In M. Carson & G. Peterson (Eds.), *Stockholm environment institute and Stockholm Resilience Centre*. http://www.arctic-council.org/arr

Bartenstein, K. (2018). Between the polar code and article 234: The balance in Canada's Arctic shipping safety and pollution prevention regulations. *Ocean Development & International Law, 15*(4), 335–362. https://doi.org/10.1080/00908320.2019.1617932

Canada. Crown-Indigenous Relations and Northern Affairs Canada. (2019). *Canada's Arctic and northern policy framework.* https://www.rcaanc-cirnac.gc.ca/eng/1560523306861/1560523330587

Coates, K., Lackenbauer, P. W., Morrison, W. R., & Poelzer, G. (2008). *Arctic front: Defending Canada in the far north.* Thomas Allen Publishers.

Dittmer, J., Moisio, S., Ingram, A., & Dodds, K. (2011). Have you heard the one about the disappearing ice? Recasting Arctic geopolitics. *Political Geography, 30*(4), 202–214.

Dodds, K. (2019). 'Real Interest'? Understanding the 2018 agreement to prevent unregulated high seas fisheries in the Central Arctic Ocean. *Global Policy, 10*(4), 542–553.

Dodds, K., & Nuttall, M. (2015). *The scramble for the poles: The geopolitics of the Arctic and Antarctic.* Polity Press.

Dremliuga, R. (2017). A note on the application of article 234 of the law of the sea convention in light of climate change: Views from Russia. *Ocean Development and International Law, 48*(2), 128–135. https://doi.org/10.1080/00908320.2017.1290486

Gjørv Hoogensen, G. (2005). Bottoms up! A toast to regional security? *International Studies Review, 7*(2), 269–274.

Gjørv Hoogensen, G., Bazley, D. R., Golviznina, M., & Tanentzap, A. J. (2013). *Environmental and human security in the Arctic.* Routledge.

Gjørv Hoogensen, G., Lanteignem, M., & Sam-Aggrey, H. (2020). *Routledge handbook of Arctic security.* Routledge.

Greaves, W. (2021). Climate change and security in Canada. *International Journal, 76*(2), 183–203.

Heininen, L., & Exner-Pirot, H. (2020). *Climate change and Arctic security: Searching for a paradigm shift.* Palgrave Macmillan.

Heininen, L., & Nicol, H. (2008). Canada and the new geopolitics of the North Pacific rim. *Northern Research Forum.* https://www.rha.is/static/files/NRF/OpenAssemblies/Anchorage2008/a/5th_nrf_anc_2008_heininen_nicol_canada_new_gepolitics_north_pacific_rim.pdf

Heininen, L., & Nicol, H. (2016). Climate change and human security from a Northern point of view. *Centre on Foreign Policy and Federalism.* https://www.sju.ca/sites/default/files/2016-heininen-nicol-climate-change-hum-sec-northern_4.pdf

Heininen, L. & Southcott, C. (2010). *Globalization and the Circumpolar North.* Fairbanks: University of Alaska Press.

Heininen, L., Everett, K., Padrtova, B., & Reissell, A. (2020). *Arctic policies and strategies: Analysis, synthesis and trends*. International Institute for Applied Systems Analysis.
Holmes, J. R. (2011). From Mahan to Corbett? The new U.S. joint operational access concept implies a shift in navy strategy in contested parts of the world. *The Diplomat, 11*. https://thediplomat.com/2011/12/from-mahan-to-corbett/
Huebert, R. (2010). The newly emerging Arctic security environment. *Canadian Defence & Foreign Affairs Institute*. https://d3n8a8pro7vhmx.cloudfront.net/cdfai/pages/41/attachments/original/1413661956/The_Newly_Emerging_Arctic_Security_Environment.pdf?1413661956.
Kee, R. (2019). Key issues to Arctic security. In D. Menezes & H. Nicol (Eds.), *The North American Arctic: Themes in regional security* (pp. 93–115). University College Press.
Krupnik, I., & Jolly, D. (2002). *The earth is faster now: Indigenous observations of Arctic environmental change*. Arctic Research Consortium of the United States.
Larsen, J. N., Schweitzer, P., Abass, K., Doloisio, N., Gartler, S., Ingeman-Nielsen, T., Ingimundarson, J. H., Jungsberg, L., Meyer, A., Rautio, A., Scheer, J., Timlin, U., Vanderlinden, J.-P., & Vullierme, M. (2021). Thawing permafrost in Arctic coastal communities: A framework for studying risks from climate change. *Sustainability, 13*(5), 26–51. https://doi.org/10.3390/su13052651
Mackinder, H. (1904). The geographical pivot of history. *The Geographical Journal, 23*, 421–437.
Mackinder, H. (1943). The round world and the winning of the peace. *Foreign Affairs, 21*, 595–605.
Matthews, E., Kee, R., Roe, J., & Keesecker, C. (2021). Increasing maritime commerce in the Arctic. Proceedings of the marine safety and security council. *Coast Guard Journal of Safety and Security at Sea, 78*(1), 56–62.
Maurer, J. H. (2016, August 11). The influence of thinkers and ideas on history: The case of Alfred Thayer Mahan. In *The American review of books, blogs and bull*. Foreign Policy Research Institute. https://www.fpri.org/article/2016/08/influence-thinkers-ideas-history-case-alfred-thayer-mahan/
Nicol, H. N. (2015). *The fence and the bridge: Geopolitics and identity along the Canada–US border*. Wilfrid Laurier University Press.
Nicol, H. (2020). The evolving north American security context: Can security be traditional? In K. Coates & C. Holroyd (Eds.), *The Palgrave handbook of Arctic policy and politics* (pp. 455–472). Palgrave Macmillan.
Nicol, H., & Chater, A. (2021). *North America's Arctic borders: A world of change?* University of Ottawa Press.
North Atlantic Treaty Organization (NATO). (2010). *Advanced research workshop (ARW) on environmental security in the Arctic Ocean*. Convened at the Scott Polar Research Institute. Workshop Summary. https://canadatogether.hypotheses.org/files/2010/12/NATO-ARW_Summary_Environmental_-Security_in_the_Arctic_-Ocean_15NOV10.pdf
Østhagen, A. (2021). The Arctic security region: Misconceptions and contradictions. *Polar Geography, 44*(1), 55–74. https://doi.org/10.1080/1088937X.2021.1881645
Pollock, T. R. (1982). The Historical Elements of Mahanian Doctrine, *Naval War College Review, 35*(4), 44–49.
Schreiber, M. (2019, April 4). Russia and Canada may lose their legal claim to Arctic seaways as ice melts, experts say. *Arctic Today*. https://www.arctictoday.com/russia-and-canada-may-lose-their-legal-claim-to-arctic-seaways-as-ice-melts-experts-say/
Sfraga, M. (2021). Navigating the Arctic's seven C's: US coast guard advancing America's interests in the Arctic. Proceedings of the marine safety and security council. *Coast Guard Journal of Safety and Security at Sea, 78*(1), 6–11.
Spykman, N. J. (1938a). Geography and foreign policy I. *American Political Science Review, 32*(1), 28–50.
Spykman, N. J. (1938b). Geography and foreign policy II. *American Political Science Review, 32*(2), 213–236.
Spykman, N. J. (1942a). *America's strategy in world politics: The United States and the balance of power*. Harcourt, Brace and Company.

Spykman, N. J. (1942b). Frontiers, security, and international organization. *Geographical Review, 32*(3), 436–447.
Spykman, N. J. (1944). *The geography of the peace*. Harcourt.
Spykman, N. J., & Rollins, A. A. (1939a). Geographical objectives in foreign policy I. *American Political Science Review, 33*(June), 391–410.
Spykman, N. J., & Rollins, A. A. (1939b). Geographical objectives in foreign policy II. *American Political Science Review, 33*(August), 591–614.
Steinberg, P. E., & Williams-Reed, E. (2017). In a world of land and water, where does ice fit in?. *A report from the ICE LAW Project*. https://lauda.ulapland.fi/bitstream/handle/10024/63201/Steinberg.Philip%3B%20Williams-Reed.Eris.pdf?sequence=1&isAllowed=y
Steinberg, P. E., Kristoffersen, B., & Shake, K. L. (2020). Edges and flows. In I. Braverman & E. R. Johnson (Eds.), *Blue legalities: The life and laws of the sea* (pp. 85–106). Duke University Press.
Stephenson, S. R., & Pincus, R. (2017). Challenges of sea ice prediction for Arctic marine policy and planning. In H. N. Nicol & P. W. Lackenbauer (Eds.), *The networked north: Borders and borderlands in the Canadian Arctic region* (pp. 77–92). Canadian Arctic Resources Committee.
Treadwell, M., & Holshouser, T. D. (2019). Bridging the gap: Fostering military-civilian collaboration to improve marine, aviation and telecommunications infrastructure in the US Arctic. In D. Menezes & H. Nicol (Eds.), *The north American Arctic: Themes in regional security* (pp. 155–167). University College Press.
UNCLOS. United Nations Convention on the Law of the Sea. (n.d.). Available online at https://www.un.org/depts/los/convention_agreements/texts/unclos/unclos_e.pdf
United States. Department of Defense. (2010). Quadrennial defense review. https://history.defense.gov/Portals/70/Documents/quadrennial/QDR2010.pdf?ver=vVJYRVwNdnGb_00ixF0UfQ%3d%3d
Vlasova, T., Petrov, A. N., & Volkov, S. (2021). Rethinking sustainability monitoring in the Arctic by linking resilience and sustainable development in socially-oriented observations: A perspective. *Sustainability, 13*(1), 177. https://doi.org/10.3390/su13010177
Vuković, N. (2020). Do we need revision of the key geopolitical paradigms?. *Medjunarodni Problemi (International Problems), 72*(1), 15–36. Institute of International Politics and Economics.
Welsh, D. A. (2020). The Arctic and geopolitics. In K. Coates & C. Holroyd (Eds.), *The Palgrave handbook of Arctic policy and politics* (pp. 473–386). Palgrave Macmillan.
Zellen, B. S. (2021). Stepping stones to a secure Arctic: Strategic importance of arctic island chains grows in lockstep with rise in polar temperatures, state rivalry. *Coast Guard Journal of Safety & Security at Sea: Proceedings of the Marine Safety & Security Council, 78*(1), 26–30.

Part III
Narrating and Visualizing Cold Waters

Chapter 9
Between Pomor Traditions and Arctic Modernities: The Northern Sea in Early Soviet Pomor Literature

Mika Perkiömäki

Abstract The Pomors are Russians living on the Arctic shores of the White Sea and the Barents Sea, and they have a distinct cultural identity of their own. A distinctive feature of Pomor identity is the dependency on northern waters. This chapter addresses meanings of the sea in the oeuvre of two key Pomor writers of the early Soviet period, Boris Shergin and Stepan Pisakhov, who have reproduced Pomor folklore in literary form. The Northern Sea is the most prominent non-human agent in their works, and the chapter studies how they depict it as the foundation of human life in Pomor areas. It further examines, how the representations and allegorical functions of northern aquatic nature in Shergin's and Pisakhov's stories and fairy tales that often rose from Pomor folk tradition respond to the positivist Soviet spirit of the times. Early Soviet discourse strove for a radically changed Soviet Arctic space by means of rapid modernization, "normalization", or "mastering" of the North, and a creation of something that has been described as Arctic modernities. Earlier studies of Shergin have explored the seas largely through the ships and the skills of the men who manned them. In the spirit of "blue ecocriticism" suggested by Sidney Dobrin, this chapter aims to shed light on whether the sea in Shergin and Pisakhov is a living thing with a history or a backdrop to the human mastery of the Arctic environment.

Keywords Arctic Ocean · Ecocriticism · Pomor folklore · Pomor identity · Soviet literature · White Sea

> We are a people of the White Sea, of the Northern shore. – Boris Shergin, "V otnose morskom", 1939.

Russia has historically been characteristically a continental country whose spatial identity has been constructed on the basis of vast terrestrial distances over

---

M. Perkiömäki (✉)
University of Helsinki, Helsinki, Finland
e-mail: mika.perkiomaki@helsinki.fi

M. Lehtimäki et al. (eds.), *Cold Waters*, Springer Polar Sciences,
https://doi.org/10.1007/978-3-031-10149-6_9

coniferous forests and wide open steppes (see Widdis, 2004). The main aquatic elements in Russian geography have been rivers that traverse these vast lands, enabling human existence on them (Perkiömäki, 2021b, pp. 40–47). One of the leading Russian historians of the nineteenth century, Vasilii Kliuchevskii (1987) gives significantly less attention to the sea when writing about the European part of Russia: "The sea forms only a small fraction of its borders, the coastline of its seas is insignificant in comparison with its continental space: for every mile of the sea coast there is 41 square miles of the mainland" (p. 64).[1]

Russia has, however, areas where the spatial significance of the sea is profound. It is most evident in the Arctic shores of the Barents Sea, which is a marginal sea of the Arctic Ocean, and its inlet, the White Sea. This is the home of the Pomors, the descendants of the Russians that have moved there since the twelfth century, at first mainly from Novgorod. The areas west of the Urals inhabited by the Pomors in the Arkhangelsk and Vologda *oblasts*, especially around the rivers flowing into the White Sea and its seashore, are usually meant when referring to the "Russian North" (*Russkii Sever*) in the Russian language (Likhachëv & Ianin, 1986, p. 115; Vlasova, 2001, p. 3). Administratively, this area also covers the Nenets Autonomous District as well as the twin islands of Novaya Zemlya and the uninhabited Franz Josef Land, which stretches beyond the 80th parallel north. The Pomors are closely associated with the concept of *Russkii Sever* (Starkov, 2012, p. 10). The centre of the Pomor area is the city of Arkhangelsk, which was founded in 1584 and lies on the Northern Dvina River, close to its mouth and its exit into the White Sea. As Russians, the Pomors are neither an ethnic group nor a tribe, but they do have a distinct cultural identity of their own (see Anufriev, 2013). This Pomor identity is central to the life of the people of the region themselves as well as to how they are perceived in the rest of Russia.

A distinctive feature of Pomor identity is the dependency on northern waters, especially the sea. This dependence on the sea and the vast distance to the more densely populated Russian regions are central to Pomor identity. The very word "Pomor" derives from the Russian word for 'sea', *more*, and refers to someone who lives by the sea. The poorly cultivable land of the Russian North has not been able to support life in the *Pomor'e*, as the area is known in Russian. Consequently, fishing and hunting of marine mammals have always been the main source of livelihood for the Pomors. The sea has been the most beneficent natural element for them, and an abundance of life-supporting meanings has been attached to it. The historian Aleksandr Okorokov (2005) compares the significance of the sea for Pomors to that of the Nile for the Egyptians and notes that if the Pomors had been pagans, they would predominantly have worshipped the god of the sea (p. 37). He also notes that the prominent nineteenth-century Russian biologist and botanist Konstantin Mereschkowski called the sea the field of the Pomors and fish their bread.

[1] Translations from the Russian are by the author unless otherwise noted.

This chapter explores representations of the Northern Seas[2] in Pomor literature written in the early Soviet period, the time when the Russian North was being rapidly modernized. The writers of the research material are two Pomor writers, Boris Shergin (1893–1973) and Stepan Pisakhov (1879–1960), who both came from Arkhangelsk. Whereas Shergin moved to Moscow upon starting his literary career in the early 1920s, Pisakhov studied in St. Petersburg, travelled extensively in Europe and the Middle East, and took part in numerous Soviet expeditions to the Arctic Ocean, but always returned to Arkhangelsk. The Northern Sea is the most prominent non-human element in the works of both writers.

The works of Shergin and Pisakhov are apt material for studying the meanings of the North Russian sea in the transitional period from imperial to Soviet Russia, because both, but especially Shergin, are considered among the most central writers of what has been described as "the northern text of Russian literature" (*severnyi tekst russkoi literatury*). According to literary scholar Elena Galimova (2013, p. 4–30), who is the central proponent of the northern text, Shergin's works form its core (p. 19). For her, the northern text means the northern Russian variant of the Russian national worldview. She has studied what is unique about the northern text and looked for answers to what it represents in literary works about the Russian North, and what kind of northern space the texts create in general. She associates the northern text with sacred and mythical meanings that connect it to the so-called "sacral geography of the Russian North" (*sakral'naia geografiia Russkogo Severa*), which is a concept coined by a researcher of Russian ethnography, culture and theology, Nikolai Terebikhin (1993). It consists of religious-mythological conceptions of the North Russian landscape and peoples, as well as concrete geographical objects that have acquired sacral meanings. Central to it is that the unpredictable and harsh climate of the Russian North has led to nature acquiring religious meanings for the people. Underlying this is the widespread "myth of the North" in Russian culture, which juxtaposes the north with the south. While in this myth Russia as a whole is opposed to southern Europe, the north of Russia represents a special kind of Russianness (Galimova, 2013, p. 15). Terebikhin (2004) writes that the mythological meanings of the sea are so central to Shergin's oeuvre that he describes it as a representative of the "text of the North Russian sea" (*severomorskii tekst*), where the life-giving and mortal qualities of the sea are central (pp. 1, 63).

Earlier studies of the natural world in Shergin and Pisakhov usually examine their works from the point of view of the northern text of Russian literature and present their works as a mythopoetical space of the sacral geography of the Russian North (e.g. Galimova, 2013, p. 31–48; Terebikhin, 2004, p. 7–19, 139–141). Such readings, which explore the mythological meanings of a Northern Sea, where the ships, their builders and the skilled seafarers appear as the most active material agents, are inadequate. As Arja Rosenholm and Elena Trubina (2021) argue, these readings that stem from Russian philosophy of religion and traditions of

[2] I use the term 'the Northern Seas' when I refer to the northern seawaters as a physical-geographical term, and the term 'the Northern Sea' when referring to the sea in the North as a concept or idea.

structuralist cultural semiotics create national and ethnic unity by understanding the Russian North as a uniform whole, and do not aim at deconstructing the myth of the North.

This chapter aims at filling this gap by concentrating on whether the Northern Sea in Shergin's and Pisakhov's works is merely a vehicle for creating a national ethos, or whether it is actually a living and agentic entity with its own history, or whether it is more ambivalent and have qualities of both of these elements. The approach is inspired by "blue ecocriticism", which Sidney Dobrin (2021) has recently suggested as an appropriate term for ecocritical studies that emerge from oceanic or aquatic frames. Ecocriticism has often been thought of as an "earth-based" approach to literary studies that is usually associated with a "green" moral or political agenda (see Garrard, 2011, p. 3). Blue ecocriticism aims to tackle this "ocean deficit" by exploring aquatic cultural imaginaries.

Blue ecocriticism is based on a new materialist framework that implies that both material and discursive practices contribute to our knowledge of reality (Oppermann, 2019, p. 446). This form of ecocriticism is critical of anthropocentric understandings of the material-semiotic (Haraway, 1991) and material-discursive (Barad, 2007) frameworks that constitute the confluence of the material and the symbolic. For the sake of aquatic environments, this means that writings about seas operate on the interface of the material and the symbolic, the physical and the discursive. What we know about the sea depends not only on our own observations of its physical appearance but also on what and how we are told about it. Because of this, the idea of the sea is partly discursively constructed. For an individual who lives far from the sea, the primary notion of the sea may be discursively constructed. Therefore, the study of writings about seas helps us understand the human experience of seas.

Blue ecocriticism is further enlightened by material ecocriticism, which treats all matter as "storied" and reads world and text as an agentic entanglement (Iovino & Oppermann, 2014, pp. 1–2, 9–10). Due to their close association with the northern text of Russian literature, Shergin's and Pisakhov's writings have storied the North-Russian waters significantly through semiotic and discursive qualities. The Arctic Ocean, the Barents Sea and the White Sea are the most prominent non-human agents in Shergin's and Pisakhov's works not only from the point of view of mythology but also materially. My aim in this chapter is to study the material-physical dimension of the Northern Sea in their works and examine the ways they depict the sea as the foundation of human life in Pomor areas. I further ask how the representations and allegorical functions of northern aquatic nature in Shergin's and Pisakhov's stories and fairy tales that often rise from Pomor folk tradition respond to the positivist Soviet spirit of the times. Early Soviet discourse strove for a radically changed Soviet Arctic space by means of rapid modernization, "normalization" (Frank, 2010), "mastering" (*osvoenie*, "deployment") of the North, and a creation of something that has been described as Arctic modernities (Hansson & Ryall, 2017; Körber et al., 2017).

After first briefly presenting my research material, I will show how the Arctic Ocean is particularly Russian in the works of the two writers. I will then handle depictions of heroic shipbuilders and seafarers in Shergin's stories. Both subchapters illustrate how the agency of the seas themselves remains in the shadows of the human, especially Russian, actors. Finally, I will investigate how the sea and the creatures living in it are represented as economic potential in Pisakhov's fairy tales. While the works idealize the Pomor lands and waters as a place where traditional Russian culture is well preserved, at the same time they also depict it as a site where Soviet modernization is spreading, which is also a positive sign. Although these two writers have received considerable critical attention, the role of Soviet modernization in Shergin and Pisakhov has not been studied before. The deployment of modern technologies in the Russian North in the early Soviet period happened so rapidly that one would suppose it to have produced resistance and confusion in Pomor communities. Shergin's and Pisakhov's works, however, imagine an early Soviet Russian North where the modern does not conflict with the traditional.

9.1 Shergin's Stories and Pisakhov's Fairy Tales, Essays and Diaries

Most of Boris Shergin's works are short stories, altogether almost 200 of them, representing his adaptations of Pomor folklore that he had largely collected himself. Shergin (1936) wrote that he had heard the stories and fairy tales in his youth in Arkhangelsk and that in his published works he repeated the story (*fabula*) but changed the form (*forma*) by, for example, adding new episodes and extending the dialogues to better suit the targeted audience and his own tastes (p. 2). Especially the first three of his nine collections, published in 1924–1936, consist of these folkloric stories. Many of the stories reappear in Shergin's later collections. Since the publication of the collection *U pesennykh rek* ("On the Song Rivers") in 1939, stories based on his own experiences of the Russian North and the *Pomor'e* also started appearing. The collection *Pomorshchina-korabel'shchina* ("Pomor Land – Land of Ships"), published in 1947, was heavily criticized by critics as reactionary, and Shergin was unable to publish again until 1957. The last three collections that came out in 1959–1971 are characterized by the increasing presence of Christian motifs (see Perkiömäki, 2021a, pp. 12–13). All the nine collections that came out during Shergin's lifetime were published by well-known Moscow-based publishers, such as Sovetskii pisatel´ and Molodaia gvardiia, except the last collection, which was published in Arkhangelsk. Shergin is also known for his visual art, and he himself drew the illustrations to some of his first story collections, although he soon concentrated on writing only.

Stepan Pisakhov likewise started as a visual artist who depicted Russian Arctic landscapes in his works, but he is best known as a writer of northern fairy tales. His works started to be published in the same year as Shergin's and as Pisakhov's whose first fairy tale, "Ne liubo – ne slushai. Morozheny pesni" ("Don't Listen If You Don't Like It. Frozen Songs") came out in the collection *Na Severnoi Dvine* ("On the Northern Dvina"), which was published in Arkhangelsk. While Shergin wrote his best-known stories between the late 1930s and the late 1950s, Pisakhov's first fairy tale perhaps remains his most popular. However, most of his fairy tales were published between 1935 and 1939, first mostly in the Moscow-based journal *30 dnei*, and later in two books, *Skazki Pisakhova* ("Pisakhov's Fairy Tales", 1938) and *Skazki Pisakhova. Kniga vtoraia* ("Pisakhov's Fairy Tales. The Second Book", 1940). On account of the publication of these collections, he was admitted to the Soviet Writers' Union. He continued to write fairy tales until the end of his life, but only a handful written after 1940 were published. The first collection of Pisakhov's fairy tales that was published in Moscow came out in 1957. Pisakhov's fairy tales have been praised for their insightful understanding of the North of Russia, the result of the writer's life-long devotion to life in the *Pomor'e* (Ponomarëva, 1985, p. 16). Pisakhov also wrote non-fiction essays and diaries about his experiences in the Russian North, especially the early Soviet expeditions to the far north. These texts are also part of my research material. Since the early 1970s, many stories by both Shergin and Pisakhov have been widely popularized by animated cartoons.

9.2 The Russian Arctic Ocean

In the following, I explore how Shergin's and Pisakhov's works build up an intrinsic connection between the Northern Sea and Pomor culture on the one hand, and Pomor culture as the archetype of Russianness on the other. By archetype I here refer to Northrop Frye's (1957) understanding of the term as a typical or recurring image, something that makes a profound imaginative impact on someone who, in my case, has never visited Russia (p. 99). The idea of Pomor culture as an archetype of Russianness has been supported by, for example, the eminent scholar of medieval Russian literary history, cultural philosophy and philology, Dmitrii Likhachëv (2004). He first visited the area in the early 1920s and was impressed by the distinctive quality of the language, songs, stories and fairy tales of the Pomors. Decades later, he described the northern region as an auxiliary nurturer of Russian folk culture that "has played an outstanding role in the history of Russian culture" (para. 5). Likhachëv equates the histories of Russia with the Russian North, and highlights the exceptional connection of the northern people with their environment. Likhachëv's view of the Russian North is idealized and leans to romantic nationalism, despite the fact that in the late 1920s and the early 1930s he spent five years in the Soviet Union's first concentration camp on the Solovetsky Islands in the *Pomor'e*, and was forced to work on the construction of the Stalin White Sea–Baltic Canal in horrible conditions.

Russia has more Arctic coastline than any other country in the world, and has played a major role in the Arctic sphere (see Hønneland, 2016). Russia has actively explored the Arctic seas between Scandinavia and Alaska for hundreds of years. Vladimir Wiese (1886–1954), a Russian explorer and scientist of Arctic waters, wrote a detailed account of this history in his two-part book, whose title, *Moria rossiiskoi Arktiki* ("The Seas of the Russian Arctic", first published in 1948), tellingly associates Arctic waters with Russia (Vize, 2008). Since the beginning of the eighteenth century, the North has been regarded as the key to the Russian national character (Boele, 1996, pp. 17–47). While the ideological significance of this was first to stress Russia's European rather than "Eastern" character, in early nineteenth-century Romantic literature it started to embody Russia's cultural and religious otherness in Europe (p. 252). In this perception of the north, Russia is both a conqueror of the hostile and barren north and an actor that identifies itself as a nation that has adapted to the special circumstances of the north (Boele, 2016). In the early Soviet Russian imagination, the Arctic was typically treated as something that needed to be rewritten as part of the Soviet Union with the help of modern technology and socialist hero figures (McCannon, 1998, pp. 81–109; Frank, 2010, 2017).

I have earlier noted that Shergin's stories are based on a presupposition that Pomor folklore reflects Russianness especially well (Perkiömäki, 2021a). The works promote an idea of the Pomor way of life as ideal, as something where traditional Russian culture has been particularly well preserved. They create an intrinsic connection between the Northern Sea and the Russian people, who are the main actors in these cold waters. The agency of indigenous people and the non-human animals is limited, and the sea itself is largely conceived of as Russian. In Shergin's works, 'Gandvik' is repeatedly used to refer to the White Sea. The title of one his collections even includes it: *Gandvik – studënoe more* ("Gandvik – Icy Sea", 1971). Shergin (1967) explains the name in the glossary attached to the end of one of his collections: "Gandvik: the song name of the White Sea, a word of Scandinavian origin" (p. 428). Indeed, the word has been used in medieval Scandinavia and appears in Norse mythology (Lundkvist, 1985, p. 90). Despite the Scandinavian origin of the word, Shergin's (1959) stories depict the White Sea as a Russian sea, as "our blessed Gandvik" or "our bright Gandvik" (pp. 21, 30). The "our" here clearly refers to the Pomors, and thus to Russians. However, Russians are not the only people who inhabit the shores of the White Sea, where also many Nenets, Sami, Karelians and Komi have lived.

The representation of the Northern Seas as Russian is most explicit in the title of Shergin's (1959) collection, *Okean-more russkoe* ("The Russian Ocean-Sea").[3] The "ocean-sea" refers to the Northern Seas generally, be it the White Sea, the Barents Sea or the Arctic Ocean. To get to Norway from Arkhangelsk, one has to enter both the White Sea and traverse the wide ocean. The idea of the Northern Sea as a Russian sea whose shores are inhabited by Russians is also present in the opening chapter, "Morskoi zazyv" ("The Call of the Sea") of Shergin's (1939) collection *U pesennykh rek* ("By the Song Rivers"): "I remembered Gandvik – the crystal sea, the Russian shores, the frequent rains, the light fogs, the flight of birds, the change of winds" (p. 3). This "gloomy, icy" Russian sea and the "grey[4] ocean" "enrich the lands of Arkhangelsk and the Dvina" (Shergin, 1959, p. 63).

---

[3] The title is sometimes referred to as *Okean – more russkoe* ("The Ocean is a Russian Sea"), but this is incorrect.

[4] '*sedoi*' – an adjective normally used to refer to the colour of one's hair.

The Russianness of the Northern Sea is further underlined by emphasizing the long history of the Pomors beside them and their navigation upon them. One example is the story "Rifmy moreplavatel'ny" ("Rhymes of the Mariners"): "Our ancestors came to the Ocean and the White Sea nine centuries ago from Novgorod. Navigation in the Northern Seas began during those days" (Shergin, 1939, p. 21). How long other peoples have inhabited the area is not relevant in Shergin's texts. The importance of the sea, "the Father sea, the supporter of life" (Shergin, 1967 ["Bratanna"], p. 237), is also connected to the Pomors, for whom it is the "giver of water and food" (Shergin, 1967 ["Matveeva radost'"], p. 279). But the sea does not give people what they need without extensive knowledge how to navigate it: "Without his vessel, even the most fragile, the Pomor is not a breadwinner, but a slave to a rich man" (Shergin, 1967, p. 279).

In Shergin's works, the North and the Northern Sea are the domain of the Pomors, who rule the northern natural world by means of the Russian language. This becomes curiously evident in the story "Russkoe slovo" ("The Russian Word"), which was published in 1957. The protagonist of the story is Captain Ust'ian the Bearded. On his way to fishing in the far north, he briefly lands on shore and meets a local reindeer herder, probably a Nenets, who laments to him that a polar bear is attacking his herd. Ust'ian is in a hurry and is unable to stop and help him, but he comes up with a solution to the problem. He tells the herder: "Go to the bear and use the Russian language to tell him: 'A Russian captain orders you to withdraw to your lair. You have no business in the reindeers' domain'" (Shergin, 1957, p. 20). This is exactly what he does, and soon enough the bear swims away "to where he was ordered". Not only does the story depict the Russians as more able to cope with the natural world than the indigenous peoples, it also sets the Russian culture hierarchically above nature, which has no other option but to obey the Russian word. It is a Russian sea, after all.

Shergin's (1967) collection includes a story titled "Po ustavu" ("By the Regulations"), where a Pomor fisherman finds a dead body in a hut in Novaya Zemlya. Due to lack of time, he cannot bury the body, so the fisherman strips and clothes the body in his own clothes. The story depicts Novaya Zemlya as something that is not Russia: the fisherman's ship is hurrying toward the "Russian side", and the captain gives orders to "get going to Rus'" (Shergin, 1967, p. 216). Clearly, even though the sea is Russian, the islands far away from the mainland are not understood to be Russian. This may be because the Pomors do not spend the winter in Novaya Zemlya, where the climate is extremely harsh. Similarly, Pisakhov's (1985) essay "On Franz Josef Land", refers to everything south of Novaya Zemlya as home: "We left Novaya Zemlya. Here it is already home" (p. 231). The Pomors are home as soon as they venture into the ocean.

In Stepan Pisakhov's fairy tales, the Russianness of the Arctic Ocean is not evident. It is, however, strongly present in his non-fiction. Pisakhov often travelled in the Arctic Ocean as part of the early twentieth-century Russian and Soviet expeditions, and wrote diaries and essays of these experiences. The purpose of these expeditions was to examine the northern waters and lands and investigate whether they had economic potential. They were an important part of the *osvoenie Severa*, "mastering of the North", whose main goal was to integrate Arctic space with Soviet space (see Widdis, 2000, p. 404, 2003, p. 7; Frank, 2010, p. 117; Bolotova, 2014, pp. 46–47). Pisakhov's writings about these journeys belong to the international literary tradition of romantic and heroic masculine conquest narratives, where the ocean acquires a frontier ethos with a sense of discovery that conceals the colonial attitude (see Dobrin, 2021, p. 16).

In the essay "Na Zemliu Frantsa-Iosifa" ("On Franz Josef Land"), Pisakhov writes about a 1928 expedition on the icebreaker *Georgii Sedov* to Franz Josef Land in the extreme north. Norwegian sealers had spotted the islands in the 1860s, and an Austro-Hungarian North Pole expedition had found and named them in the 1870s (Vize, 2008, pp. 172–181). The Soviet Union annexed Franz Josef Land in 1926, and at the time of Pisakhov's expedition its ownership was very much disputed, especially by Norway (p. 270).

Pisakhov's essay reflects the patriotic mood of the expedition. On the way, the ship visits the town of Yokanga[5] on the remote northern shore of the Kola Peninsula, where a monument has been erected to commemorate the Allied intervention of foreign troops – mostly British and American – in the Russian Civil War. After visiting the monument, Pisakhov (1985) writes: "Foreigners passing by can read the sign and they will know that WE [Soviet Russians] REMEMBER!"[6] (p. 223; emphasis in the original). Later, news arrives that Umberto Nobile's flight over the North Pole in the airship *Italia* had ended in a crash on the way back that left ten men of the crew trapped on the ice. An international rescue operation ensued. Pisakhov's ship also considered taking part, but it was too far away. Subsequently, news arrives that the Soviet icebreaker *Krassin* had rescued the survivors of the crash. Again, patriotic feelings are palpable in Pisakhov's diary: "It felt that we were all proud that *Russians saved* them" (p. 225; emphasis in the original). Curiously, Pisakhov underlines the Russianness of the rescuers rather than their Sovietness.

Similar remarks continue to fill the pages upon arrival at the destination: "We are eager to set foot on Franz Josef Land to raise *our flag*!" (Pisakhov, 1985, p. 226; emphasis in the original), as well as after disembarking: "I painted the hammer and sickle of the USSR in red paint on a large stone" (p. 227). Later, an American ship *Hobby* arrives, and the Soviet crew visits it. Pisakhov seems to be genuinely impressed by the technical sophistication of the American ship, but he feels it lacks the more human scale of the Soviet ship: "It looks like everything on their ship is well equipped, but ours is somehow warmer, simpler and cosier." Unlike the rescue operation of Nobile and his crew, Pisakhov here makes it explicitly clear that the expedition's achievements are milestones carved by the Soviet Union. Thus, this essay depicts the Arctic waters not only as Russian but also as a site where Soviet modernity acquires a positive label.

---

[5] Renamed Gremikha in 1938, and known as Ostrovnoi since 1981. It is a closed town that has hosted a naval base since 1915.

[6] Pisakhov himself seemed to support the interventionists and the Whites in the Russian Civil War, however (Galimova, 2012). Being critical of them in his essays might be a tactic that he was forced to adopt to compensate for this.

Both Shergin's stories and Pisakhov's essays are explicit about marking the Northern Sea as Russian. With the help of the Soviet Union's investments in Arctic modernities, Russians also rule these waters. It is remarkable how the connection of Russianness and northern waters is created with semiotic and discursive practices, with a material agency that is reserved exclusively for Russian rulers. I will now look closer at the main actors in these Russian seas.

9.3 Heroic Shipbuilders and Seafarers

In addition to the Russianness of the Northern Sea, Shergin's stories highlight the heroism of the northern shipbuilders and seafarers on its waters. In the collection *Okean-more russkoe*, these kinds of stories are under the section titled "Ottsovo znan'e", which translates to "Knowledge of the Fathers". The idea is that the mastering of shipbuilding and seafaring skills is passed on from fathers to sons and that these skills are essential for the Pomor way of life. Both shipbuilding and seafaring are male professions, and thus in Shergin's works the Northern Sea is an exclusively masculine space.

Although Shergin wrote his stories in the early twentieth century, the prominence of shipbuilders and seafarers in them is strikingly similar to how John Gillis (2013), para. 5, see also 2012, pp. 68–98) describes sea fiction before the nineteenth century, including his note how the sea itself is represented almost as an afterthought. In this way, Shergin's stories follow a very traditional way of depicting the sea, and they highlight a connection of Pomor life in the early Soviet period to premodern times. My reading of Shergin's cold waters, inspired by blue ecocriticism, shows that there is little room for non-human, indigenous or female actors in them. The stories follow the old view, where skilful shipbuilders are the backbone of the relationship between the people and the sea. At the same time, while they depict shipbuilding as a long-established skill that the Pomors have mastered for centuries, they connect it to traits that have to do with modernity. A great shipbuilder must have extensive education in many subjects, and one has to travel overseas to receive this education. Additionally, shipbuilding is associated with the visual arts.

A prime example of the prominence of shipbuilders is the story "Rozhdenie korablia" ("Birth of a Ship", 1939), where three nineteenth-century shipbuilders are described as "famous shipbuilders of the Arkhangelsk Admiralty that spread glory all over the world on account of the ships built in Arkhangelsk" (Shergin, 1959, p. 5). The story depicts shipbuilding as the backbone of Pomor life. The variety of different ships that are built in the *Pomor'e* is stunning: "shkuny, boty, gal'oty, likhtery, kutera, ëly, murmanskie, shnëki, karbasa" (Shergin, 1959, p. 6). Additionally, "lod'i, brigantiny, kochi and barki" were built earlier but have become outdated. The narrator explains how his father used to sail on these ships to the Netherlands and Norway, and how local shipbuilding masters drew pictures of the ships to be able to explain their construction details to their apprentices.

The central hero of the story is, however, a later shipbuilder, whose work the narrator was able to follow in the late nineteenth century. His name is Konon Vtoroushin, he comes from the shores of the White Sea, and he is an educated man. He started to learn shipbuilding in the *Pomor'e*, but received further education in Norway and Denmark in languages, mathematics, navigation, astronomy and drawing. Later, Konon returned to the *Pomor'e* and worked as a highly regarded shipbuilder. His extensive experience abroad as well as his education are clearly markers of his fame. When in the story a new ship needs to be built, it is only Konon who can be trusted with the job. In this way, northern shipbuilding skills are connected to modernity, because extensive technical knowledge is needed in order for one to be a great builder of ships that traverse the northern waters. Konon is also familiar with the techniques of some of the most famous Western European artists, such as da Vinci and Michelangelo (Shergin, 1959, p. 13). Hence, shipbuilding is considered a form of art in the story.

Shergin's stories are explicit about the significance of scientific education, received overseas, for shipbuilding skills. Practical experience in shipbuilding, however, gets much less attention. Even though Shergin's stories often narrate the Northern Seas in connection with men who have great shipbuilding skills, and people's connection to the sea is associated with a centuries-long history of fathers passing on their skills to their sons, at the same time, they also intriguingly connect shipbuilding skills to modernity. There does not seem to be any conflict between the traditional and the modern. Even though modern technology is essential for shipbuilding, it seamlessly blends with the traditional way of life.

The skills that Shergin's stories attach to seafaring are quite different from those that are associated with shipbuilding. An example of a formidable seafarer in Shergin's stories is Pafnutii Ankudinov, who is one of the central recurring figures in several of Shergin's stories, such as "Dvinskaia zemlia" ("The Dvina Land"), "Novozemel'skoe znanie" ("The Knowledge of Novaya Zemlya"), "Zapechatlennaia slava" ("Imprinted Glory"), and "Ded Pafnutii Ankudinov" ("Grandfather Pafnutii Ankudinov"). Like the narrator's father, who "travelled on the Arctic Ocean throughout his life" (Shergin, 1959, p. 25), Pafnutii did the same but for an even longer period of time; the father considered Pafnutii to be his teacher and was the person the narrator most revered in his childhood. The narrator's father explained to him how practical experience was essential for learning the necessary seafaring skills on the difficult northern waters: "Listen son, at that time, everything was learned without books. Practice taught us the knowledge needed for seafaring" (Shergin, 1959, p. 59).

To illustrate his point, the father tells a story of a trip with Pafnutii to Novaya Zemlya, when he saved a boat from a disaster in difficult weather with the help of his superb practical knowledge of the area. Based on his seafaring experience, he calculates where the other boat had drifted. The captain of that ship, for his part, uses his experience to estimate where Pafnutii will most likely be searching for him: "The day and the place of the meeting we determined by the knowledge of the wind, the knowledge of the sea, the knowledge of the shores – not by fortune-telling or witchcraft" (Shergin, 1959, p. 62). The idea is that only with this practical knowledge are the Pomors able to fish and hunt in the treacherous northern waters. Since fishing and hunting are essential for the Pomor way of life, excellent seafaring skills are also required.

Pafnutii's skills do not end here for he is also described as a master of words and song: "Pafnutii Ankudinov was known as an excellent storyteller and singer of bylinas both during sea voyages and hunting trips. He also knew how to sing in the ancient way," the narrator of the story "Zapechatlennaia slava" describes (Shergin, 1959, p. 315).[7] The stories "Dvinskaia Zemlia" and "Ded Pafnutii Ankudinov" include similar remarks about Pafnutii's gift of speech despite him not performing officially. The idea is that storytelling skills are as important for Pomor identity as seafaring skills, and Boris Shergin sees himself as part of the northern storytelling tradition. Indeed, he is predominantly still known as such today in the *Pomor'e*.

Traits that connect seafaring skills to modernity are also present in Shergin's stories, although not as much as those that connect them to old traditions. This is most evident in the story "U korabelnogo pristanishcha" ("At the Haven of Ships", 1939), which praises the extent of Soviet sea exploration:

> My city, my homeland, you are the door, you are the gateway to unknown polar lands. Scientists arrive in Arkhangelsk and are equipped to map the depths and distances of the Northern Ocean. Ships in all directions of the world are constantly sailing from the piers of Arkhangelsk. To the West – to Norway, Sweden, Denmark, Germany, England and America; to the North – to Novaya Zemlya, to Svalbard, to Franz Josef Land. In our days, the people's rule[8] has opened the gates to the East, and it has shown the great northern path. The rule of the Soviets[9] has equipped ships for long voyages that were previously only dreamed of. The rule of the Soviets looked with an inquiring eye and stepped with firm foot on such shores and lands and islands, where before the gull has not flown nor the halibut fish visited. The Soviet rule[10] has shaken up the polar bears, seals and walruses, brisked up the lazy and raised the drowsy to their feet. (Shergin, 1939, p. 15).

The same passage, with only the last sentence missing, is also present in the story "Dvinskaia zemlia", which was originally published in 1947. The story was republished in 1957, 1959 and 1967, and in these publications, the text is identical to the 1939 version. Another update appears in the last version, which was published in 1971. In it, the Soviet rule sends to the north not only sea vessels but also airplanes, and it sets foot not only on previously unknown northern islands and shores but extends its territory all the way to the North Pole. Thus, the 1971 version advocates even more modern technology than the earlier versions.

In addition to the praising of modern technology in sea exploration, the story "U korabelnogo pristanishcha" also repeatedly underlines how the Pomors have always

---

[7] In the forewords of the second part of his book, *U pesennykh rek*, Shergin (1939) also associates the skilful use of words with ordinary Pomors from whom he had collected Pomor folklore: "Northern fishermen, hunters, lumberjacks, factory workers do not realize that they are word artists even though they are uneducated" (p. 83).

[8] *narodnaia vlast´*

[9] *Vlast´ sovetov*

[10] *sovetskaia vlast´*

lived by the sea and off the sea, despite the challenges the sea has posed. One of the most often cited passages is as follows:

> And in our land, water is the beginning and water is the end. Waters give birth, and waters bury. The sea waters and feeds... But who knows the sea? We do not walk on land, but in the depth of the sea. And our destiny is something we have in common [with the sea]. (Shergin, 1939, pp. 7–8).

This passage does not explicitly refer to sea voyages, but it exemplifies how the same story that praises modern exploration of the northern waters at the same time advocates the idea of an age-old connection of the people with the sea. Just as the modern and the traditional blend together with shipbuilding in Shergin's stories, there is no apparent conflict between the two in connection to seafaring either. The mythical sea, the "grey-haired ocean" and the age-old companion of the Pomors, merges together with the technological progress that was pivotal for the early Soviet state, similarly to how Rolf Hellebust (2003) describes the spiritualization of matter and the materialization of spirit in what he calls the "flesh-to-metal myth" in early Soviet literature, where machinery is often mystified (pp. 144–151). In the case of Shergin's stories of the Russian North, machines are not assimilated into the human body, as in Hellebust's material that is obsessed with metallization of human flesh, but into the Northern Sea, bringing it under the rule of magic and myth. In this way northern myths and mysteries do not have to give way to modernity, because they are incorporated into it, even though modernity is at the same time solving the mysteries of the far North by extending its influence still further toward the North Pole.

9.4 The Northern Sea as an Inexhaustible Source of Natural Resources

Marine animals often appear in both Shergin's and Pisakhov's stories. Below I concentrate on two of Pisakhov's fairy tales, where schools of fish convey an understanding of the Northern Sea as an inexhaustible source of natural resources. Many scholars write that the main protagonist of Pisakhov's fairy tales, Senia Malina, embodies an equal relationship to the non-human environment and is a "man of nature" (see Ponomarëva, 1985, p. 16; Kamalova, 2012, p. 68; Johansen, 2017, p. 49). However, despite his superficial connection to natural elements, Senia's relationship to nature is highly instrumental. His motivation to benefit from fish is evident in the two stories that I analyse below. In literature and culture, the sea often functions as a trope for the boundlessness, which, as Patricia Yaeger (2010) notes, may contribute to people treating it as an inexhaustible storehouse of goods (p. 535). The trope of the boundless sea is clear in Senia's adventures in the Northern Seas with their endless supplies of fish, which are represented as collective bodies rather than as individual organisms with agency (cf. Dobrin, 2021, p. 151).

In Pisakhov's (1985, pp. 145–146) fairy tale, "Na treske guliali" ("A Walk on Cod"), a shipmaster collides with an immense shoal of cod in the Arctic Ocean.

There are so many fish that one could walk on them. The attention of the crew is on the economic potential of the event. The crew beg the captain to let them harvest the fish, because otherwise "so much wealth goes to waste" (Pisakhov, 1985, p. 146). This is a prime example of a typical human approach to the creatures of the sea: if you can catch them and make use of them, do so or they will be wasted; if they can possibly be used by humans but people do not exploit them, it is wasteful. In the story, the cod are caught and the crew live on them for a whole winter.

Later in the story, the same happens to Senia Malina. He is alone in the Barents Sea in a small boat, taking a nap, when he runs into a shoal of cod that is many kilometres long and extends to twenty-five metres in depth. Senia starts travelling on the shoal. He builds a hut on top of the fish and is overnight there for several days, eating fish whenever he is hungry: "The cod both feeds and carries me," Senia notes (Pisakhov, 1985, p. 146). The shoal carries him north past Novaya Zemlya, into the Arctic Ocean. However, he works out how to harness the leader of the fish and leads the shoal south towards his hometown. Once he arrives there, he sells the cod for a good price. Once again, the fish have only instrumental value; they can be used to help sail in the ocean and when the trip is over they can be exchanged for money.

A similar fairy tale by Pisakhov is "Sauna on the Sea" ("Bania v more"), where Senia improvises a boat out of his sauna and goes to sea to fish from it. It appears that his sauna attracts the fish, who are "a curious lot, they need to know everything, and in the sauna the news is always fresh and new" (Pisakhov, 1985, p. 36). Senia tricks the fish into thinking that he is inviting them to the sauna to hear the latest gossip, and they start flocking in. Senia catches more fish than his sauna can take, and shares the catch with the other fishermen. This companionship with men who appear to be the masters of the sea is enhanced when Senia lets the others bathe in his sauna. Again, a sense of losing the potential of natural resources that could be used by humans is present: "To prevent letting out smoke in vain, they converted the chimney into a smokehouse" (Pisakhov, 1985, p. 36). Senia and the other fishermen also take great pains to collect as many fish as they possibly can, even though the number of fish clearly exceeds their needs.

Senia's extravagant spectacle with the cod takes place again in winter, when the source of the fish is inexhaustible: "We fished the whole winter. It is impossible to catch all the fish in the sea" (Pisakhov, 1985, p. 37). Using a sauna is yet another way to fish, and the fish are nothing more than riches lurking in the sea, waiting to be exploited by humans. This is completely in line with the 1930s Soviet state's aspirations, as it was a time when it sought rapid exploitation of Arctic natural resources with little thought for the vulnerability of the northern environment, and when the number of fish trawled in the Barents Sea grew exponentially (Bruno, 2016; Vize, 2008, p. 184). It reflects a typical idea of the sea as an immensely vast storehouse of unlimited resources, which it is not, of course.

9.5 Conclusion

This chapter shows that while early Soviet Pomor fiction idealizes the traditional Pomor life that is dependent on the Northern Sea, at the same time it imagines the arrival of Soviet Arctic modernity as something that coexists without trouble with the traditional way of life. Even though the Russian North is represented as an ancient living place of the Pomors, as the readings based on "the northern text of Russian literature" emphasize, modernization has reached it too. The modern features of the North do not appear, however, to be in conflict with traditional ways of life. Instead, they complement it. In this way, Boris Shergin's and Stepan Pisakhov's works contrast with late Soviet "village prose", which also emphasized the long coexistence of people with non-human nature, but was critical of modern ways of exploiting the environment. The idea of the conflict-free existence of traditions and modernity probably contributes to the better reception of the emergence of modern technology in the *Pomor'e*. My reading, however, cannot clarify whether this is intentional or not.

Reading representations of the Northern Seas in Shergin's and Pisakhov's works from the point of view of blue ecocriticism shows that the material qualities of these seas are not prominent. Instead, they usually discuss the sea through allegories, where the sea is a semiotic space rather than an agentic actor. Skilful shipbuilders and experienced seafarers are the main agents in the Northern waters, and value is attached to the non-human world mainly via its material and economic potential for humans. The Northern ocean is prominently a Russian sea in the works of Shergin and Pisakhov. The other peoples that inhibit the northern shores of the White Sea and the Arctic Ocean are rarely mentioned, and when they are, they are subordinate to Russian command. This is logical, because since the Pomor way of life is imagined as an especially genuine variant of Russianness, and the Northern Sea as an integral part of the Pomor culture, then it must be a Russian sea.

References

Anufriev, V. (2013). *Kul'tura russkikh pomorov: Istoriko-kul'turologicheskii analiz*. Forum.

Barad, K. (2007). *Meeting the universe halfway: Quantum physics and the entanglement of matter and meaning*. Duke University Press.

Boele, O. (1996). *The north in Russian romantic literature*. Rodopi.

Boele, O. (2016). The north as a symbol of national identity in Russian culture. In K. A. Myklebost, J. P. Nielsen, V. V. Tevlina, & A. A. Komarov (Eds.), *Net Severa, a est' Severá: o mnogoobrazii poniatiia "Sever" v Norvegii i Rossii* (pp. 114–131). Lenand.

Bolotova, A. (2014). *Conquering nature and engaging with the environment in the Russian industrialized North*. Doctoral dissertation, University of Lapland. https://urn.fi/URN:ISBN:978-952-484-780-3

Bruno, A. (2016). *The nature of soviet power: An Arctic environmental history*. Cambridge University Press.

Dobrin, S. I. (2021). *Blue ecocriticism and the oceanic imperative*. Routledge.

Frank, S. K. (2010). City of the sun on ice: The soviet (counter-)discourse of the Arctic in the 1930s. In A. Ryall, J. Schimanski, & H. H. Wærp (Eds.), *Arctic discourses* (pp. 106–131). Cambridge Scholars Publishing.
Frank, S. K. (2017). Ice as a literary motive in soviet Arctic modernities. In H. Hansson & A. Ryall (Eds.), *Arctic modernities: The environmental, the exotic, and the everyday* (pp. 16–38). Cambridge Scholars Publishing.
Frye, N. (1957). *Anatomy of criticism: Four essays*. Princeton University Press.
Galimova, E. (2012). Genii mesta. Stepan Grigor'evich Pisakhov. In N. A. Korolkova, L. A. Tolochin, & L. E. Karshina (Eds.), *Stepan Grigor'evich Pisakhov. Bibliograficheskii ukazatel´* (pp. 7–32). Arkhang. obl. nauch. b-ka im. N. A. Dobroliubova.
Galimova, E. (2013). *Poeziia prostranstva: obrazy moria, reki, lesa, bolota, tundry i motiv puti v Severnom tekste russkoi literatury*. KIRA.
Garrard, G. (2011). *Ecocriticism*. Routledge.
Gillis, J. R. (2012). *The human shore*. University of Chicago Press.
Gillis, J. R. (2013). The blue humanities. *Humanities, 34*(3) https://www.neh.gov/humanities/2013/mayjune/feature/the-blue-humanities
Hansson, H., & Ryall, A. (2017). Introduction. Environmental, exotic and everyday Arctic. In H. Hansson & A. Ryall (Eds.), *Arctic modernities: The environmental, the exotic and the everyday* (pp. 1–13). Cambridge Scholars Publishing.
Haraway, D. (1991). *Simians, cyborgs, and women: The reinvention of nature*. Free Association Books.
Hellebust, R. (2003). *Flesh to metal: Soviet literature and the alchemy of revolution*. Cornell University Press.
Hønneland, G. (2016). *Russia and the Arctic: Environment, identity and foreign policy*. I. B. Taurus.
Iovino, S., & Oppermann, S. (2014). Stories come to matter. In S. Iovino & S. Oppermann (Eds.), *Material ecocriticism* (pp. 1–17). Indiana University Press.
Johansen, L. A. (2017). *Tradition and novelty in Stepan Pisakhov's works*. The Arctic University of Norway. https://hdl.handle.net/10037/11204
Kamalova, A. (2012). Khudozhestvennaia kartina mira skazok S. Pisakhova. *Vestnik Buriatskogo Gosudarstvennogo universiteta, 10*, 66–70.
Kliuchevskii, V. (1987). Lektsiia III. In *Sochineniia v deviati tomakh. Kurs russkoi istorii. Chast´ I*. Mysl´.
Körber, L.-A., MacKenzie, S., & Westerståhl Stenport, A. (2017). Arctic modernities, environmental politics, and the era of the Anthropocene. Introduction. In L. A. Körber, S. MacKenzie, & A. W. Stenport (Eds.), *Arctic environmental modernities: From the age of polar exploration to the era of the Anthropocene* (pp. 1–20). Springer.
Likhachëv, D. (2004). Russkii sever. In K. P. Kemp (Ed.), *Skaz o Belomor'e. Slovar´ pomorskikh rechenii* (pp. 3–4) http://likhachev.lfond.spb.ru/articl100/Russia/rus_sever.pdf
Likhachëv, D., & Ianin, V. (1986). Russkii Sever kak pamiatnik otechestvennoi i mirovoi kultury. *Kommunist, 1*, 115–119.
Lundkvist, S. (1985). Northern Scandinavia during the middle ages. In M. Backe, I. Bergman Hennix, L. Forsberg, L. Holm, L. Liedgren, A.-K. Lindqvist, I.-M. Mulk, M. Nejati, P. Perstrand, & P. H. Raqvist (Eds.), *In honorem Evert Baudou* (pp. 89–97). Department of Archaeology. University of Umeå.
McCannon, J. (1998). *Red Arctic: Polar exploration and the myth of the north in the Soviet Union, 1932–1939*. Oxford University Press.
Okorokov, A. V. (2005). Sistema ėkspluatatsii prirodnykh resursov. In Ė. L. Bazarova, N. V. Bitsadze, A. V. Okorokov, E. N. Seleznëva, & P. I. Chernosvitov (Eds.), *Kul'tura russkikh pomorov* (pp. 35–44). Nauchnyi mir.
Oppermann, S. (2019). Storied seas and living metaphors in the blue humanities. *Configurations, 27*(4), 443–462. https://doi.org/10.1353/con.2019.0030
Perkiömäki, M. (2021a). Boris Šerginin pohjoinen. *Idäntutkimus, 28*(1), 3–19. https://doi.org/10.33345/idantutkimus.107838

Perkiömäki, M. (2021b). *Imagined riverography of late twentieth-century Russian prose*. Doctoral dissertation, Tampere University. https://urn.fi/URN:ISBN:978-952-03-2012-6
Pisakhov, S. (1985). *Skazki, ocherki, pis'ma*. Severo-zapadnoe knizhnoe izdatel'stvo.
Ponomarëva, I. (1985). Stepan Pisakhov (1879–1960). In S. Pisakhov, *Skazki, ocherki, pis'ma* (pp. 5–20). Severo-zapadnoe knizhnoe izdatel'stvo.
Rosenholm, A., & Trubina, E. (2021). "Pohjoisen teksti" – kirjallisuudentutkimus kansallisen eetoksen rakentajana. *Idäntutkimus, 28*(1), 36–49. https://doi.org/10.33345/idantutkimus.107840
Shergin, B. (1924). *U arkhangel'skogo goroda, u korabel'nogo pristanishcha. Sbornik starin*. Gosudarstvennoe izdatel'stvo.
Shergin, B. (1936). *Arkhangel'skie novelly*. Sovetskii pisatel´. https://bookscafe.net/read/shergin_boris-arhangelskie_novelly-246797.html#p1
Shergin, B. (1939). *U pesennykh rek*. Gosudarstvennoe izdatel'stvo "Khudozhestvennaia literatura".
Shergin, B. (1947). *Pomorshchina-korabel'shchina*. Sovetskii pisatel´.
Shergin, B. (1957). *Pomorskie byli i skazaniia*. Detgiz. https://www.litmir.me/bd/?b=184491
Shergin, B. (1959). *Okean-more russkoe. Pomorskie rasskazy*. Molodaia gvardiia.
Shergin, B. (1967). *Zapechatlennaia slava. Pomorskie byli i skazaniia*. Sovetskii pisatel´.
Shergin, B. (1971). *Gandvik – studënoe more*. Severo-zapadnoe knizhnoe izdatel'stvo.
Starkov, V. (2012). *Ocherki istorii osvoeniia Arktiki. Tom III: Russkii Sever v kartografii XVI–XVII vv*. Professional.
Terebikhin, N. (1993). *Sakral'naia geografiia Russkogo Severa (Religiozno-mifologicheskoe prostranstvo severnorusskoi kul'tury)*. Izdatel'stvo pomorskogo mezhdunarodnogo pedagogicheskogo universiteta imeni M. V. Lomonosova.
Terebikhin, N. (2004). *Metafizika Severa*. Pomorskii universitet.
Vize, V. (2008). *Moria rossiiskoi Arktiki: Ocherki po istorii issledovaniia*. Evropeiskie izdaniia.
Vlasova, I. (2001). *Russkii Sever: Ėtnicheskaia istoriia i narodnaia kul'tura: XI-XX veka*. Nauka.
Widdis, E. (2000). Borders: The aesthetic of conquest in soviet cinema of the 1930s. *Journal of European Studies, 30*(120), 401–411. https://doi.org/10.1177/004724410003012004
Widdis, E. (2003). *Visions of a new land: Soviet film from the revolution to the second world war*. Yale University Press.
Widdis, E. (2004). Russia as space. In S. Franklin & E. Widdis (Eds.), *National identity in Russian culture: An introduction* (pp. 30–50). Cambridge University Press. https://doi.org/10.1017/CBO9780511720116
Yaeger, P. (2010). Editor's column: Sea trash, dark pools, and the tragedy of the commons. *PMLA, 125*(3), 523–545. https://doi.org/10.1632/pmla.2010.125.3.523

Chapter 10
Water, Oil and Spirits: Liquid Maps of the Taiga in Eremei Aipin's Novel *Khanty, or the Star of the Dawn*

Eeva Kuikka

Abstract In this chapter, I present a reading of Khanty author Eremei Aipin's novel *Khanty, or the Star of the Dawn* through its main liquid elements: water, oil and spirits. I approach the novel through the theoretical framework of postcolonial ecocriticism. I argue that all these three fluids participate in the imagining of the literary space of the novel by creating their own liquid maps that reflect rather different ways of relating to the surrounding environment. The cold, northern river is depicted as a watery map that connects the Indigenous Khanty people on a geographical and temporal level as well as transmits the Indigenous epistemology, underlining the inseparability of people and non-human nature. This map of the river is juxtaposed with maps of oil and spirits that represent the colonial mapping of the settler state and demonstrate an instrumentalised view of non-human nature. Re-mapping the area through oil and spirits leads to devastating consequences for the taiga environment and the Indigenous people whose epistemology and cultural memory are entangled in it.

Keywords Indigenous people · Postcolonialism · Postcolonial ecocriticism · Soviet Union

From the 1930s onwards, the need for electricity for building the communist society and its industries led to massive hydroelectric projects, such as the Dnieper Hydroelectric Station, that were meant to harness Russia's largest rivers for electricity production (Josephson et al., 2013, pp. 76, 81–82). After World War II, the oil production of the country shifted from faraway Caucasus to the Volga-Ural region and, later in the 1960s, to West Siberia (Moser, 2018, pp. 55, 59–60). River-based electricity was needed in order to secure the extractive industries (Josephson et al., 2013, pp. 128–130), which in return polluted the local river basins with oil and petroleum runoffs (Weiner, 1999, pp. 382–383). Moving the emphasis of the

E. Kuikka (✉)
Tampere University, Tampere, Finland
e-mail: eeva.kuikka@tuni.fi

M. Lehtimäki et al. (eds.), *Cold Waters*, Springer Polar Sciences,
https://doi.org/10.1007/978-3-031-10149-6_10

resource extraction to West Siberia also had a massive impact on the Indigenous peoples of the area, whose livelihoods depended on the reindeer pastures, rivers and lakes that were turned into extraction sites or contaminated as an effect of the mining and extraction industries.

The focus of this chapter is on the interconnectedness of water and oil. I discuss how Khanty author Eremei[1] Aipin's novel *Khanty, or the Star of the Dawn* (*Khanty, ili zvezda utrennei zari*, 2014) depicts the northern taiga as intersected by three different flowing substances: water, oil and spirits. I ask what meanings these liquids are given in the novel, from what kinds of epistemologies these meanings emerge and how they participate in drawing a fluid map – or maps – of Aipin's home region. I argue that comparing these different maps brings out the conflict between Khanty people's relationship with the non-human environment of the region and the Soviet Union's settler-colonial claim on the natural resources in the area. For the Khanty, the river appears as an organic watery map that connects Khanty people on a geographical level, as well as a temporal level, whereas oil and alcohol are defined by the settler-colonial logic of conquering the region through its natural resources. If the map of the river emerges in the entanglement of the river and the Khanty people, the oil stream and alcohol are related to the state's understanding of the non-human environment and people as hierarchically separated from one another. Even if Aipin's novel was published in 1990, the juxtaposition of these flowing substances continues to demonstrate the tension between Indigenous people and the nation state today as oil companies aim at expanding their extraction sites near Lake Numto, a sacred place for many West-Siberian Indigenous peoples (Golovnev & Golovneva, 2021).

I approach these questions through the theoretical perspective of postcolonial ecocriticism. As I read the different liquids of the novel through the notion of mapping, I pay special attention to the intersections of maps and cartographic theory, ecocriticism and postcolonial theories. Because the liquids not only flow through space but also through time and because they interconnect with questions of remembering and forgetting, my analysis also draws from memory studies. I start this chapter with short introductions of Khanty people in general, Eremei Aipin in particular and his novel, as well as the above-mentioned theoretical frames. Then I move onto analysing the river in Aipin's novel as a spatial actor that reflects Khanty people's relationship with their home region and maps the geography and different temporalities of the area. After that, I look at oil and its related mapping practices as antithetic to the flowing river. Finally, I discuss how the settler-colonial oil extraction, which destroys the northern environment, becomes transcribed in Khanty people's bodies through alcohol and hence underlines the embodied entanglement of the Khanty people and the environment they inhabit and upon which they depend.

[1] Depending on transliteration system, the author's name is sometimes transliterated as Yeremei Aipin.

10.1 *Khanty, or the Star of the Dawn* as a Native Historiography

Eremei Aipin belongs to the Khanty people, an Ugrian Indigenous people living in western Siberia. Khanty's traditional livelihoods include fishing, hunting and reindeer herding. They live a semi-nomadic life as they change their habitation according to the season: they change from permanent winter huts to temporary summer shelters in their fishing and hunting grounds. The Khanty were subdued under the reign of the Muscovite empire in the late sixteenth century, and like other Native peoples in Siberia, the Khanty had a *yasak* (a fur tribute) imposed on them (Forsyth, 1992, pp. 36–37; Gurvich, 1982, p. 11). The Bolshevik revolution and the change in policies concerning Indigenous peoples affected the Khanty people as well. At first, the communist rule aimed at ameliorating Indigenous peoples' lives and elevating them from what were perceived to be backward conditions to the same level as the rest of the country, yet the philanthropic spirit disappeared with Stalin's seizure of power. The urge to "civilize" the Natives grew more pressing in the rush to realise similar collectivisation in Siberia that was seen in other parts of the country (Forsyth, 1992; Slezkine, 1994). Soviet policies were not met without resistance. Among the Khanty, they led to the Kazym rebellion that started in 1931 with an uprising which grew into a series of short violent conflicts in 1933 and the demand for Russians to, for example, close the cultural bases, stop fishing in the sacred Lake Numto and leave the area. The revolt was supressed by Soviet forces in a short confrontation (Golovnëv, 1995, 165–178).

One of the Bolsheviks' main tasks among the Indigenous peoples was promoting literacy, which necessitated developing written forms for Indigenous languages. The uniform language policy in the Soviet Union led to a similar development of written languages and literary tradition among all the Indigenous peoples of the North. Along with literacy and education, a group of northern national intellectuals emerged, Native people who were educated at the Institute of the Peoples of the North in Leningrad State University. This group was needed by the Soviet authorities in order for them to serve as mediators between the new power and the local Indigenous communities (Toulouze, 2005, p. 141). Nevertheless, the first Khanty authors and translators emerged from a literary circle at the Ostiako-Vogul'sk pedagogical institute (Ogryzko, 2002, pp. 11–14). Regardless of the early names in Khanty literature, Viacheslav Ogryzko (2002) primarily connects the rise of the prose tradition in Khanty culture with Eremei Aipin, whose first book came out in 1979 (p. 23). The main conflict in Aipin's works is the threat of oil extraction to the Khanty culture, which is dependent on the surrounding natural environment. Ogryzko (2002) compares Aipin to Russian village prosaists, especially Valentin Rasputin, whose *Farewell to Matyora* (*Proshchanie s Matëroi*, 1976) presents a similar tragedy of a traditional culture being swept away by modernity to that which Aipin presents in his works on Khanty people and oil extraction (pp. 23–24). As the most prominent author in Khanty literature and with a political career at the state level, it is no wonder that Aipin's works have provoked scholarly interest (Ogryzko,

2006; Lagunova, 2007; Miliakhova, 2009). The critical tone of his works has also enabled a discussion of his literature from a postcolonial point of view, even though the postcolonial paradigm remains generally underdiscussed in the Soviet context (Smola, 2017; Boliachevets & Sablin, 2016).

Critique towards Soviet power is present in Aipin's *Khanty, or the Star of the Dawn,* which deals with the dreadful consequences of oil drilling in the home area of the Khanty people. The first version of the novel was published as a short story in 1980, in the journal *Ural*, but the much more critical and pessimistic novel came out only at the dawn of the collapse of the Soviet Union. Since the publication of *Khanty, or the Star of the Dawn*, Aipin has published short stories and a novel, as well as worked in state politics, but nevertheless, the tragic novel about a Khanty hunter, Dem'ian, and his three-day journey to a local village and back remains his most notable work. The temporally leaping narration and multiple focalisers[2] from different decades turn Dem'ian's journey in the novel into a historical journey whose timespan stretches from the purges of the 1930s to the late 1980s. In addition to the main character's life story and journey, the novel entails stories of his relatives, other kinsmen and that of his son in the future. These embedded narratives take the reader on a journey through the Khanty people's history at the same time as Dem'ian travels across the taiga. Besides historical events, the novel is rich in allusions to Khanty myths and other folklore, which are transmitted to the reader in the narration as part of the Khanty people's thoughts or communication (Miliakhova, 2009, pp. 15–16). Even if previous studies on Aipin and his novel acknowledge the criticism of Soviet practices and the importance of the non-human environment in his works, a more elaborate analysis of the interconnectedness of these themes is still to be made. The present study strives to fill this void.

10.2 Postcolonial Ecocriticism, Mapping and Memory

As stated, the main conflict in *Khanty, or the Star of the Dawn* is caused by state-controlled natural resource extraction that harms both the non-human environment and Indigenous people of the area. The interrelatedness of environmental damage and settler-colonialist practices makes postcolonial ecocriticism a suitable and natural theoretical framework for approaching the novel. Postcolonial ecocriticism is a field that discusses the intersections of postcolonial and ecocritical theories, a field that emerged in the second half of the 2000s, predominantly in the writings of Graham Huggan and Helen Tiffin (2007, 2010). Bringing together postcolonialism and ecocriticism is not unproblematic as these two frameworks have mutual tensions: first world nations' environmentalism may risk stepping over marginalised people's and cultures' interests and adapt a universalist white gaze on the (formerly)

[2] The concepts of an 'anachronism' as a temporal discrepancy between the story and the narrative and 'focalization' as the point of view through which the story is narrated are both developed by Gérard Genette in his book *Narrative Discourse* (Genette, 1983).

colonised countries, which consequently reinforces imperial and colonial hierarchies (Cilano & DeLoughrey, 2007, pp. 72–73; Huggan & Tiffin, 2010, pp. 2–5).

However, in Aipin's novel, questions of environmentalism and Native people's rights to their land are not conflicted, they are closely interlinked as both non-human nature and Indigenous people appear in a marginalised and oppressed position in relation to the settler-colonial Soviet power. The consequences of this uneven power hierarchy lead to environmental damage and the degradation of the Khanty people. As they mostly occur gradually, in the long term, I approach them through Rob Nixon's concept of slow violence. According to Nixon (2011) slow violence is "a violence that occurs gradually and out of sight, a violence of delayed destruction that is dispersed across time and space, an attritional violence that is typically not viewed as violence at all" (p. 2). Even though Nixon's focus lies on environmental humanities, the concept has been adapted in order to describe the consequences of settler colonialism (Wakeham, 2021). It also fits a reading of Aipin's novel in which environmental and settler-colonial slow violence cannot be separated from one another. As Aipin's novel underlines the embodied character of Khanty people's interrelatedness with the non-human environment, another relevant concept in my analysis is Stacy Alaimo's concept of trans-corporeality. With trans-corporeality Alaimo (2008, 2018) refers to the enmeshment of the (human) body in the environment, which not only surrounds the body but also permeates the body and diffuses the boundary between the physical self and the material world. Understanding the Khanty people as trans-corporeally entangled in the surrounding nature makes it possible to see how the consequences of slow violence in the non-human environment are reflected in Khanty people's bodies.

In this chapter, postcolonial and ecocritical theories also intersect in the concept of mapping that is central in my reading of Aipin's novel. In postcolonial theory in particular, the role of maps and mapping practices has been vastly discussed as maps have been seen both as tools for legitimising colonial conquest and as demonstrations of colonialist discourses. As John Harley (1988) pointed out, maps have formed nation states' and empires' main tool for executing territorial power by demarcating territorial claims on spaces that have been detached from their social dimension through the cartographic representation. The imperialist function of maps is emphasised by the fact that the map is far from a neutral and objective geographical image, it is laden with power structures and projective distortions that easily remain unrecognised (Harley, 1988, pp. 287–294; Huggan, 2008, pp. 23–26). As a means of knowledge production with a certain epistemological basis, maps not only reflect the relationship with human others, they also reflect the relationship with non-human ones as well. For example, according to Leila Harris and Helen Hazen (2009), maps not only "privilege certain social groups over others", "similar hierarchies of values operate to privilege some species or non-living natures over others" and they "naturalize contingent links between spaces and territories, and constellations of human–environment relations" (pp. 55, 57). Both postcolonial and non-human perspectives on cartographies are visible in Aipin's novel where the map used by oil-seeking geologists both disregards the Indigenous presence and habitats in the area by labelling their habitat as drilling fields and naturalises environmental damage through regarding non-human nature as a natural resource.

The last important conceptual tool for my analysis is memory. Even if such an abstract concept might seem unrelated to a postcolonial and ecocritical reading that emphasises materiality and embodiment, in line with the suggestion of Hartmut Böhme (1988, pp. 24–25), I understand the liquid elements, especially the river, to simultaneously be both material and symbolic. Therefore, discussing water from the perspective of cultural studies necessitates considering the multiple discursive, cultural and symbolic meanings embedded in this material entity. In *Khanty, or the Star of the Dawn,* the flow of the river is elementarily linked with flows of time and life that are manifested through memory. My discussion on memory relies on the notion of cultural memory elaborated by Jan Assmann (1992) and Aleida Assmann (1999, 2011). With cultural memory they refer to memory that is stored outside of the human mind in rituals, myths and cultural artefacts. As it passes from one generation to another, it transfers cultural and social knowledge and creates cohesion between the members of a community. If cultural memory creates a sense of togetherness and social belonging through a set of cultural formations, it is communicative memory that passes on social memory between the individual members of a community (Welzer, 2005, pp. 13–14). Whereas cultural memory maintains the grand narrative of a social group, Harald Welzer (2005, p. 15) points out that communicative memory defines one's personal belonging to that group by determining one's individual history in relation to the collective past.

10.3 Mapping with the River

Even though Dem'ian's journey in Aipin's *Khanty, or the Star of the Dawn* takes place on a road, it is not the terrestrial road but the river alongside which he travels that constitutes as the central element in the literary space of the novel. In fact, the terrestrial road has formed in order to travel alongside the river as both "the winter road and the summertime 'road', the water",[3] serve as the route between villages and smaller settlements around the taiga (Aipin, 2014, p. 55). Therefore, it is specifically the river that defines Dem'ian's movement "up the river". Moreover, rivers and river systems designate the larger geographical context of Dem'ian's home region near the Arctic as up in the North "begins the Nenets tundra, near the river basin" and the rivers "flow into the Arctic Ocean whose breath he sometimes feels in autumn and spring" (Aipin, 2014, pp. 42–43). The river also appears as the defining feature when describing the Khanty people's settlements around the northern taiga as different Khanty families are not only defined through their clan's names but also through their clan's residence in relation to the river, whether they are upstream or downstream and along which tributaries the families have spread. The

[3] Translations from the Russian sources are my own.

significance of the river for the local Khanty families also becomes evident in the way the narrator only rarely refers to the Khanty people as "the Khanty people" and rather identifies them through the river as the "river people" (Aipin, 2014, p. 127).

The river's importance in depicting Dem'ian's home region turns the waterscape into a watery map through which the reader perceives the northern taiga and its inhabitants. Even though the spatial description of the novel is not a map in a traditional sense, it enables the reader to form a map-like image of the literary space in which the work takes place, and hence, it may be understood "as a form of mapping or a cartographic activity" (Tally, 2013, p. 45). Whereas in 2013 Robert Tally (pp. 49–50, 54) saw the writer or the narrator as the active mapmaker who chooses what to include in the literary cartography of the work, in Aipin's novel, the description of the Khanty's home region is given in terms of the river. The narrator may travel with the characters up and down the river, but it is the flow of the water, the riverscape and the appearance of tributaries, lakes or swamps that lead the narration of the space. In this sense, the narrator does not merely map the river but rather maps *with* the river, which makes the river an active participant in the production of the literary cartography of the area. This idea of mapping with a river resonates with Cecilia Chen, Janine MacLeod and Astrida Neimanis' (2013) idea of thinking with water rather than "*of* or *about*" it (p. 3) in order to dismantle the idea of water as a resource and to emphasise our own material, embodied and discursive entanglement with it. Similarly, the way in which Aipin's narrator engages in the creation of the literary cartography *with* the river questions the distinction between the observing subject and the passive observed object typical for cartographic activity (cf. Kitchin et al., 2009, pp. 2–3).

The entangled nature of the Khanty people and the river is visible in the way the river is used in the novel as a metonym for the whole Khanty community that lives along it. It is the river (instead of people) that turns into a subject who "remembers" (Aipin, 2014, p. 271) or "loses its last soldier and last singer" (p. 334). The narrator's way of dissolving individual human subjects with the use of the river metonym questions divisions between human and non-human, as well as a milieu and the persons or beings dwelling in it. Instead of underlining the Khanty people as individual subjects living along the river, the novel depicts both the rivery space and its inhabitants as an entity with agentic capacities. In this way, the description of the river concretises Mielle Chandler and Astrida Neimanis's (2013, p. 76) idea of how watery milieus challenge the common way of understanding a milieu as "a passive backdrop" by simultaneously being an inextricable part of people's material being and providing the conditions for existing and operating. Acknowledging the watery nature of the milieu necessitates the "dissolution of oneself as a sovereign subject" that, at the same time, dissolves the dichotomy between human and non-human by underlining the interconnectedness between human individuals and different non-human entities around us (Chandler & Neimanis, 2013, p. 75).

This dissolution of human subjects and the human/non-human dichotomy by means of the watery milieu becomes specifically apparent when the narrator tells how the information of a death of a Khanty person travels among Khanty families on the riverside: "The piece of news spread slowly, stayed overnight at smoky campfires, travelled along over grassed crossing places, leaning on a flint oar in a furious flow, stayed few days in fishermen's and hunters' huts, climbed up the riverside. Finally, it reached those to whom it was primarily meant" (Aipin, 2014, pp. 76–77). The agentic subject of the passage is not limited to the people carrying the message, it is the message itself, as well as the whole riverscape that consists of people, water, and other non-human entities. Such a mode of narration takes a step away from human-centeredness towards more all-encompassing modes of imagination that recognise and give space to a more nuanced understanding of agency in which it is not only understood as a property of human subjects but also as emerging in the entanglements and networks of different entities. At the same time, it points to the trans-corporeal relationship between the Khanty people and the river as it underlines how the people and other entities "are intermeshed with the dynamic, material world" (Alaimo, 2018, p. 435). Mika Perkiömäki (2017) also noted the trans-corporeal connectedness of people and a riverine ecosystem in his analysis of Viktor Astaf'ev's *Queen Fish* (*Tsar'-ryba*, 1976). According to his interpretation, Astaf'ev's way of describing the river's food chain – which consists of gnats, fish, and people – underlines how they all depend on the river and become one through the flow of matter and substances between human and non-human bodies (Perkiömäki, 2017, pp. 158–159). In a similar way, Aipin's narration points to the oneness of the river and the beings depending on it.

10.4 Mapping Temporalities

In addition to the river's geographical map-making agency, it also appears as a temporal map that maps the past of the Khanty people. By mapping the physical locations of the Khanty families, whose members of different generations characterise the history and flow of time, the river turns into a metaphorical "river of life and time" (Lagunova, 2007, p. 68). On the other hand, as a physical watercourse, the river evokes memories in individual Khanty characters with its material composition. For example, as Dem'ian travels alongside the river, memories of a boat journey in the opposite direction with a Russian doctor, Marina, "flow along the way towards him" (Aipin, 2014, p. 55). The memory of Marina becomes inscribed in the riverscape, and the shapes of "every springlike swamp, every autumn colour, every curve in the river, and every lake shore" remind Dem'ian of Marina who is "everywhere around him: in colours, voices, the contours of woods, and his land" (Aipin, 2014, p. 281). At the same time as Dem'ian's journey proceeds, the narrator takes the reader to Dem'ian's memories of the same journey from a long time ago. For one of Dem'ian's kinsmen, the river gives rise to a memory of his son's fall in the Second World War, and for him, the physical river turns into a symbolical "river of eternity" along which the lost son "sails away" instead of "sailing across these waters" (Aipin, 2014, p. 49). The "river of time" eventually becomes a "river of eternity" in Dem'ian's case too as the novel ends in his death, and the last journey he makes to the village alongside the river is also his last journey into his personal memories and back.

The memories that the river gives rise to in the Khanty characters are extremely personal, those of desire, loss and grief. Yet these individual memories are connected with the broader Soviet history and society. The Second World War, in which the old Khanty man's son fell, and the fraternity of peoples are important symbols and building material for the Soviet identity and a sense of Soviet history. However, the memories that arise from the river do not follow the national discourses of heroic deeds, victory over fascism, a multi-ethnic Soviet people or proletarian internationalism that dominate the discussion on these themes. Jane Costlow and Arja Rosenholm (2007, p. 5) pointed to water's interconnectedness with the subconscious and repressed; in Aipin's novel, the flowing water also brings back marginalised and even suppressed memories: personal grief instead of national celebration, sexual desire instead of platonic comradeship. In this sense, the river in Aipin's novel shares a similar destabilising potential as that which Withold Bonner (2017) reads into the sea in Anna Seghers' novella *Überfahrt* (1971), in which the fluidity and ambiguity of the water undermines the binary oppositions of the time in which the work was written. Just as the ever-flowing water of the river raises buried objects to its surface, the constant movement of the river in Aipin's novel cannot let the bottled-up memories lie at the bottom of the mind.

Because of this capacity to bring out suppressed memories, it is no wonder that Dem'ian shares his memories of Khanty's bear feast specifically during the river journey with Marina. This memory is narrated in a detailed, nine-page-long description of a five-day feast held in order to return a killed bear's soul back to the forest. If the river evokes in Dem'ian a personal and silenced memory of a women he once desired, the moment of remembering and recounting a feast of great importance to his family and people reveals the river's ability to also surface supressed (or at least marginalised) collective memories. As the bear feast that Dem'ian recounts step by step is a piece of cultural memory (Assmann, 1992; Assmann, 1999) the river not only evokes individual supressed memories but also communal ones and followingly the communal identity of the whole Khanty community. This capacity to strengthen the Khanty people's self-identity demonstrates how "[r]iver narratives […] are essential parts of cultural history" and how "the perception and representation of a river" not only create cultural narratives for empires such as Russia but also for marginalised and subjugated communities (Costlow & Rosenholm, 2007, p. 3).

Within the novel's context, in which the surrounding society favours a national Soviet identity instead of minor ones, the attachment to a local river as the basis of one's identity becomes a way of resisting the superimposed, unifying master narratives of the state. This becomes evident when Dem'ian's sons, Mikul' and Iuvan, visit a monument for the fallen soldiers of their home region. The monument represents Pierre Nora's (1996) *lieu de mémoire* as it is a symbolic site to which a memory or a history is attached. However, instead of being a real and lived environment of that memory, a setting "in which memory is a real part of everyday experience" (Nora, 1996, p. 1), the *lieu de mémoire* is rather a reminder of the past. According to Nora (1996) it serves a collective purpose by participating in the reconstruction of a "commemorative consciousness" (p. 6), a sense of the shared history of a community that would be swept away if the material symbols were to disappear. Correspondingly, the monument of the Second World War is meant to reconstruct a memory of the war as a part of Soviet history, a unified narrative of the past of the Soviet people. However, in Dem'ian's sons the monument leads them to recall different memories to those of the official history writing as they both connect the names of fallen kinsmen on the obelisk with the rivers of their home region. When Mikul' reads the list of names, he complements each name with the name of the river along which the person's family originates. He constantly asks: "From which river? From which family? Did they leave behind any relatives on this land? Do they remember them?" (Aipin, 2014, p. 333). Also, Iuvan recalls a story of some parents who escorted their son to the war front down the river by boat. In their minds, the river preserves the intimacy and the communicativity of the memory that the monument ineluctably loses according to Nora. Whereas the monument symbolizes a historical memory belonging to no specific individual but to the abstract yet discursively dominant nation, the flowing rivers entail local memories of the fallen soldiers that their relatives keep alive by continuing their lives besides the river.

10.5 Settler-Colonial Mapping and Oil

As pointed out above, the river in Aipin's novel can be read as a natural map that both shows the geographical distribution of the Khanty families in the region and reflects the symbolic and embodied interconnectedness of the Khanty people with the river and their riverine home area. In the novel this organic map that underlines the entanglement of humans and their non-human environment is juxtaposed with another kind of a map, one made by Russian oil seekers to locate oil deposits and mark oil pipes across the taiga: "They go according to a paper that has the earth's face drawn on it: lakes and rivers, forests, swamps and streams, hills and valleys of the taiga. That drawing they call a map." (Aipin, 2014, p. 41.) Several scholars have pointed out how – especially in a Canadian context (see, e.g. Hoogeveen, 2015; Simpson, 2019; Schmidt, 2020) – natural resource extraction and the geological search for new resource deposits are settler-colonial techniques that lead to the dispossession and destruction of Indigenous lands and that naturalise the settler state's claim for the territory. In this sense, the oil seekers' search for new oil deposits in Aipin's novel appears as a similar settler-colonial technique as it leads to corresponding results in Khantys' home area. Therefore, the oil seekers' map that Dem'ian recalls is not a mere representation of the region but a piece of settler-colonial apparatus that aims at taking over the region and the natural resources it contains.

If the novel's literary cartography of the river offers an example of mapping with a non-human environment rather than considering the river as a mere object of mapping, the oil seekers' map of the area resonates with Cecilia Chen's (2013, p. 280) reading of oil rigs. She understands them as apparatuses that mirror an instrumentalised view of the non-human world as they concentrate on viewing the ocean solely from the perspective of exploitation opportunities. In a similar way, the maps of the Russian oil seekers reflect a corresponding instrumentalist thinking that disregards the interests of both non-human entities and Indigenous people. The geologists move straight forward without considering any of the elements that constitute the space outside of its cartographic representation and occupy the space through a linear, superimposed motion. A line on the map becomes a sign of destruction as the oil seekers 'drill along it deep into the soil and place, in 20 meters deep, something like gun powder that bangs and shoots deep into the ground' (Aipin, 2014, p. 41). However, the environmental damage appears as a sign of power in the eyes of oil seekers and drillers as they "brag about how nothing stays alive in a ten-kilometre-wide area on their route" (Aipin, 2014, p. 42).

The difference in the ways in which Khanty people and the oil seekers see the region and its non-human environment becomes evident through Dem'ian's thoughts on the oil seekers' map. Whereas the oil seekers see the map as a description of a formerly valueless taiga that attains value through the oil deposits, Dem'ian sees the region and Khanty life as interconnected with each other:

> Probably Dem'ian's winter dwelling was marked in it with a pen as well as a cone-shaped granary made of laths, a fish house meant for storing fish, meat, and hunting equipment. [...] On the right shore of Iukhan-Iagun was the spring dwelling of the family – huts carved out of poles with shingle roofs. [...] The map did not lack the family's summer house on the Woodside river either. [...] Also, the family's autumn house on the swamp area of the great river, in the middle of centuries-old pines on the edge of a sand cliff fit on the map. (Aipin, 2014, p. 41)

Dem'ian's way of engaging with the space follows the seasonal changes of the non-human environment and is based on his movement across the region. The dwellings of his family are built to fit in the non-human surroundings, their locations and descriptions reflect the knowledge and respect for different watersheds or "centuries old pines". They mirror the way in which Khanty life is rooted in the environment. This rootedness appears as an embodied experience as Dem'ian senses the threat that the oil seekers pose to his home physically as "sharp furrows scored onto his heart" (Aipin, 2014, p. 43). The fear and pain that the oil seekers awoke in Dem'ian's mind reveal his feeling of powerlessness in front of the Russian oil seekers and demonstrates how he understands his claim to the land as subordinate in relation to the settler power. Even if Dem'ian refers to himself as the "host" of his land, he is aware of the marginal authority he has over the land when it is dispossessed by the settler state in its search for natural resources. As the embodiedness of Dem'ian's fears suggests, the destruction of Indigenous peoples' lands and homes destroys the very basis of their life.

The contrast between the oil seeker's map and the liquid map of the river creates a juxtaposition between the oil and the river, the two liquids flowing on and under the ground. These concrete and material fluids are entangled in manifold webs of discourses, desires and imaginaries. The inseparability of what is imagined and what is material in Aipin's descriptions of flowing water and oil bring forth Gaston Bachelard's (1983) argument on using the interconnectedness of imaginary concepts and material entities to express abstract ideas. In the novel, the river appears as an element that is ultimately connected with life as its physical flow is parallelised with the flows of life, time and memories and, at the same time, it sustains the organic life of the Khanty people by providing them with the necessary living conditions. Mika Perkiömäki (2021) has noticed a similar tendency to imagine and present the major Siberian rivers in Russian village prose texts as "rivers of life". The metaphor "underline[s] the common history, interdependency and fate of the river and the people' and 'connects to the active, agentic river, whose actions have implications for both human culture and the more-than-human world" (Perkiömäki, 2021, p. 110). He has also noticed a prevailing metaphor of the "river of death" that presents the river as "dangerous and lethal" or as related to environmental damage and contamination (p. 107). In contrast to Perkiömäki's observations, the river and the water in Aipin's novel do not bear any lethal connotations as the deadly characteristics are attached to the oil. The oil, depicted as "the black, hot grease" (Aipin, 2014, p. 24) that flows in the black pipes, becomes a literal stream of death as oil pumping destroys the environment for both human and non-human entities. The notion of death is drawn on the geologists' maps as well as the lines marking out possible oil deposits signify destruction for everything that gets in the way of those lines. If the river serves as a live, flowing and constantly transforming map of the watery milieu that connects the lives of the Khanty people with the non-human entities with which they share and co-constitute the space they inhabit, the streams of the oil draw a map that enforces settler-colonial power and an instrumentalised view of the non-human world. Whereas the river adapts to the relief of the environment and sustains the ecosystems and living entities to which it is connected, the oil reserves lead to constructions of straight railway- and pipelines whose builders "cut down pine woods, covered swamps, built bridges over big and small rivers" (Aipin, 2014, p. 24).

The fear that the oil seekers provoke in Dem'ian is based on what he has heard from the regions where oil is already pumped. The novel's temporally non-linear narrative structure reveals to the reader that his concern is not in vain by revealing the destructive development of the region to the reader through the narrative line that follows Dem'ian's son Mikul'. Even if Aipin's novel entails depictions of brute acts of violence, especially in the flashbacks depicting the purges of the 1930s, the main narrative strategy of presenting colonial violence in the novel is to underline the long-term impacts of the settlers and the environmental colonialism experienced by the Khanty people. In the next section of this chapter, I argue that the depiction of the settler-colonial violence in the novel can be read through Rob Nixon's (2011) concept of slow violence. In my view, Aipin's novel offers a demonstrative example of how slow violence can constitute different interrelated and cumulative processes that have a destructive effect on both subaltern groups of people and the non-human world.

10.6 Flows of Slow Violence

As presented earlier in this chapter, Nixon's slow violence is a concept that is used to describe gradual violence, the effects of which are only visible in the long term or/and in the distance, effects that may not be even recognised as violence. This graduality and temporal dispersal is visible in Aipin's novel, and whereas Nixon (2011, p. 3) pointed out what challenges presenting slow violence might face, Aipin's work resolves this problem through its narrative structure. When the narration of the novel leaps back and forth between the past, present and the future, violent causal connections whose consequences are not revealed during one's lifetime or during a generation are made visible. In particular, the narrator's tendency to move into the future and narrate Mikul's experiences in adulthood sheds light on the disastrous results of the Soviet policies and resource extraction.

Despite the novel's anachronism, the reader can distinguish the development of the region since the establishment of the Soviet power and the arrival of oil seekers. In the beginning of Dem'ian's journey, he recalls a time "when he did not think of anything other than hunting and fishing" and "a peace lived in his soul". That peace is ended when "new people appeared on the Vat'ëgan river to build a metal road" (Aipin, 2014, p. 23), and at the time of Dem'ian's journey, "[t]he bottomlands in the East were split in two by a pipe; there was oil in the South and a railway line in the West" (p. 42). The consequences of Soviet policies for the Khanty's home region are already present by the time Dem'ian makes his journey: the hunters' hunting areas of the "lowest river" have been destroyed, alcohol consumption is slowly increasing, and Khanty families suffer from a lack of reindeer, the most important companion of Indigenous hunters moving across the taiga. However, only the narrator's leaps to the1980s with Dem'ian's grown-up son Mikul' reveals the tragic development of the situation.

Mikul's experiences and observations in his home region underline the interconnectedness of the resource extraction, environmental degradation and wilting of the Indigenous community:

> Under the pressure of geologists and oil men the village slowly fell in onto economic depression. First, the cooperative lost its reindeer herd of 300 reindeer, then its cows and horses. Growing vegetables ended due to a lack of fertilizers. Game and fish disappeared due to contamination of lakes and rivers and unreasonable logging. That forced people to close down the fox farm. (Aipin, 2014, p. 265)

The environmental degradation and animal loss that bring an end to the traditional livelihoods is followed by alienation of Native children from their parents' culture as boarding schools move further away from the taiga and into urban centres. All these factors are depicted as interrelated and leading to the gradual decline of the Native people. Therefore, Aipin's description of the consequences of settler colonialism is in line with Wakeham's (2021) notes on slow, attritional violence leading to the annihilation of Indigenous societies, social practices and cultures with a genocidal effect (cf. Wolfe, 2006).

The correlation between environmental damage and social decline is illustrated through the interconnectedness of the settler-colonial flows of death, oil and spirits. In the same way that the people who map the region and extract its oil come to the taiga from outside, spirits are also brought there from elsewhere. Yet, it is the river along which alcohol is transported: "The river did its job: it fed and watered. But it did not only give clear water to drink. A regular transportation of boxes with alcohol bottles to the village began ten years later" (Aipin, 2014, p. 125). The same river that connects the Khanty families with each other and connects the Khanty people with the region also connects the area with other parts of the country, and the river flow carries the alcohol that then flows down the Khanty people's throats. However, the "era of the bottle" (Aipin, 2014, pp. 125–126) only starts after the arrival of the oil seekers and the village they establish, where "the shops are open without a break and during the weekends" (p. 125). The intersections of these different flows bring out the embodied character of the relationship between the Khanty people and their home region. As these liquids flow between different regions, as well as between human bodies and the non-human environment, they point to the porosity of their interfaces in a trans-corporeal way and show how inseparable human bodies are from their material surrounding. When the oil extraction and consequent environmental damage are approached through the oil's interconnectedness with spirits and how it spills over the region with the geologists, environmental problems and the alcoholism among the Khanty people cannot be seen as separate phenomena. Instead, they appear as a continuation of the same "flow of death". Whereas the geologists following the flowing oil "dig up the forests and the endless swamps" (Aipin, 2014, p. 42), the spirits flowing with the geologists "takes" one Khanty person after another. The rampant alcoholism in the region is not only a social consequence of the natural resource extraction but also a continuum of the settler-colonial slow violence in different human and non-human bodies.

After all, in the novel the alcohol-related deaths of the Khanty people seem almost less tragic than the break of the communicative and cultural memories, which not only destroys individuals but also the Indigenous people as a nation. The Russian literary scholar Ol'ga Lagunova (2007, p. 123) pointed out the importance of memory for the survival of the people by comparing Dem'ian's recount of the bear feast to an event taking place in the future, after Dem'ian's journey. In this event, Dem'ian's son, Mikul', contacts his old relative Galaktion Kurpelak in order to hear old songs and stories, but the old man, who has lost all his relatives to alcohol, states: "My mind is empty, my words have run off. What stories there were … In this century of sufferings" (Aipin, 2014, p. 260). In the contrast between these scenes, Lagunova (2007, p. 123) sees the end of an era in the history of the Khanty people, a change from the time when stories, myths and rituals are carried in people's mind to one when they have disappeared from memory:

> The space of the feast, in which everyone knew when and what they were supposed to do and say, and the space of the seemingly boundless fairy tale that joins together everything alive, entails all times and unites generations, have fallen apart: the elderly people have left and the young ones do not remember. […] And so comes the end, the end for all: for songs, fairy tales, feasts, family, home, Land [*sic*], life. (Lagunova, 2007, p. 123)

As Lagunova points out, when Galaktion Kurpelak lets go of the tales and stories he used to tell, the cultural memory of the Khanty people is not transmitted from one generation to another. Whereas Mikul's father has participated in the bear feast himself and is therefore able to recall its course and related songs, for Mikul' the oral tradition of the Khanty remains a vague recollection from childhood. The following generation will have no personal experience of the national feasts or folklore. The break in the communicative chain of memory between different Khanty generations stifles the live and natural flow of the cultural memory and solidifies it into Nora's *lieux de mémoire*, forming lifeless symbols that merely represent the Indigenous community rather than reconnect to it. This is already the case with one distant relative of Mikul' who "collects exhibits of Khanty's applied arts for an exhibition of the Achievements of National Economy" in Moscow (Aipin, 2014, p. 262). Gradually, the vivid Indigenous culture turns into a curiosity and a symbolic demonstration of the multi-ethnicity of the Soviet Union until it becomes a remnant of an extinct community. As a result of alcohol abuse, the break of the communicative and cultural memories cannot be separated from oil drilling and its environmental consequences, and hence, the thoroughness of the following devastation of the Khanty people is comparable to the destruction of the taiga caused by the natural resource extraction.

10.7 Conclusion

The interconnectedness of Indigenous peoples and their Native land, as well as the devastating consequences of settler colonialism and resource extraction, are widely recognised in the North American context. However, in Russian studies, postcolonialism as a theoretical framework has not yet reached its full potential when discussing the fates of Indigenous peoples in Russia and the former Soviet Union. In my reading of Eremei Aipin's *Khanty, or the Star of the Dawn*, I wished to show how Aipin addresses the nexus of environmental destruction and the gradual degradation of his people in such a way that underlines the mutual correlation between the two. In Aipin's novel, this degradation of his people is both physical and cultural, and it is parallel to the devastation of the non-human environment with which the Indigenous life is entangled.

The non-human environment does not only provide the material basis for the Native people, it is also the source of the living collective memory that passes on from one generation to another and keeps the Indigenous community alive. In Aipin's novel, this entanglement of the non-human environment and Indigenous culture is most apparent in the narration about the river on which the Khanty people's lives depend, the river that connects the Khanty people with each other as well as with other places. The river is a piece of cultural memory that serves as a collective symbol for the northern Indigenous people, but even more, it illustrates the Khanty people's epistemological views and understanding of themselves in relation to the surrounding world. The river's role in transmitting the Indigenous epistemology becomes evident in its cartographic function as a watery map that, like any other map, is not only a spatial representation but a means of knowledge production as well. Hence, as the narrator creates the novel's literary cartography using the cold northern water of the river, it reflects an epistemological view of both the non-human environment and the human. According to this view, they are not divided into a subject who maps and an object that is being mapped, they are fundamentally entangled.

In this sense, the river and its water are both a geographical denominator and a means of thinking that represent the relationship between a local Indigenous people and their native northern region. However, in a similar way to the way that the North, and especially the Arctic, are often taken as objects of Western (or, geographically speaking, southern) "control and knowledge" (Huggan, 2016, p. 7), the Khanty people's home region is taken over by the oil seekers from southern parts of the country. They arrive with maps that stem from a completely opposite epistemological basis than that upon which the map that is drawn with the river as their maps demonstrate a radical separation from and an instrumentalised view of the non-human environment. If it is the cold water of the river that is emblematic of the northern Indigenous epistemology and way of thinking, the flowing "hot grease" (oil) and alcohol represent the colonial, southern/Western epistemology in which the non-human environment is subdued to serve human interests. As the region is mapped anew according to these "hot" liquids, instead of being mapped according to the cold water, the supersession of local epistemology appears as one of the forms of settler-colonial violence that eventually leads to the annihilation of Indigenous Khanty culture that the novel depicts.

References

Aipin, E. (2014). *Khanty, ili zvezda utrennei zari* [Khanty, or the star of the dawn]. Amfora.

Alaimo, S. (2008). Trans-corporeal feminisms and the ethical apace of nature. In S. Alaimo & S. Hekman (Eds.), *Material feminisms* (pp. 237–264). Indiana University Press.

Alaimo, S. (2018). Trans-corporeality. In R. Braidotti & M. Hlavajova (Eds.), *The posthuman glossary* (pp. 435–438) Bloomsbury Academic.

Assmann, J. (1992). *Das kulturelle Gedächtnis: Schrift, Erinnerung und politische Identität in frühen Hochkulturen*. C. H.

Assmann, A. (1999). *Erinnerungsräume: Formen und Wandlungen des kulturellen Gedächtnisses*. C. H.

Assmann, J. (2011). Communicative and cultural memory. In P. Meusburger, M. Heffernan, & E. Wunder (Eds.), *Cultural memories* (pp. 15–27). Springer.

Böhme, H. (1988). Umriss einer Kulturgeschichte des Wassers: Eine Einleitung. In H. Böhme (Ed.), *Kulturgeschichte des Wassers* (pp. 7–44) Suhrkamp Verlag.

Boliachevets, L., & Sablin, I. (2016). The second and the fourth world: Critique of communism and colonialism in contemporary north Asian literature. *Ab Imperio, 2*, 385–425. https://doi.org/10.1353/imp.2016.0038

Bonner, W. (2017). The interplay of water, home, and narration in *Überfahrt* by Anna Seghers. In J. T. Costlow, Y. Haila, & A. Rosenholm (Eds.), *Water in social imagination: From technological optimism to contemporary environmentalism* (pp. 100–118) Brill Rodopi.

Chandler, M., & Neimanis, A. (2013). Water and gestationality: What flows beneath ethics. In C. Chen, J. MacLeod, & A. Neimanis (Eds.), *Thinking with water* (pp. 61–83). McGill-Queen's University Press.

Chen, C. (2013). Mapping waters: Thinking with watery places. In C. Chen, J. MacLeod, & A. Neimanis (Eds.), *Thinking with water* (pp. 274–298). McGill-Queen's University Press.

Cilano, C., & DeLoughrey, E. (2007). Against authenticity: Global knowledges and postcolonial ecocriticism. *ISLE, 14*(1), 71–87.

Costlow, J., & Rosenholm, A. (2007). Introduction. In J. Costlow & A. Rosenholm (Eds.), *Meanings and values of water in Russian culture* (pp. 1–12). Routledge.

Forsyth, J. (1992). *A history of the peoples of Siberia: Russia's north Asian colony 1581–1990.* Cambridge University Press.

Genette, G. (1980). *Narrative discourse: An essay in method* (trans: J. Lewin). Cornell University Press. (Original work published 1972).

Golovnev, I., & Golovneva, E. (2021). Traditional ethno-cultural communities in the modern Russian north: Oil field as a documentary film case. In M. Lehtimäki, A. Rosenholm, & V. Strukov (Eds.), *Visual representations of the Arctic: Imagining shimmering worlds in culture, literature and politics* (pp. 279–294). Routledge.

Golovnёv, A. V. (1995). Govoriashchie kul'tury: Traditsii samodiitsev i ugrov. UrO RAN.

Gurvich, I. S. (1982). *Etnicheskaia istoriia narodov severa.* Nauka.

Harley, J. B. (1988). Maps, knowledge and power. In D. Cosgrove & S. Daniels (Eds.), *The iconography of landscape* (pp. 277–312). Cambridge University Press.

Harris, L., & Hazen, H. (2009). Rethinking maps from a more-than-human perspective: Nature-society, mapping and conservation territories. In M. Dodge, R. Kitchin, & C. Perkins (Eds.), *Rethinking maps: New frontiers in cartographic theory* (pp. 50–67). Routledge.

Hoogeveen, D. (2015). Sub-surface property, free-entry mineral staking and settler colonialism in Canada. *Antipode, 47*(1), 121–138. https://doi.org/10.1111/anti.12095

Huggan, G. (2008). *Interdisciplinary measures: Literature and the future of postcolonial studies.* Liverpool University Press.

Huggan, G. (2016). Introduction: Unscrambling the Arctic. In G. Huggan & L. Jensen (Eds.), *Postcolonial perspectives on the European high north: Unscrambling the Arctic* (pp. 1–29). Palgrave Macmillan.

Huggan, G., & Tiffin, H. (2007). Green postcolonialism. *Interventions, 9*(1), 1–11. https://doi.org/10.1080/13698010601173783

Huggan, G., & Tiffin, H. (2010). *Postcolonial ecocriticism.* Taylor and Francis.

Josephson, P., Dronin, N., Mnatsakanian, R., Cherp, A., Efremenko, D., & Larin, V. (2013). *An environmental history of Russia.* Cambridge University Press.

Kitchin, R., Perkins, C., & Dodge, M. (2009). Thinking about maps. In M. Dodge, R. Kitchin, & C. Perkins (Eds.), *Rethinking maps: New frontiers in cartographic theory* (pp. 1–25). Routledge.

Lagunova, O. (2007). *Fenomen tvorchestva russkoiazychnykh pisatelei nentsev i khantov poslednei treti XX veka (E. Aipin, Iu. Vella, A. Nerkagi): Monografiia.* Izdatel'stvo Tiumenskogo gosudarstvennogo universiteta.

Miliakhova, L. P. (2009). *Roman Eremeia Aipina "Khanty, ili zvezda utrennei zari": Genezis, obraznyi stroi, kontekst poetiki.* Avtoreferat dissertatsii na soiskanie uchenoi stepeni kandidata filologicheskikh nauk. Rossiiskii gosudarstvennyi pedagogicheskii universitet imeni A. I. Gertsena.

Moser, N. (2018). *Oil and the economy of Russia: From the late-tsarist to the post-soviet period.* Routledge.

Nixon, R. (2011). *Slow violence and the environmentalism of the poor.* Harvard University Press.

Nora, P. (1996). *Realms of memory: The construction of the French past* (trans: A. Goldhammer). Columbia University press. (original work published 1984–1992).

Ogryzko, V. (2002). Literatura otchaianiia i nadezhdy. In V. Ogryzko (Ed.), *Khantyiskaia literature* (pp. 3–29) Literaturnaia Rossiia.

Ogryzko, V. (2006). V szhimaiushchemsia prostranstve: Portret na fone bezumnoi epokhi. Literaturnaia Rossiia.

Perkiömäki, M. (2017). 'The sovereign of the river and the sovereign of all nature – In the same trap': Viktor Astafiev's *queen fish*. In J. Costlow, Y. Haila, & A. Rosenholm (Eds.), *Water in social imagination: From technological optimism to contemporary environmentalism* (pp. 145–166). Brill Rodopi). https://doi.org/10.1163/9789004333444_009

Perkiömäki, M. (2021). *Imagined riverography of late twentieth-century Russian prose*. Tampere University.

Schmidt, J. J. (2020). Settler geology: Earth's deep history and the governance of in situ oil spills in Alberta. *Political Geography, 78*, 1–11. https://doi.org/10.1016/j.polgeo.2019.102132

Simpson, M. (2019). Resource desiring machines: The production of settler colonial space, violence, and the making of a resource in the Athabasca tar sands. *Political Geography, 74*. https://doi.org/10.1016/j.polgeo.2019.102044

Slezkine, Y. (1994). *Arctic mirrors: Russia and the small peoples of the north*. Cornell University Press.

Smola, K. (2017). Postkolonial'nye literatury Severa: avtoetnografiia i etnopoetika. *Novoe literaturnoe obozrenie, 144*(2), 429–447.

Tally, R. T., Jr. (2013). *Spatiality*. Routledge.

Toulouze, E. (2005). The intellectuals from Russia's peoples of the north: From obedience to resistance. In M. Mandelstam Balzer (Ed.), *The northern peoples and states: Changing relationships* (pp. 140–164). University of Tartu Press.

Wakeham, P. (2021). The slow violence of settler colonialism: Genocide, attrition, and the long emergency of invasion. *Journal of Genocide Research*. https://doi.org/10.1080/14623528.2021.1885571

Welzer, H. (2005). *Das kommunikative Gedächtnis: Eine Theorie der Erinnerung*. Verlag C. H. Beck.

Weiner, D. (1999). *A little corner of freedom: Russian nature protection from Stalin to Gorbachëv*. University of California Press.

Wolfe, P. (2006). Settler colonialism and the elimination of the native. *Journal of Genocide Research, 8*(4), 387–409.

Chapter 11
The Ambiguity of the Arctic Littoral: Changing Perspectives of Chukchi Communities in Two Russian Films

Tatu Laukkanen and Arja Rosenholm

Abstract This chapter discusses two Russian-language feature films about the Chukchi, the Soviet film *The Most Beautiful Ships (Samye krasivye korabli*, 1972) and the recent *The Whaler Boy* (*Kitoboi*, 2020). Using narrative and formal analysis, the study focuses on the cinematic representation of the liminal and highly complex space between the sea and the land. The chapter examines the role of the marine shoreline of the Arctic Ocean and how it becomes an agentic figure in the narrative and subjectification process of the characters. The shore between the water and earth acts as a topos of negotiation and the conflict of belonging experienced by the indigenous people coping with the changing world on the Arctic coastline. The multidimensional qualities of the shore as conceptualized by liminality, instability, and transitoriness allow the advancement of the idea of correspondence between littoral and psychosocially engaged cultural identities. Utilizing concepts from ocean and littoral studies, social psychoanalysis and comparative literature, as well as film and cultural analysis, the chapter indicates how the shore's epistemological and aesthetic qualities shape the identity formation of the protagonists in their social growth. The study also charts social and ideological changes in the representation of the indigenous Chukchi, the Arctic environment, the state, and gender in these two films, which are separated by almost 50 years.

Keywords Chukchi · Film · Russia · Littoral studies · Belonging

The Soviet film *The Most Beautiful Ships (Samye krasivye korabli*) opens with an evocative scene with a little boy running along the shoreline towards an older man

---

T. Laukkanen (✉)
Independent Scholar, Helsinki, Finland

A. Rosenholm
Faculty of Information Technology and Communication Sciences, Tampere University, Tampere, Finland
e-mail: arja.rosenholm@tuni.fi

M. Lehtimäki et al. (eds.), *Cold Waters*, Springer Polar Sciences,
https://doi.org/10.1007/978-3-031-10149-6_11

standing on the shore. The young and old encounter, hug, and the grandfather hands the boy an amulet-like gift. The boy then climbs up to the mountain cliff to leave his Chukchi community for a boarding school while the grandfather remains on the shore, contemplating the infinite ocean.

The scene is not only significant in this particular film but emblematic of the use of the marine shoreline in several Soviet-Russian fictional film narratives about indigenous communities living on the marine coastline of the Arctic Ocean. This northern coastline area is a backdrop for transformation where environmental and socio-political changes "have an enduring impact on a collective sense of home and belonging", as Christensen (2013) has noted in her study on the spiritual and immediate homelessness experienced by the indigenous people in the Canadian North. Our film analysis examines the Arctic shoreline as a topos of negotiation and the conflict of belonging, as experienced by the northern indigenous people, the Chukchi. This chapter will pay attention to the alternating water-land images running through the films' narratives, which, we suggest, become integral to the conflicts of belonging and the characters' subjectification process.

The focus of our interest is on this liminal and highly complex space in between the sea and the land, a prime example of a contact zone (Pratt, 1992, p. 4). As a contact zone, the littoral space of the shore is a palimpsest site, constantly reconfiguring in a material as well as in a symbolic sense (Kluwick & Richter, 2016, p. 2; see also Corbin, 1994; Anderson & Tabb, 2002). Accordingly, the emphasis in our film analysis is placed on the ways in which both case films encode the Arctic coastline and how the films – with an almost five-decade time gap – interpret the identity transformations of the Chukchi living in this littoral space.

The chapter discusses two Russian-made films that both highlight the shore's epistemological and aesthetic qualities, shaping it as a complex space that structures the identity formations of the protagonists in their social growth. The films are *The Most Beautiful Ships,* from the Soviet period (1972), and *The Whaler Boy* (*Kitoboi*),[1] which was released in 2020. Both films are set in Chukotka in the Russian Far East, the north-eastern region of the country less than a hundred kilometres from US Alaska, which lies across the icy waters of the Bering Strait. Located along the Bering Strait, Chukotka is home to the Chukchi who dwell both in the coastal areas and the inner tundra (Vdovin, 1965). The littoral in these films is related to the representation of confrontations within the community and between Chukchi culture and colonial forces – the Soviet paternal "rescue" policy regarding the "small peoples" (Slezkine, 1994; Hirsch, 2005) and Americanized global culture. Accordingly, the comparative analysis also investigates how the colonized history of the Chukchi people is covered in the Soviet and Post-Soviet frame.

While focusing on the interconnected space on both sides of the shoreline, we draw on spatial and littoral studies that conceptualize the shore as "a polyvalent site from the point of view of the humanities", following Kluwick and Richter (2016,

[1] The translated and Russian titles carry different tones as the original title is directly translated as *Whaler*, the English title adds *Boy,* highlighting the themes of the ambiguousness and subject-in-process of the protagonist in the film.

p. 2; see also Gillis, 2012; Rostek, 2011; Cohen, 2010; Denning, 2004). Of importance for our analysis is also Richter's (2016, p. 155) observation that littoral sites "are constituted both materially and symbolically through the interactions of various human and non-human inhabitants of littoral space". The chronotope of the shore, which involves multiple imaginative and spatial forms, is "a creative trope and a socio-cultural site, as well as an aesthetically productive topography […], [with] a multiplicity of meanings and functions" (Kluwick & Richter, 2016, p. 2). We concur with Cohen's (2006, p. 649) view that the topos of the shore – as one of the remarkably constant waterside chronotopes – not only features in multiple artforms (including film and literature in Western European contexts), but as this chapter shows, it also has a strong symbolical resonance among (Russian) narratives engaged with indigenous communities located by the shore of the Arctic Ocean.

The multidimensional qualities of the shore – as conceptualized by liminality, instability, and transitoriness – allow us to forward an idea of correspondence between the littoral and the psychosocially engaged cultural identities. As Mack (2011, p. 165) noted, the shoreline, as "an ambiguous place, an in-between place", coincides with processes of transformation deeply affecting also the indigenous lifestyle. Both films adapt the motif of disrupted tradition and fractures in both individuals and communities. Problems underlying the personal narratives are associated with the Soviet (Russian) modernization and articulated as detachment from family, community, and cultural identity; inter-generational distance; and the vulnerability of existence under the environmental threats affecting the indigenous homeland (Forsyth, 2000; Zamyatin, 2021; Golovnev & Golovneva, 2021).

As the seashore is open and porous, the young Chukchi protagonists' transformation is correspondingly conflictual and uncertain; their cultural heritage and the continuity of tradition are threatened by Soviet-Russian and US hegemony. What we aim to illuminate is this "fluid" situation that throws the young men into doubt and challenges their self-realization and future outcomes, which they had taken for granted. To accentuate the corresponding linkage between spatial liminality and psychosocial transgression, we draw on identity studies, according to which cultural identity undergoes constant transformation while taking shape in a specific social, cultural, and historical context (Hall, 1994). A parallel linkage is drawn between the ambiguity arising out of the shoreline as a topos of "mutation and perpetual motion; of fluidity and flux" (Samuelson, 2016, p. 122) and the ambivalence in the way in which the films signify the changing world and the subject-in-process in it. The ambiguities connecting the social and the subject are visualized as a communication between different but interdependent modalities of the visual imagery and the use of sound. This oscillation is in line with the concept of signification as a continuous process (Kristeva, 1990, pp. 90–136; Moi, 1990, p. 13) occurring between the symbolic order that controls it, and the semiotic, which "speaks" through the symbolic with its pre-linguistic shifts, impulses, and metaphorical and metonymical transgressions. The films show how the symbolic order of coherent identities and stable social systems turn fluid when an individual transgresses the collective truth or, in the terms of Victor Turner (2008), when the participants are in the liminal stage of their rite of passage.

11.1 The Most Beautiful Ships

In January 1973, Soviet Central Television broadcast a two-part television film, *The Most Beautiful Ships,* based on a screenplay by the well-known Chukchi and Soviet writer, Yuri Rytkheu[2] (1930–2008) and directed by the Russian director, Anatolii Nitochkin (1932–2001). The film is an adaptation of both Rytkheu's novel of the same title (Rytkheu, 2016/1967) and his short story *Veket and Agnes* (*Veket i Agnes*) (Rytkheu, 1972). Being long almost forgotten, the film has recently experienced a revaluation since, according to Damiens (2020, p. 17), "it is related to the possible emergence of an indigenous, or 'Fourth cinema' in the USSR". The film tells the story of a young Chukchi man, Veket, the boy on the shore in the opening scene described above. Veket successfully finishes his residential school in Chukotka. Contrary to his fellow students, Veket does not leave to study in a big Soviet city but stays in Chukotka as a reindeer herder. In the tundra he meets Vaike, a young art student from the Estonian minority, studying in Leningrad, who comes to Chukotka to paint. They fall in love. Vaike returns to Leningrad, and Veket follows her. However, once there, he finds that he cannot live anywhere but in Chukotka and returns to the Arctic.

While Damiens focuses attention on the genesis of the film, its alternative visual regime of the colonial-backed ambivalence, and the film's reception, our analysis demonstrates how the littoral as a particular landscape enables psycho-social and cultural expressions of ambiguity. The key scenes we analyse demonstrate how the inconstant shoreline allows the characters to encounter otherness and is a location where the traditions of one's "native land" ("*rodniaia zemlia*") are both called into question and reaffirmed (Fig. 11.1).

The opening scene on the shoreline is repeated and modified as a key motif of ambivalence. As a liminal site, the shore is associated with the assessment and revaluation of indigenous tradition and cultural heritage of backwardness, memory, and respect for ancestors. The scene where the little Veket joyfully runs along the narrow strip between water and land towards his grandfather raises feelings of love and community. However, the old man's words for Veket, with which he advises him "not to forget the tundra", evoke feelings of loss. The scene harbours mixed emotions related to the anticipation of the residential school and mourning over the break with his ancestors and local tradition.

The scene engenders an ambivalent perspective toward Veket's Chukchi identity as he oscillates between staying and leaving, as well as among past, present, and future. The next scene shows Veket and other Chukchi children leaving their mothers and being sent to boarding school. We see the little boy travelling away in a snowmobile, without family, sobbing. The next sequence takes place years after, in a classroom, with the teenage Veket proudly reciting poems of the Russian poet, Aleksandr Blok. Suddenly, a telegram arrives and informs him of the death of his

---

[2] We use this version of his name as it is likely to be well known to the Western readers in the commonly used form.

Fig. 11.1 Veket and grandfather taking farewell

mother whose funeral he then misses as he arrives to Chukotka too late. The death of his mother, which interrupts Veket's successfully indicated acculturation in the Russian majority culture, makes Veket conscious of an existential loss: "I thought mother and sea would be there for ever". The loss of his mother unexpectedly threatens his symbolic rite of passage. The loss is significant, reminding him of the trauma caused by the forced separation from his mother and the Chukchi home shore. The missed funeral makes the loss turn into feelings of abandonment and guilt that affect Veket's further development.

11.2 Flashbacks of Trauma

Veket's trauma is not explicated in verbal words, quite the contrary; the split in his illusory unity is mediated by his speechless contemplation of the sorrow. His tears over the loss of his mother express the bodily memory, which intrudes into Veket's current world in images and voices of the past that displace him back to the seashore and his mother. In flashbacks we see little Veket happily walking along the shoreline with his mother who sings the family song, which she had composed, about the ships, of which those passing by are "the most beautiful". In Veket's memory, that which is maternal is associated with music and water as the affective mediators of the impulsive semiotic activity tied to instincts that escape the denotative meanings of words. The flashbacks induce the past separation from the maternal shore as if it is happening in the present, and the memories flow over Veket's conscious self and give rise to his fragmented ambiguity. He is caught in between the fixed and "land-locked" Soviet habitus on the one hand and his nomadic and maritime Chukchi heritage on the other. The maternal shore keeps a hold on him with conflicting memories of tradition, cultural heritage, and traumatic departure that Veket tries to compensate for by imitating Russian high culture and becoming the Soviet "cultured person" (*kul'turnyi chelovek*).[3] Mimicry as a problematic means of surviving and resisting colonial power (Bhabha, 1984, 1994) does not liberate him from the "black sorrow" of melancholy (Kristeva, 1992), which keeps a hold on him, causing a yearning to return to the maternal shore.

Due to the uncertainty about whether to leave or to stay in Chukotka after returning from the boarding school, Veket ends up walking along the shore from where he can see the ships tempting him to depart. Ships become the very metaphor of existence, of "a kind of reality model" (Vidauskyte, 2017), and at the harbour Veket makes his decision not to leave. The ships remind him of the alternatives and the visions of future professionalism that departing would offer him. But Veket's visions of the future are as vague as the ships as they fade into the fog. Veket does not board the boat with his fellow students in order to sail away to the big Soviet cities and institutes. The ambivalence of his decision to stay in Chukotka is mediated by a dual perspective: The vocal departure of his fellow youths is contrasted by shots of Veket and the families staying behind, the sorrow of the loss reflected on their mute and motionless faces. Once again, Veket stands on the shore, indicating his alienation in both worlds; he does not quite fit in either. Among his fellow Chukchi, Veket is an odd "*akademik*" due to his background in the residential school and his vague future plans, while the modern world outside is alien but desirable with its technology, the high culture of fine art, classical music, and the Russian literary canon. Nonetheless, Veket is in danger of being late: The Russian motorist, as one of the symbolic mediator-fathers who separate the young Veket from the maternal community, repeatedly hurries up Veket, saying that the "motor is already running". The untimely haste of Soviet modernization forced the indigenous community to catch up with the benefits of modernity, and merging into the Soviet homogenic space brings about a "break" in (or even "a termination" of) the minority culture; both the material and the spiritual indigenous world are at stake (Morch, 2017, p. 229).

11.3 The Potential of Art and its Imaginary Powers

The longed-for family community is brought back to Veket in flashbacks full of maternal images and the affective pre-linguistic voices of his mother's song. Throughout the film, Veket is in an uneasy emotional process that is encoded in aquatic fluidities and music; while the Arctic Ocean and the shore belong to the maternal and the family song, Leningrad and its river embankments signify his love

[3] Kelly and Volkov (1998, p. 295) compare the concept of *kul'turnost'* with the process described by Norbert Elias's classic analysis in *The Civilizing Process*, "which described the evolution of a concept of the individual needed in a modern society".

to Vaike and the attraction of classical music. Veket leaves the shores of Chukotka for Leningrad to visit Vaike, his beloved mirroring other (Damiens, 2020, p. 26), who is exhibiting her paintings of Chukchi culture. They meet in a symphony concert which re-mobilizes the semiotic processes of the unconscious sorrow constantly affecting Veket's adult mind. In the associative flashbacks evoked by the concert, Veket moves back to the maternal shore of his childhood where the Soviet-Russian symphony orchestra had once given a classic concert to the Chukchi people who listened to the orchestra by the sea (Fig. 11.2).

Both water and the art of music as metaphors for fluidity and indeterminacy communicate Veket's experience of self-reflection. According to musical semiology, music is a specific modality of signification and can enter realms that language does not enter, or dares not enter, indicating a preoccupation with underlying prelinguistic affects (Szekely, 2006; Barthes, 1998; Välimäki, 1998, p. 373; Välimäki, 2005, pp. 16, 63, 138, 164 ff.). Musical pleasure as *jouissance* resonates with the prelinguistic state of unpredictability, providing emotional bliss that transgresses stable or pre-defined symbolic codes and allows the listener to break out of his or her subject position. Music offers a moment of merging when the listener "gathers" himself or herself in his or her ear (Barthes, 1985, p. 252). Therefore, music, as a link to mother and childhood, serves as an essential mediator in Veket's identity process. It fuses the ephemeral, affective, and nonsignifying moments of communication, which are articulated by the non-linguistic semiotic agency, the technical competence, and the performance of the symphony (Barthes, 1977, p. 182). It is obvious that the maternal song of the "most beautiful ships" is a "a key element that Rytkheu promotes in his scenario", as Damiens (2020, p. 23) also stated. Veket's

Fig. 11.2 Veket and mother

identity is not only reinforced by the perfect musical performance of the orchestra but also by music as something deeply imbued in the materiality of maternal singing. The reception of the bodily basis of singing and affective manipulations happening within singing are incorporated in the auditive expression of music that allows a temporary regression since "[m]usic may be one way to 'purify the abject' and 'name the un-nameable' – to ameliorate, through sublimation, the rejection of and separation from the m/other's body" (Välimäki, 2005, p. 141). Veket can regain what once was lost – his mother, and his Chukchi community and culture. The cathartic moment of reconciliation is fantasized as an illusion of an idyllic reintegration into the Arctic shore: An imaginary flashback depicts Veket simultaneously as a little boy again, joyfully running on the shoreline and boisterously playing with splashing waves, and as an adult self, dressed in European attire and in a happy encounter with his laughing grandfather who has been waiting for him on the shore.

Both music and the imagery of water as recurrent motifs inspire the aesthetics of the film and underline the fluidity of life. Contrary to the dynamic, bright, and tumultuous Arctic Ocean, Leningrad is a rainy place; the harnessed and channelled Neva becomes a synecdoche to what happens to the love story set on its granite embankments. The motionless and dammed river, with autumn leaves silently floating on its surface, indicates what is going to happen: Veket leaves Leningrad and returns to his Arctic shore as someone he was not before.

The name Veket means "moving forward" ("*idushchii*"). His name connotes with multiple balancing acts taking place not only in between indigenous culture and Russification, but also within his own culture in relation to his "more indigenous" ancestors and fellows. This "double scale" (Mathisen, 2004, p. 155) is illustrated by the contradictory image of how Veket sees his "most beautiful ship": While the "most beautiful ships" for the mother were those passing by the shore, Veket states that his ship stands "in the tundra" (*v tundre*).[4] The idea of a land-locked ship points to Veket's position in the cross pressures of hegemonic Soviet and Chukchi culture. This ambivalent position of psychological and ideological struggling that Barker (1993) has called a "dual life" was also well-known to Rytkheu as an indigenous writer within the Soviet intelligentsia.

The potential of art and its imaginary powers, as imbued both in Vaike's paintings and the affective powers of music, unsettle both the linear development of Veket as a subject-in-process and also the classic continuity of film narration unsettled with flashbacks. Music and water, both continually changing entities, underline the ambivalence of Veket's development that, despite the emphasis on indigenous heritage, is caught in a trap of the Soviet master narrative with its accent on Russian high culture (Clark, 1981). Classical music brought from the Soviet metropole and played on the shore of the Arctic Ocean is a way of encoding the Arctic with Soviet symbols and making the indigenous Arctic become part of the "civilized" Soviet culture. The film ends on the very same shore with a Chukchi artist group dancing

[4] In Veket's words: "*moi korabl' stoit v tundre i zhdët menia*" ("my vessel stands in the tundra and waits for me"). Translation by Arja Rosenholm.

with accompanying local music. Having both (Western) classical music and ethnic folklore on the same littoral "stage" denotes a hybrid existence or even equality, even if idealized, among multiple national and cultural agencies within the boundaries of the Soviet nationality policy. The location of the dancing group on the shore at the end of the film indicates hope for future unity in cultural differences and makes the viewer, at least for a moment, lose sight of power disparities.

11.4 *The Whaler Boy*, a Post-Soviet Return to the Arctic Shore

Like *The Most Beautiful Ships*, *The Whaler Boy* represents the shoreline of the Arctic Ocean as a location of migration, transaction, and return, following Cohen's (2006, pp. 649, 662) marine chronotope of the beach as a location of encounters. Correspondingly, the unstable geology of littoral areas again forms the material surroundings as an individual entering an intermediate state, returning to his community after change, and moving on to a new stage in life. The sea promises a possibility of change and, above all, romance overseas to the main protagonist, Lëshka, who lives in a small Chukchi community near the Russian coast of the Bering Strait. The naval journey, the sea, and the littoral again serve as the backdrop of a subject-in-process narrative that involves a dissipation of stable subjectivity with a negotiation between the symbolic and semiotic, culminating in a metaphysical death and rebirth. The semiotic bursts forth in the naval sequence that immerses the main protagonist, and the viewer, in both narrative and formal oneiric and even surreal scenarios. The cold and unforgiving real ocean from which the community has to wrest its living, turns into a "hot" entity; charged emotions and desire are projected onto the sea and the opposite shore, upon which Lëshka escapes his home village. The Arctic shore has evolved from the place in *The Most Beautiful Ships* in which domestic Soviet cultural contradictions are problematized and played out into the location in *The Whaler Boy* in which cultural, ecological, and geopolitical shifts and fears, driven by globalization, are narrated through the development of a young Chukchi man.

The Whaler Boy is produced at a time when climate change, which particularly affects the Arctic, has become a reality. Water and the ocean become symbols of anxiety with rising sea levels and dwindling fishing resources, not least for coastal indigenous populations such as the Chukchi, and they raise spectres of ecological disaster and even extinction. As the shore is framed as a location of desire and becoming, this atmosphere of anxiousness can be seen to also contribute to the framing of the shore as a Darwinist battleground upon which whales are cut up and the protagonist robbed of his boat and belongings simultaneously. Physical (not only cultural) deaths become an acute theme in the film in which Lëshka almost perishes on the barren shore.

The whale is a central entity in Chukchi lore and economy, and whaling is represented to largely define the community. The impetus of the story is that in addition to the decidedly male activity of whaling, Lëshka also spends time on the internet enthralled with the cybersex web model "Hollysweet_999" with whom he subsequently falls in love. Lëshka gets into a fight with his best friend Kolia over Hollysweet and, thinking he has killed him, decides to steal a whaling motorboat and go to Detroit via Alaska to meet her. However, Lëshka's journey becomes unmoored drifting over which he has no control. The menacing awareness of being at the mercy of forces beyond one's control correlates with the subject position of Lëshka being from a marginalized community in the global economy.

11.5 Spectres of Extinction

The impoverished Chukchi whaling community is portrayed as almost devoid of women. The men discuss bringing in prostitutes and watch net porn, the foregrounded commodification of sex and masturbation underscoring their lack of sexual initiative and appeal. The absence of women also ties in with the spectre of extinction that is connected to the overt presence of whales, and their hunting and slaughter. As Richter (2016, p. 158) noted, as "[o]ne of the iconic animals of conservationism", whales have been a symbol of endangered species for decades. Chukchi culture is similarly threatened.[5]

Littoral space is represented as a hotbed of interaction between humans and whales, this relation culminating in non-lingual communication and bonding. The whales in the film are not only reminders of the possibility of extinction but become markers of Chukchi masculinity as well. As if paying homage to the American cultural hegemony the film sets out to criticize, the film starts in the U.S., the camera moving into a cybersex production centre, via a sleazy alley, and then concentrating on Hollysweet doing her labour in front of a webcam. Then, in a displacing move, the camera zooms in on Hollysweet's laptop screen but zooms out in Chukotka where a group of Chukchi men watch her on a computer screen. By her name, Hollysweet is metonymic to the larger dream machine of Hollywood, the central motor of cultural Americanization. The character sexualizes the U.S. as a target for desire, but, at the same time, desire is depicted as a constructed illusion as it guides the viewer through the cybersex production centre. After the men have watched net porn, we see documentary-like footage of whaling, filmed through a cinematic form that connects and objectifies women and whales. The whale hunt sequence ends in a long drone shot of the beached, bleeding whale being cut up by the men. At one point in the film, Kolia feels around what can be called a vaginal slit in a beached whale with a fishhook. In her analysis of the Pacific literature of whaling populations, Virginia Richter (2016) noted that the shore in such narratives is at the same

[5] Director Filipp Yuryev commented upon the situation in an interview: "I discovered that there are 15,000 whales [referring to the bowhead whales that the men of the village hunt] left and about 13,000 Chukchi, so they are also a disappearing species" (as cited in Beumers, 2020). We use the name Yuryev that is likely to be well known to Western readers.

time a space of regeneration and a place of masculine violence against animals. This dual character of the shore is replicated in *The Whaler Boy*.

11.6 Dissolving Borders and Identities

Lëshka's border crossing can be tied to many Soviet-Russian narratives in which the Arctic and its shores have been represented as a space of masculine liminal testing and even reconstruction (Sarkisova, 2015, p. 231; Bugaeva, 2015, p. 323; Rosenholm 2021, pp. 217–237). However, *The Whaler Boy*'s parallels with past narrations of the Arctic frontier are adjusted to a re-evaluation of the idea of borders in the era of globalization and a critique of global digital culture and Americanization. In notable contradiction to the tradition of Russian films about the Arctic (films such as *The Most Beautiful Ships*) that display ample majestic vistas of the shoreline and mountains in their establishing shots, *The Whaler Boy* starts with mundane roadside imagery of the US (such as gas stations, diners, and a motel) before moving to Chukotka via the cybersex centre in Detroit. The cultural consequences of a shrinking world are foregrounded in the film, but globalization's narrative of the weakening of the nation state – and the implied geopolitical fears resulting from it – can be read as a subtext of the film as well. If the state plays a great (paternal) role in *The Most Beautiful Ships,* it is absent in *The Whaler Boy*. Post-Soviet films set on the Arctic Ocean, such as *Leviathan* (2014), represent the area as a post-Soviet *terra nullius,* a grey area of jurisdiction and sovereignty. *The Whaler Boy* follows suit in this trend. Lëshka's "global" outlook is represented to be driven by digitality as he explores imagery and maps of the US on the internet before his breakout. Simultaneously, the transnational outlook indicated by global cultural currents and digital mass media recalls the state borders between Russia and the North American continent. The film does not touch on border-crossing from the indigenous point of view or pay attention to the mobile history of the Chukchi and their Inuit neighbours across the northern waters of the Bering and Chukchi seas (Huntington, et al. 2020).

If the cultural boundaries seem to fade, state borders stand firm. Earlier in the film a Chukchi is killed by U.S. border guards as he tried to cross to Alaska. The long burial sequence, one of the rare glimpses of traditional Chukchi ways, is another allegory of the extinction of the community. Later, Lëshka himself is almost killed by a U.S. border guard himself. The guard – who perhaps misleadingly takes Lëshka under his wing, like the earlier shady poachers – becomes a symbolic father figure for fatherless and motherless Lëshka. The poachers and the policeman, here allegorically, can be read as the market and the state, both tied to authority and, following Kristeva, the socio-economical realm of the symbolic powers to which Lëshka and his marginal community are subjugated.

Unlike *The Most Beautiful Ships,* which features both Chukchi history and particularly Veket's personal past, *The Whaler Boy* is strikingly ahistorical. This is obvious both in the depiction of the Chukchi community and in the private temporality of Lëshka himself, exemplified by his orphanhood. The history of Russian imperialism and the colonization of the Chukchi is not touched upon. The erasure of the past becomes more problematic as the filmmakers are of the Russian majority, leaving the Chukchi themselves to blame for their sorry predicament. Ahistoricity, the bane of postmodern texts (Jameson, 1991), has also been a trend in films about "natives" and the "cultureless" Arctic (MacKenzie & Westerståhl Stenport, 2016, pp. 15–18). In terms of a post-colonial reading, the film is rife with the visual and narrative "tropes of imperialist cinema" (Shohat & Stam, 1994, pp. 137–150) when representing the Chukchi men as feminized and exoticized. In the case of Lëshka, these tropes include animalization, as he is narratively and cinematographically bonded to whales, infantilization in his naïve approach to digital relationships and geography, and even feminization as he promises to be the webcam model's "wife". The representation becomes even more problematic as the Chukchi and their marine-oriented mythology are drafted and geopoliticized into a narrative of the debunking of illusions diffused by U.S. cultural hegemony.

If Veket tried to compensate for the traumas of leaving his traditional home and separation by embracing the high culture of the Soviet metropole in *The Most Beautiful Ships*, in the case of *The Whaler Boy* and the era of globalization, the object of imitation is American (pop) culture, to the extent that Lëshka painfully tries to learn English. Despite some references to ancient customs, the Chukchi are depicted as thoroughly immersed in American culture, as even the older folks play instrumental rock. In another metonymic reference to cultural extinction, Lëshka's grandfather brings up the fact that he feels that his end is near. The depiction of the comic rock-and-rolling grandfather, who in the end cannot even die according to Chukchi customs, significantly differs from the grandfather who reminds Veket "not to forget" tradition in *The Most Beautiful Ships*. The Chukchi traditional songs – an important cultural attribute (not only as artistic expression but also having connections to Chukchi spiritual life) that is foregrounded in *The Most Beautiful Ships* – have been taken over by rock and roll in *The Whaler Boy*. Still, even though imported, like in the *Most Beautiful Ships,* the music in *The Whaler Boy* on the soundtrack is tied in with scenery of water and the sea, again expressing the characters' ambivalent position and the liminal subject-in-process.

Even though the idea of an authentic and untouched Chukchi culture was problematized in *The Most Beautiful Ships*, not least by the in-between character of Veket, the film still celebrated traditional dances, opulent fur garments, cosy *iarangas*,[6] and so on. *The Whaler Boy* has none of this: The Chukchi live in barrack-like dilapidated housing with ship containers and rusting oil barrels thrown into the scenery; the few Chukchi rituals shown, such as the funeral, are almost comical residues. The film paints a picture of a lost authentic culture, as well of a separation from nature and the "spirit world" of the Chukchi, that Lëshka seems to reconnect with at the end of his journey.

[6] Chukchi tent-like dwellings.

11.7 Littoral Dreaming and Awakenings

The final sequence starts after the policeman drops Lëshka off on the beach and he heads off to the "Motor City" via the tundra and the coast. At first Lëshka walks with confidence, his journey depicted with wide shots of the landscape that dwarf him in the best formal tradition of wilderness and adventure films. Gradually Lëshka's strength withers as he runs out of water. Soon he collapses and falls asleep. When curling up he chants "*Amerika, Amerika*", connecting the U.S. to dreams and a fantasy land or illusion.

Along with Lëshka being cast off into the sea, a strong oneiric component is introduced to the film that adds an ambivalence to the narrative and enables various ways of reading it. Falling asleep and waking up at sea or on the shore becomes a recurring motif in the film. This dialectic follows the subject-in-process as he oscillates between the symbolic world of the knowing self and the prelinguistic fluidities and desire flows of the semiotic. On the limitless and changing ocean and beach, Lëshka's subjectivity and ego are faced with dissolution. The focus on falling asleep and waking up raises questions about whether the narrative or parts of it are just a dream – the film becomes more surreal as the journey progresses; for example, the poachers on the island start shapeshifting into animal-like entities by the end of the night. Besides the littoral liminality, the oneiric quality of the film strengthens its ambiguity while connecting Lëshka's dreams with his struggle to find a place between two continents, the past, and the future.

Freud (1920/2010) and his later interlocutors, such as Bachelard (1999) and Theweleit (2003), have forged a strong connection with the aquatic and birth as well as death. This connection is also made in the shot that initiates the scene in which Lëshka is "re-born". Lëshka is shown to sleep and shiver in a foetal position next to the beach with the flow of water on the soundtrack, the visuals and sounds creating a womblike ambience. Lëshka then continues his trek, walking on marshland, the focus on his boots on the wetland, the water/land environment adding another hybrid element to the diegesis. There he finds a discarded plastic water bottle from which he, parched, tries to drink. The plastic bottle in the middle of nowhere is one of the environmental commentaries in the film. As Dodds and Nuttall (2019, p. 203) noted, the Arctic Ocean has become a repository of plastic waste and microplastics from around the planet, one of the downsides of the "globalized Arctic" (Fig. 11.3).

Blowing across the bottle top, the wind starts to make a tone in the empty bottle that grows into a howling sound as if a genie of past ancestors had been released. The organic sound is akin to whale song, and it beckons Lëshka to continue into what could be called a whale graveyard or shrine.[7] There, a whale's skull is stuck into the ground as if a totem to which he is drawn. The totem has been seen as a nexus of nature and culture (Lévi-Strauss, 1991), as well as kinship, and the whale skull serves both these functions in the film. An exchange and connection between

[7] Places like these are found on the shores of the Bering Strait, such as Whale Bone Alley on Yttygran Island.

Fig. 11.3 Lëshka falling asleep on the beach

Lëshka and the skull is created with the use of cinematic techniques, such as focus, close-ups, and POV shots. This passage of non-lingual bonding between human and whale culminates in an edit of shots of the skull and Lëshka's face that are similarly composed as if they were the same entity. The scene reminds us of a special relation, a sense of kinship, that the Chukchi share with these mammals that are of existential importance in their life (see Rytkheu, 2010, pp. 133–138).[8] The scene cinematically obscures the great divide between human and non-human and brings about the abject experience threatening the divide between the self and the other (Kristeva, 1982, pp. 12–13). The meeting of a Chukchi boy and whale bones taking place on the liminal shore demonstrates the parallel history of "the endangered species" – the Chukchi and the whales (see Slezkine, 1994, pp. 337ff).

Exhausted, on the threshold of consciousness or half asleep, Lëshka returns to having a pre-objectual relationship with his surroundings in the highly non-symbolic surroundings of the whale burial ground, reinforced by the non-verbal whale calls running throughout the scene. Lamenting near the whale totem, with a sharp snap on the soundtrack, Lëshka opens his eyes. As if woken from a dream to reality, Lëshka sees that the whale bones around him have turned into rusting oil cans. The scene also evokes Kristeva's (1982, p. 4) notion of abjection with the residual whale skull literalizing the breakdown of the distinction between subject and object, and

[8] This component of Chukchi culture, as well as the act of whaling itself, is left out of *The Most Beautiful Ships* as Chukchi traditional whaling was put an end to during Soviet times when commercial and logistical-type vessels were used for whaling. The practice was resuscitated in the post-Soviet era (IWC website, 2021).

between conscious and unconscious that are essential for the establishment of identity and entering the symbolic order. The sequence also becomes exemplary of the Darwinist qualities of the shore in Lëshka's struggle for survival. Kluwick and Richter (2016, pp. 5–6) noted that the changing shore is also an amalgamation of life and death with the laws of natural selection foregrounded in this environment. At the beginning of the journey, Lëshka is still equipped, both technologically and economically (he has his boat, provisions, etc.). After his drifting encounters and wandering down the shoreline, he is stripped to the point that he struggles to find food, drink, and warmth. Lëshka's status is equated with that of other animals, this levelling coming to a culmination with the symmetrical identification with the whale remains. Lëshka's encounter and revelation can thus be read as a dual narrative of re- and degeneration, correlating with the ambiguity of the coast (Fig. 11.4).

The scene becomes the accomplishment of Lëshka's rite of passage that started with his ritual fight with Kolia. Lëshka's bonding with the whales' spirits is literally an eye-opening experience, and he sheds his illusions of America and abruptly seems content with what he has. Lëshka laments to the totem, "I have had enough of this America" as if the fantasies of love and the U.S. no longer delude him, and he returns home to his village – incidentally, he is right next to it. Read literally, the film and its ending could be seen as an articulation of the colonization of the unconscious with mass media imagery, the field of the semiotic and its flows of desire shaped by the *Amerika*-nization that Lëshka finally sees as unattainable. However, the journey and the subject-in-process narrative arguably define the film and house its transformative potential.

Fig. 11.4 The whale graveyard

11.8 Conclusions

The film analysis shows how the coastal shore between water and land, as one of the major imaginary tropes, is "the point of departure, of immersion into turbulence" (Doody, 1996, pp. 321, 326). In addition, both films align with the Soviet/Russian cinematic tradition of representing the Arctic as "entangled in connotations of identity and belonging" and the coastal area as a frontier justifying territorial control and further expansion (Sarkisova, 2019, pp. 231, 232). Furthermore, we show how two films with decades and great politico-economic upheaval between them thematize identity and belonging by utilizing attributes of liminality, both geographical and social. Both films' stories are shaped by their setting, the Arctic coastal area, following Moretti's statement, that "[s]pace is not the 'outsider' of narrative, then, but an internal force, that shapes it from within" (Moretti, 1998, p. 70). The topos of the coastline becomes ingrained in these films of transformation in which the central values of the Chukchi communities are at stake. The shore, being between solid and fluid, is represented as a terrain for liminal periods and becomes intrinsic to the expression of how the continuity of tradition and taken-for-granted futures turn suspect. Collective indigenous history is translocated into an individual initiation narrative with both men becoming subjected to cultural and psychosocial processes involving loyalty, intercultural love, and its loss.

The shore becomes witness to and part of communal trauma in both films. The transfers, conflicts, mergers, exits, and returns give rise to palimpsest identities and keep the subjects in an in-between state of engagement with the past and present, as well as with different cultures. If in *The Most Beautiful Ships* the Chukchi community is represented as vital, even though it is changing, in *The Whaler Boy,* the community is reduced to a fragile settlement struggling for its survival with the spectre of physical and cultural extinction hovering over it. While Veket still has a community to negotiate with, Lëshka lives as an orphan in a world of digital illusion. The personal oscillation between indigenous tradition and an imitation of Soviet-Russian or American pop culture is commensurate with the community's vulnerability.

Even though the shoreline persists as an ambivalent contact zone in these films, as they are separated by over four decades of great changes, they show considerable shifts in the ideological constellation and aesthetics. Both films deal with cultural hybridity differently: The screenplay written by Yuri Rytkheu, a Chukchi himself, provides *The Most Beautiful Ships* with an insider perspective. The film draws vigorously from its contemporary Soviet society: The (Soviet) state is seen as benevolent, even though Veket does not approve all the "developmental" policies, and the Chukchi are tightly knit with the state through cultural interaction and technological innovations. The indigenous community is not a monolithic unity defined against the Russian majority but portrayed as complex and tangible due to contemporary references. Simultaneously, while cultural hybridity is conceived of as beneficial for the "small peoples", the film also succeeds in communicating issues of traumatic experiences, such as resettlements of indigenous children in remote boarding schools, cultural interruptions, and changes of lifestyles along with inter- and transgenerational conflicts. In comparison, the Chukchi in *The Whaler Boy,* almost 50 years later, are residual and endangered, even self-destructive. In the films the state pivots between paternal control, and retreat and abandonment. Also, the perspective on colonial agency is different: while the Soviet film still aims to tackle the total assimilation of the indigenous Chukchi culture into Soviet hegemony, the perspective of the recent film is of an outsider being as much interested in the global competition for cultural hegemony between Russia and the US as in the history of the indigenous people. In *The Most Beautiful Ships*, the balancing act between land and sea is imagined within Soviet borders and in accordance with the ideology of Soviet multinational equality policy. In *The Whaler Boy*, the transmarine dreams, coined as "America", indicate the fluidity of cultural borders, and the rivalry of ideas and bodies in intense overseas competition.

One clear difference is also in how future expectations are represented. Veket chooses to return, giving up a career in the Soviet metropole. Lëshka, by contrast, is caught between a dying Chukchi male society and global culture. Lëshka's precarious and powerless position in the film can be seen as allegorical of the marginal position of his community in the global economy. For most of the film, he is drifting and at the mercy of powers beyond his command. The rather tacked-on happy ending of *The Whaler Boy* offers few solutions to the problems that caused him to take off in the first place.

Lëshka's drifting and *The Whaler Boy*'s lack of historical depth contribute to the film's parable-like timelessness. The film's ahistoricity leaves aside the histories of Chukchi oppression, which makes the representation problematic. Furthermore, the film uses the dangerously porous shore of the Arctic Ocean, and the Chukchi as an appendage, for the displacement of both the desire for and fears of U.S. cultural hegemony and environmental anxieties. *The Whaler Boy* is conscious of its character as a cultural text. The "ironic quote marks" (Hutcheon, 1989, p. 102) of "Hollysweet_999" refer to Hollywood, the global hegemonic "dream factory". The film becomes a story not only about the Chukchi but a commentary on the illusive nature and power relations behind the medium itself; respectively, the Russian–U.S. coastline is both a geopolitical and an aesthetic borderline that resonates with the story's ambivalent plot. The moral of the story is that Lëshka may be a victim in a world of confrontations between major cultures, but he may get his reward in achieving some kind of inner peace by stating, "I have had enough of this America".

Lëshka's statement reveals what is common in both films. The Arctic Ocean is not only a remote locality, impacted on and defined by climate change, but – like more southern shores with their dynamic social and cultural histories – the Arctic coastline is also equally, and perhaps even more so today, central for the historical changes of people living in these areas. Moreover, the shore is an affective terrain that mobilizes intensive passions, a place where emotions play a role. The balanced life between land- and liquid-based elements is threatened by (feminized) waters that can dissolve both communities and solid male subjects since, as Cohen (2006, p. 652) noted, "shore encounters test boundaries by mixing danger and desire".

Respectively, both Veket and Lëshka have to decide how to deal with the unknown, symbolized by the ocean and the ambivalence of the shore. They both leave but come back to stand on the terrestrial side of the shore. Going out to the ocean provides positive possibilities, offered by the Soviet state to the "small nations" of the north, even if the voyage out would steer away from the traditional Chukchi way of life. For Veket, the voyage is made under state-guided control, strictly keeping the sailors within the geographic and imaginary borders of the Soviet empire. Lёshka, instead, faces the challenges of the late-modern global world where fantasies flow through digital networks as well as oceans.

References

Filmography

Nitochkin, A. (Director). (1973). *The most beautiful ships (Samye krasivye korabli).* [A two-part television film based on a screenplay by Yuri Rytkheu]. Soviet Central Television, 2 hr., 22 https://www.youtube.com/watch?v=DgVoMnefwec&t=15s

Uchitel, A. (Producer), & Yuryev, F. (Director). (2020). *The whaler boy (Kitoboi).* [Film]. Rock Films, Orka, Man's Film Productions 1 hr., 33 min. DVD.

Literature

Alexander, N. (2016). Shorelines: Littoral landscapes in the poetry of Michael Longley and Robert Minhinnick. In U. Kluwick & V. Richter (Eds.), *The beach in anglophone literatures and cultures: Reading littoral space* (pp. 71–85). Routledge. https://doi-org.libproxy.tuni.fi/10.4324/9781315613932

Anderson, S. C., & Tabb, B. H. (2002). Water, leisure and culture: European historical perspectives. .

Bachelard, G. (1999). *Water and dreams: An essay on the imagination of matter* (trans. E. R. Farrell). Dallas Institute of the Humanities and Culture. (Original work published 1942).

Barker, A. (1993). The divided self: Yuri Rytkheu and contemporary Chukchi literature. In G. Diment & Y. Slezkine (Eds.), *Between heaven and hell: The myth of Siberia in Russian culture* (pp. 215–226). St. Martin's Press.

Barthes, R. (1977). The grain of the voice. In R. Barthes, *Image, music, text* (trans. S. Heath) (pp. 179–189). Fontana Press.

Barthes, R. (1985). *The responsibility of forms: Critical essays on music, art and representation* (trans. R. Howard). University of California Press. (Original work published 1982).

Barthes, R. (1998). *The pleasure of the text* (trans. R. Miller). With a note on the text by R. Howard. Hill and Wang. (Original work published 1973).

Beumers, B. (2020). Interview with Filipp Yuriev (in English), Venice, 8. September. *Kinokultura* (70). http://www.kinokultura.com/2020/70r-kitoboi.shtml

Bhabha, H. (1984). Of mimicry and man: The ambivalence of colonial discourse. *October, 28* (Spring 1984), 125–133. Discipleship: A special issue of psychoanalysis. The MIT Press. https://doi-org.libproxy.tuni.fi/10.2307/778467

Bhabha, H. K. (1994). *The location of culture*. Routledge.

Blumenberg, H. (1997). *Shipwreck with spectator* (trans. S. Rendall). MIT Press.

Bugaeva, L. (2015). 'Here will be a garden city.' Soviet man on an Arctic construction site. In S. MacKenzie & A. Westerståhl Stenport (Eds.), *Films on ice: Cinemas of the Arctic* (pp. 310–324). Edinburgh University Press.

Carson, R. (1955). *The edge of the sea*. Staples.
Christensen, J. (2013). 'Our home, our way of life': Spiritual homelessness and the sociocultural dimensions of indigenous homelessness in the northwest territories (NWT), Canada. *Social & Cultural Geography, 14*(7), 804–828. https://www.tandfonline.com/doi/abs/10.1080/14649365.2013.822089
Clark, K. (1981). *The Soviet novel: History as ritual*. Indiana University Press.
Cohen, M. (2006). The chronotopes of the sea. In F. Moretti (Ed.), *The novel. Volume 2: Forms and themes* (pp. 647–667). Princeton University Press.
Cohen, M. (2010). *The novel and the sea*. Princeton University Press.
Corbin, A. (1994). *The lure of the sea: The discovery of the seaside in the western world 1750–1840* (trans. J. Phelps) University of California Press.
Damiens, C. (2020). The birth of a Soviet 'fourth cinema' on prime time television: Iurii Rytkheu and *The most beautiful ships*. *Studies in Russian and Soviet Cinema, 14*(1), 17–36. https://www.tandfonline.com/doi/pdf/10.1080/17503132.2020.1711607?
Denning, G. (2004). *Beach crossings: Voyaging across times, cultures, and self*. University of Pennsylvania Press.
Dodds, K., & Nuttall, M. (2019). *The Arctic: What everyone needs to know*. Oxford University Press.
Doody, M. (1996). *The true story of the novel*. Fontana Press.
Forsyth, J. (2000). *A history of the peoples of Siberia: Russia's north Asian colony 1581–1990*. Cambridge University Press.
Freud, S. (2010). *A general introduction to psychoanalysis* (trans. G. Stanley Hall). Boni and Liveright. (Original work published 1920) www.bartleby.com/283/
Gillis, J. R. (2012). *The human shore: Seacoasts in history*. University of Chicago Press.
Golovnev, I., & Golovneva, E. (2021). Traditional ethno-cultural communities in the modern Russian north: *Oil field* as a documentary film case. In M. Lehtimäki, A. Rosenholm, & V. Strukov (Eds.), *Visual representations of the Arctic: Imagining shimmering worlds in culture, literature and politics* (pp. 279–295). Routledge.
Hall, S. (1994). Cultural identity and diaspora. In P. Williams & L. Chrisman (Eds.), *Colonial discourse and post-colonial theory: A reader* (pp. 227–237). Harvester Wheatsheaf.
Hirsch, F. (2005). *Empire of nations: Ethnographic knowledge and the making of the Soviet Union*. Cornell University Press.
Huntington, H. P., Binder, R., Sr., Comeau, R., Kielsen Holm, L., Metcalf, V., Oshima, T., SimsKayotuk, C., & Zdor, E. (2020). Crossroads of continents and modern boundaries: An introduction to Inuit and Chukchi experiences in the Bering Strait, Beaufort Sea, and Baffin Bay. *Water, 12*(6) https://www.mdpi.com/2073-4441/12/6/1808
Hutcheon, L. (1989). *The politics of postmodernism*. Routledge.
International whaling commission. (2021, November 9). *Website description of the aboriginal subsistence hunt in Chukotka, Russian Federation*. https://iwc.int/russian-federation
Jameson, F. (1991). *Postmodernism, or the cultural logic of late capitalism*. Verso.
Kelly, C., & Volkov, V. (1998). Directed desires: *Kul'turnost'* and consumption. In C. Kelly & D. Shepherd (Eds.), *Constructing Russian culture in the age of revolution, 1881–1940* (pp. 291–313). Oxford University Press.
Kluwick, U., & Richter, V. (2016). Introduction. 'Twixt land and sea': Approaches to littoral studies. In U. Kluwick & V. Richter (Eds.), *The beach in anglophone literatures and cultures: Reading littoral space* (pp. 1–20). Routledge. https://doi-org.libproxy.tuni.fi/10.4324/9781315613932
Kristeva, J. (1982). *Powers of horror: An essay on abjection* (trans. L. S. Roudiez). Columbia University Press.
Kristeva, J. (1990). Revolution in poetic language: The semiotic and the symbolic. In T. Moi (Ed.), *The Kristeva reader* (pp. 89–136). Basil Blackwell.
Kristeva, J. (1992). *Black sun: Depression and melancholia*. Columbia University Press.
Lehtimäki, M., Meretoja, H., & Rosenholm, A. (2018). Veteen kirjoitettu – veden kirjoittamia: Kulttuurisen vedentutkimuksen suuntaviivoja. In M. Lehtimäki, H. Meretoja, & A. Rosenholm (Eds.), *Veteen kirjoitettu: Veden merkitykset kirjallisuudessa* (pp. 7–23). Suomalaisen Kirjallisuuden Seura.

Lévi-Strauss, C. (1991). *Totemism* (trans. R. Needham). Merlin Press. (Original work published 1962).
Mack, J. (2011). *The sea: A cultural history*. Reaktion.
MacKenzie, S., & Westerståhl Stenport, A. (2016). Introduction: What are Arctic cinemas. In S. MacKenzie & A. Westerståhl Stenport (Eds.), *Films on ice: Cinemas of the Arctic* (pp. 1–28). Edinburgh University Press.
Mathisen, S. (2004). Ethnic identities in global and local discourses: Contested narratives of Sámi ethnic heritage. In J. Kupiainen, E. Sevänen, & J. A. Stotesbury (Eds.), *Cultural identity in transition* (pp. 141–157). Atlantic Publishers & Distributors.
Moi, T. (1990). Introduction: The semiotic project. In T. Moi (Ed.), *The Kristeva reader* (pp. 1–22). Basil Blackwell.
Morch, A. J. (2017). Between the primordial, the modern and the post-modern: The Chukchi writer Yuri Rytkheu. In H. Hansson & A. Ryall (Eds.), *Arctic modernities: The environmental, the exotic and the everyday* (pp. 213–229). Cambridge Scholars Publishing.
Moretti, F. (1998). *Atlas of the European novel, 1800–1900*. Verso.
Payne, C. (2007). *Where the sea meets the land: Artists on the coast in nineteenth-century Britain*. Sansom.
Pratt, M. L. (1992). *Imperial eyes: Travel writing and transculturation*. Routledge.
Richter, V. (2016). Where things meet in the world between sea and land: Human-whale encounters in littoral space. In U. Kluwick & V. Richter (Eds.), *The beach in anglophone literatures and cultures: Reading littoral space* (pp. 155–173). Routledge. https://doi-org.libproxy.tuni.fi/10.4324/9781315613932
Rosenholm, A. (2021). The masculine north in popular Russian films: *Territoriia* as a case study. In M. Lehtimäki, A. Rosenholm, & V. Strukov (Eds.), *Visual representations of the Arctic: Imagining shimmering worlds in culture, literature and politics* (pp. 217–237). Routledge.
Rostek, J. (2011). *Seaing through the past: Postmodern histories and the maritime metaphor in contemporary anglophone fiction*. Rodopi.
Rytkheu, Y. (1972). *Veket i Agnes. Povesti* [Veket and Agnes. Novels]. Khudozhestvennaia literatura.
Rytkheu, Y. (2010). *Dorozhnii leksikon* [A traveller's manual] (pp.133–138). Zhurnal "Zvezda".
Rytkheu, Y. (2016). *Samye krasivye korabli* [The most beautiful ships]. Veche. (Original work published 1967).
Samuelson, M. (2016). Literary inscriptions on the South African beach: Ambiguous settings, ambivalent textualities. In U. Kluwick & V. Richter (Eds.), *The beach in anglophone literatures and cultures: Reading littoral space* (pp. 121–138). Routledge. https://doi-org.libproxy.tuni.fi/10.4324/9781315613932
Sarkisova, O. (2015). Arctic travelogues: Conquering the Soviet North. In S. MacKenzie & A. Westerståhl Stenport (Eds.), *Films on ice: Cinemas of the Arctic* (pp. 222–234). Edinburgh University Press.
Sarkisova, O. (2019). Moving through the century: The far north in Soviet and contemporary Russian nonfiction. In L. Kaganovsky, S. MacKenzie, & A. Westerståhl Stenport (Eds.), *Arctic cinemas and the documentary ethos* (pp. 231–253). Indiana University Press.
Shohat, E., & Stam, R. (1994). *Unthinking eurocentrism: Multiculturalism and the media*. Routledge.
Slezkine, Y. (1994). *Arctic mirrors: Russia and the small peoples of the north*. Cornell University Press.
Szekely, M. (2006). Gesture, pulsion, grain: Barthes' musical semiology. *Contemporary aesthetics, 4*. https://digitalcommons.risd.edu/liberalarts_contempaesthetics/vol4/iss1/5
Theweleit, K. (2003). *Male fantasies. Volume 1: Women, floods, bodies, history* (trans. C. Turner et al.). University of Minnesota Press. (Original work published 1977–1978).
Turner, V. (2008). *The ritual process: Structure and anti-structure*. Aldine Transaction Press. (Original work published1969).
Välimäki, S. (1998). Musiikin ei-kielellinen merkitysmaailma. *Musiikki, 4*, 371–393.

Välimäki, S. (2005). *Subject strategies in music: A psychoanalytic approach to musical signification*. The International Semiotics Institute.

Vdovin, I.S. (1965). *Ocherki po istorii i etnografii chukchei*. [Survey of Chukchi history and etnography]. Izd. "Nauka".

Vidauskyte, L. (2017). Metaphor of existence: Seafaring and shipwreck. *Filosofija. Sociologija, 28*(1), 11–19. https://www.researchgate.net/publication/315806087_Metaphor_of_Existence_Seafaring_and_Shipwreck

Zamyatin, D. (2021). Geo-cultural space of the Arctic: Landscape visualization and ontological models of imagination. In M. Lehtimäki, A. Rosenholm, & V. Strukov (Eds.), *Visual representations of the Arctic: Imagining shimmering worlds in culture, literature and politics* (pp. 295–312). Routledge.

Chapter 12
"The Silvery Song of Water": Nature, Experience, and Time in Paul Harding's Fiction

Markku Lehtimäki

Abstract In Paul Harding's novels *Tinkers* and *Enon*, cold water functions as a kind of transmission medium, bringing forth memories and even mystical connections to the past. As it is argued in this chapter, water is connected to consciousness and knowledge, as well as experiencing time, in these narratives. Water with its complex material-symbolic entanglements also connects Harding's fiction to the Puritan and Transcendentalist tradition of New England literature, while his fiction's naturalist and materialist emphasis simultaneously goes beyond that tradition. Therefore, my reading of Harding's two novels emphasizes the physical and material presence of water in addition to its symbolic and spiritual meanings. The narratives make achingly clear that human lives are fragile and fleeting, but they also emphasize the ways in which the natural environment is vulnerable and subject to change. The chapter especially focuses on the ways in which fictional minds work in natural environments, and consequently the analysis requires integrating perspectives on nonhuman nature with research on consciousness presentation. According to my reading, while non-human nature is cruel and indifferent to human strivings, the characters try also to find some meaning to their lives from the harsh beauty of nature. Cold water, especially, functions as a portal to the past and to the dead, also giving hope.

Keywords Consciousness · Experience · Memory · Time · New England · Paul Harding

> There are very few things that modern science does not yet understand. One of them is consciousness; another is water (Henry, 2020, p. 23).

M. Lehtimäki (✉)
Faculty of Humanities, University of Turku, Turku, Finland
e-mail: markku.lehtimaki@utu.fi

M. Lehtimäki et al. (eds.), *Cold Waters*, Springer Polar Sciences,
https://doi.org/10.1007/978-3-031-10149-6_12

The rich literary tradition as well as the diversity of the natural world of New England still resonates in contemporary fiction, as in the work of Paul Harding.[1] It has been suggested that the literature of this specific northeastern area of the United States has had an enduring influence on American culture in general, beginning with the oral traditions of Native American tribes and continuing through the Puritan sermons and Transcendentalist nature writing to the modes of naturalist fiction or modernist poetry (see, e.g., Westbrook, 1988; Ruland & Bradbury, 1991). The manifold literary genres of New England, such as Ralph Waldo Emerson's philosophical essays, Henry David Thoreau's life writings, or Nathaniel Hawthorne's romance novels, also appear to be inspired by the abundantly rich natural history of the land – most famously in the case of Thoreau's Walden Pond. It is especially this tradition (both Puritan and Transcendentalist) that informs my reading of water in the twenty-first century New England author Paul Harding's fiction, but, at the same time, my reading aims to emphasize the physical and material presence of water in addition to its symbolic and spiritual meanings.[2] Therefore, even as Harding situates his writing in the Puritan and Transcendentalist traditions of New England literature, his novels bring this heritage into a critical discussion with the subsequent naturalism and modernism. Still, the fundamental question shines through these literary and philosophical modes of thinking, idealist and materialist alike: how can we humans know nature, or, more specifically in the context of this chapter and this volume, how can we know water?

As I argue in what follows, in Harding's novels *Tinkers* (2009) and *Enon* (2012)[3] water – and especially cold water – functions as a kind of transmission medium, bringing forth memories and even mystical connections to the past. In these narratives, water is connected to consciousness and knowledge; while human beings cannot know water in itself, their sense of its material presence helps them to recall things and find missing links to the past experiences. While I will also discuss these aspects of the narratives in my analysis, I also aim to read the human stories against the grain, as it were, emphasizing the ways in which human needs and concerns are always also surrounded and enveloped by the harsh reality of the natural environment, in this case especially the winter and its accompanying phenomena (frozen water and abundance of snow). It appears, however, that in prevailing readings of Harding's novels, non-human nature is in the background as human consciousness is foregrounded.

---

[1] Paul Harding (b. 1967) was previously the drummer in the appropriately titled rock band Cold Water Flat. *Tinkers*, his first novel, was a surprise winner of the Pulitzer Prize after having been rejected by several publishers. *Enon* is Harding's second (and as of writing this chapter, the latest) novel.

[2] This kind of symbolic-material reading of water, including cold and Arctic waters, has been recently practiced and developed in literary and cultural studies (see, e.g., Costlow et al., 2017; Lehtimäki et al., 2021).

[3] When referring to the novels in in-text citations, I will use the following abbreviations: *Tinkers* = Harding (2010) and *Enon* = Harding (2014).

Both knowing and narrating water therefore forms a central question in Harding's fiction.[4] However, water and other non-human nature are obviously not for human needs alone; water and nature – in this context, water and nature of Maine and Massachusetts – have their own reality and history as well. In fact, inside the human stories in *Tinkers* and *Enon* there runs a parallel, or hidden, story about the changing environment of New England, its nature and landscape, which are gradually being transformed and even destroyed by the human habitation and construction. The narratives make achingly clear that human lives are fragile and fleeting, but they also emphasize the ways in which the natural environment is vulnerable and subject to change. The Puritan and Transcendentalist traditions of New England, with their emphasis on a spiritual understanding of nature, are gradually transforming into a more secular and material approach to nature. This development is not altogether negative, obviously; in the materialist and naturalist conception of nature, water is not only seen as a mirror reflecting the human mind but also as a non-human agency of its own. The tension between consciousness and materiality – that is, seeing water as both responding to human hopes and resisting them – also structures the narrative practice of Harding's fiction.

12.1 Consciousness and Water

The first sentence of *Tinkers* states: "George Washington Crosby began to hallucinate eight days before he died" (T, p. 7). *Tinkers*, which is mainly situated in Maine, in some of the northernmost parts of the United States (excluding Alaska), tells about the fragmented memories and gradually weakening consciousness of its protagonist, George Washington Crosby, lying on his death-bed and surrounded by his family and relatives. George, suffering a dementia, is gradually moving into a free-floating recollection of separate images of his father and his childhood in the Maine woods. The main story appears to concern itself with George's memories of his father – and his father's memories of his own father inside George's memory – although George's grandson Charlie, who eventually becomes the narrator of Harding's next novel, *Enon*, also exists as an external character outside of George's mind. Another, as if hidden, story in *Tinkers* is the reality and history of winters in Maine, with their cold waters, frozen lakes, freezing wind, and piles of snow. As the narrative seems to suggest, the characters' minds and memories are solidly linked to this natural environment. Indeed, non-human nature, including water, plays a crucial role in the narrative. Yet in most reviews of the novel – as well as in the few scholarly studies published of it so far – the focus is on the human interest: language, religion, consciousness, memory, and trauma (see Bouchard, 2014; Çirakli,

[4] Etymologically, narrative derives from the Latin word *narrare*, which means "telling," as well as from the Latin word *gnarus*, which means "being acquainted with something" or "having knowledge of something" (Kvernbekk & Frimansson, 2013, p. 571).

2014; Sykes, 2018). This betrays a strongly anthropomorphic reading of Harding's novel, as water as an agency of its own is not really touched upon in these studies.[5]

The novel itself transforms into a stream of consciousness, often resembling more poetry than narrative. In her reading of *Tinkers*, Rachel Sykes speaks of "quiet" novels, in which "the interior lives of introverted, scholarly, and often reclusive characters are prioritized" and in which a narrative proceeds at a slow pace and lacks in action, instead foregrounding the movement of thoughts and consciousness of the narrator or the protagonist (Sykes, 2018, pp. 3, 91). In his classic study of the technique of stream of consciousness as a continuous flow of sense perceptions and memories in the human mind, Robert Humphrey suggests that "[t]he word 'stream' need not concern us immediately, for the representation of the flow of consciousness is [...] entirely a matter of technique" (Humphrey, 1954, p. 6). We may note, then, that Humphrey, in his definition, argues for the technical aspects of stream of consciousness and is less interested in the ways in which fictional minds function in natural environments – or indeed, in the ways in which water provides the technique with its governing metaphor. Yet, in William James's original coinage of the concept ("stream of thought"), as in James's pragmatic philosophy in the larger sense, the natural world is a springboard for human thinking and experience.[6]

Some of the well-known concepts related to the consciousness-reading techniques – "stream of consciousness" or "transparent minds" (Cohn, 1978) – interestingly evoke watery metaphors, even though in classical narratology the focus is solely on fictional minds and not on natural worlds. What I especially emphasize in my own reading of Harding's fiction, however, is the ways in which fictional minds work in natural, including nonhuman, environments. Accordingly, as Thomas Polger argues in his *Natural Minds*, "the problem is not *whether* minds are part of the natural world, but *how* they are" (Polger, 2004, p. 2). Humanist or anthropomorphic understandings of narrative have been recently challenged and questioned in posthumanism and material ecocriticism, which see certain narrative techniques and literary conventions as hermetically sealed textuality. Accordingly, Stacy Alaimo suggests that "the psychological interiority stressed by the stream of consciousness form isolates the characters, not only from each other, dramatizing a familiar modernist sense of alienation, but also from the wider material world" (Alaimo, 2012, p. 478). In Harding's fiction, modernist techniques of consciousness presentation and inner monologue are connected to the wider material world, and in what follows, I aim to read these fictional minds in the context of the watery and wintry environment of New England.

---

[5] A considerable exception is Marco Caracciolo's (2020) reading of *Tinkers* as a "we"-narrative, in which the presence of various nonhuman actors (such as animals and material objects) problematize and complicate the narrative's – and narrative theory's – anthropomorphic assumptions. In Caracciolo's analysis of the novel, the significance of water is not the main point, however.

[6] In *The Principles of Psychology* (1890), James writes: "Consciousness, then, does not appear to itself as chopped up in bits [...], it is nothing joined; it flows. A 'river' or a 'stream' are the metaphors by which it is most naturally described. In talking of it hereafter, let's call it the stream of thought, consciousness, or subjective life" (James, 1950, p. 239).

In *Tinkers*, George's floating mind creates the very style of narration:

> George Crosby remembered many things as he died, but in an order he could not control. To look at his life, to take the stock he always imagined a man would at his end, was to witness a shifting mass, the tiles of a mosaic spinning, swirling, report raying, always in recognizable swaths of colors, familiar elements, molecular units, intimate currents, but also independent now of his will, showing him a different self every time he tried to make an assessment (T, p. 18).

There is both physical pain and mental breakdown in the main character George's mind, which also explains the fragmented nature of his memories. In one of the few analyses of the novel, it is argued that "the parallel narratives in *Tinkers* come together in George's gradually dwindling but still questing mental functioning that orients his consciousness and the narrative progression, to the far past in order to reconcile him with his father and his father with his grandfather at the same time" (Çirakli, 2014, p. 228). Thus, George's memorizing is a frame narrative in which his father's memory of his father is embedded; the three generations of men construct a continuity in which their memories and experiences flow together like water.

In my view, water is a governing element in the narrative, even though water is hardly even mentioned in the existing research of the novel. Furthermore, water is not only a natural phenomenon outside the human mind, but it also constitutes the human body. George's gradually weakening body as if betrays its watery nature and, finally, its lack of water: "George had the watery, raw feeling" (T, p. 10); "George was dehydrated ninety-six hours before he died" (T, p. 57). George, lying on his bed and sensing his end approaching, feels like he was in the midst of water or walking on thin ice or eventually drowning to cold waters: "His lungs were full of liquid and he felt like he was drowning" (T, p. 178); "It was as if he lay faceup just beneath the surface of water" (T, p. 179). There is also a traumatic experience somewhere hidden in George's mind, so that his repressed memories about his father appear in his consciousness in a fragmented but also flowing style. Howard, the father, unexpectedly left George and his mother Kathleen and was totally absent for twenty-five years, during which time George got married and his mother died. The absence of a father therefore constitutes the prevailing trauma of George's subsequent life. He simply tries to forget his father, but the figure of the father is constantly moving on the margins of George's consciousness. As George's fragmented mind floats between life and death, he is able to see the meaning of the moments of his life. What follows is that George imagines the experiences of his father Howard, who was a tinker and travelling salesman of various mismatched items and who suffered from epileptic seizures. George's imagining of Howard's life evokes senses and smells of the lost life, sensing even "frozen wood so brittle that it rang when you split it" (T, p. 23). Here Howard's experience as if becomes George's own through his active imagination. But George also feels that Howard "leaked out of the world gradually" and that he "seemed merely vague or peripheral" (T, p. 134). Howard becomes only a glimpse in the margins of memory and vision, since "we could no longer even see him, but felt him in brief disturbances of shadow or light [...] such as the snow melting into the wool of his winter coat" (T, p. 134).

In *Tinkers*, the main characters are clock-makers and they are keen to measure nature scientifically, finding "methods of capturing time more precisely" (T, p. 55). As he is in the process of dying, George Crosby's inner thoughts and floating memories represent a subjective sense of time as juxtaposed with the external narrator's deliberately detached measurements of clock time: "One hundred and thirty-two hours before he died, George awoke from the racket of the collapsing universe to the darkness of night and silence" (T, p. 28). The narrative corresponds to a well-known distinction in modern philosophy between "the uniform, objective time of clocks and the subjective temporality of human experience" (Hogan, 2011, p. 29). Clocks and calendars are the characters' way of controlling nature and organizing it to scientifically measurable parts, even though it is emphasized in the novel that time functions subjectively and in a layered or flowing way. There is a sense of simultaneity in the events, as Howard's experiences are vivid in George's memory: "The person contained hundreds of years, but they overlapped, as if the person experienced any number of times at once" (T, p. 158). Howard's actions inside George's mind are even narrated in the present tense: "Howard eventually comes to the outlet at Tagg Pond. […] The silt and water combine in an element that is half earth and half liquid" (T, pp. 143–144).

Therefore, from a static space of George's bedside the narrative flows – in the style and technique of stream of consciousness – into his fragmented, floating memories of his father in a different time and space in the wintry Maine woods, in the midst of which his father Howard recalls the incidents of the life of his *own* father (see also Çirakli, 2014, p. 228). Indeed, there is a juxtaposition between mechanical and experiential understandings of time in the narrative. As natural science informs George, "our universe is a mechanism consisting of celestial gears, spinning ball bearings, solar furnaces" (T, p. 179). What is, however, relevant is that the ticking of the clocks is associated to the beating of the human heart. Thus, George feels his "faintly ticking heart" (T, p. 34) and senses the "clocks, which did not seem to him to tick but to breathe" (T, p. 35). The passing of time, and the human experience of it, is captured in the phrases such as "the calming, reassuring tick-tocks of a winter's night" (T, p. 162). It is especially this "wintery" experience in the cold and dark natural environment – with its icy surroundings and frozen waters – which constructs a specific kind of human minds in *Tinkers*.

12.2 Transcendentalism and Materialism

In Harding's fiction, an oscillation between Transcendentalism as spiritual idealism and naturalism as a materialist conception of non-human nature is the very basis of the author's vision, even as it is a poetic basis of many other American writers' work. Transcendentalism has often been regarded as an essential characteristic and a specific trait of American literature. As Roger Assilineau has it, with a reference to Henry David Thoreau's *Walden* (1854), American writers are "Narcissuses looking at their own changing reflections in the Walden Ponds of their minds" (1980,

p. 9). In his classic essay "Nature" (1836), Ralph Waldo Emerson believes that natural things signify spiritual phenomena and that nature is a metaphor of the human mind. According to Emerson's Transcendentalist vision of nature, a true lover of nature is the one whose inward and outward senses are adjusted to each other; one becomes a "transparent eyeball," viewing nature from the vantage point of the disembodied eye (Emerson, 2003, pp. 39–40; see also Porter, 1981, pp. 94–106; Person, 2008, pp. 147–148). However, while "Nature" is an example of Emerson's idealism, the subsequent essay "Experience" (1844) is an example of his realism, presenting "an unbridgeable chasm" between the human and the non-human (see Lundin, 2005, p. 61). This very brief foray into the origins of American Transcendentalism in my chapter is meant to emphasize the ways in which the Transcendentalist heritage in American writing also paves a way to a more naturalist and materialist experience and understanding of non-human nature.[7]

As Assilineau argues, American Transcendentalists "see the whole and perceive the design of the mosaic" as well as see nature "as made up of hieroglyphics to be deciphered, [...] recorded, and interpreted" (Asselineau, 1980, pp. 9–10). In its partial adherence to the Transcendentalist tradition in American literature, *Tinkers* suggests how non-human nature is sometimes a text to be read and interpreted by humans:

> We kicked the bark off of dead trees and the soft wood beneath was as pale as sawdust and sometimes covered with strange designs that looked like writing that had been drawn into the wood with a stylus or fine-carving tool and the bark then fitted back over the trunk—a rough skin, a splintery hide that protected the secret language. These hieroglyphs we discovered like revelations, like messages someone had left for us to discover and to ponder [...]. [T]he people who were supposed to read these messages had to go through a very long and difficult process of deciphering obscure clues and directions [...] (T, p. 188).

In nature, there are, therefore, designs, writing, language, and messaged carved by a stylus (i.e., style) to be read and interpret by the human mind. Thus even natural formations and shapes are seen as linguistic signs to be read and interpreted. Words are also a human way of penetrating the surface of water:

> Light skin of sky and cloud and mountain on the still pond. Water body beneath teeming with reeds and silt and trout (sealed in a day skin and night skin and ice lids), which we draw out with silk threads, fitted with snags of fur or bright feathers. Skin like glass like liquid like skin; our words scrieved the slick surface (reflecting risen moon, spinning stars, flitting bats), so that we had only to whisper across the wide plate. Green drakes blossomed powder dry among the stars, glowing white, out of pods, which rose from the muck at the bottom of the pond and broke open on the skin of the water (T, p. 45).

[7] More systematically, Richard M. Gale defines that in the American pragmatic philosophy "anthropomorphic or humanistic naturalisms [...] depict nature as made to order for us human beings because it answers back to our deepest feelings and aspirations," whereas "reductive materialistic naturalisms [...] strip nature of all the qualities that give it human meaning and purpose" (Gale, 2010, p. 55). This kind of tension between anthropomorphic and materialistic naturalisms can be seen constituting Paul Harding's fiction as well.

Here, the seemingly objective description "light skin of sky and cloud and mountain on the still pond" appears as a lyrical image in its ungrammatical sentence, so that the omission of any action emphasizes the stillness of the pond. However, human beings soon disrupt the natural surroundings with their language. As the narrator says, "like glass like liquid like skin; our words scrieved the slick surface"; here people are trying to understand what water is with the help of similes – like glass, like liquid, like skin – and finally the stillness of the pond is disturbed by the use of words, human language. This attempt to penetrate water through language is, obviously, one example of *aquagraphy* or water writing[8]; but simultaneously phrases like "water body" and "the skin of the water" emphasize the material quality of water, even so that "the skin of the water offered [...] resistance" (T, p. 80).

As suggested previously, the story of the natural environment is a kind of a hidden narrative in Harding's fiction. *Tinkers* includes in itself a brief history of the violent transformation of the non-human nature of Maine into human habitats, so that under the surface of the human story there lives and breathes a larger ecological story about "the raw earth" (T, p. 61) and its domination. According to this history, "people arrived with their catastrophic voices and saws and plows and began to sing and hammer and carve and erect" (T, p. 61), and they "pounded, carved, blasted, and peeled the earth open through forests, rivers, and gorges, mountains and swamps" (T, p. 185).[9] Yet, there is something in *Tinkers* that suggests the ways in which nature welcomes such human acts, which are not too disturbing or destructive. The intrusion into nature is clear when divers in George's childhood memory enter the sea and disrupt the ocean waters: "The green sea turned gray and its surface rolled like a membrane. When we dived for shells, it parted for us without resistance, and sealed itself behind our up-pointed toes. We felt around, blind, in its slick graphite body" (T, p. 189). The collective "we" character (see Caracciolo, 2020) infiltrates the natural body of the sea and affects everything around it. Perhaps, in this sense, nature and water are open to such "souls", who themselves have "a heart open to nature" (T, p. 55). Generally, in *Tinkers* anthropomorphic language is a characteristically human way to think about non-human nature, to transform wilderness into human concepts in order to survive in it:

> The reasonable, sensitive soul who perhaps one day while taking his rest along the banks of a bubbling brook came to hear [...] a regularity in the silvery song of water over pebbles,

---

[8] According to Isabel Capeloa Gil's definition, "*scriptura aquae* [...] can either be read as a writing about water or as the writing modes produced by water and the liquid imaginary" (Capeloa Gin, 2008, p. 10).

[9] The feminization of the land – as that which receives and nurtures (Mother Nature), but that can also be dominated (Virgin Territory) – is a central theme in ecofeminist criticism (see, e.g., Westling, 1996, p. 45; Jarvis, 1998, p. 7). It is of course possible to interpret that in Harding's masculine fiction the feminized water receives and nurtures the male protagonists while at the same time the history of New England (originally named as "Virginia" in colonial times) is that of domination by men who "pounded, carved, blasted, and peeled the earth." It is as if after centuries of nature destruction, men seek for Mother Nature's nurturing and healing power. In Gaston Bachelard's poetic – and somewhat gendered – definition, water is "the mistress of liquid language" (Bachelard, 1983, p. 187).

> that soul is unknown to us. [...] [A] heart open to nature [...], he arrives at a device which marks time by a steady flow of water through its guts (T, pp. 55–56).

On the one hand, the "silvery" in the quotation above is a sign of the human mind's way of conceptualizing the material presence of water; elsewhere in the narrative, water is haptic ("the skin of the water" [T, p. 45]) or mineral ("the sharp mineral cold" [T, p. 61]); "metallic water" [T, p. 142]; "seamless, molecular, slick, atomic" [T, p. 190]). On the other hand, the "song" in the quotation implies the human mind's characteristic way of anthropomorphizing water. Here, sounds made by water resemble the flow of language. Or, at least, this would be a human interpretation of the language of water: a gently flowing stream has a quiet babble of its own ("the silvery song of water over pebbles").

In *Tinkers* and *Enon*, water may be comforting even in its coldness, but this is an especially human interpretation of its meaning. According to the concept of "surface reading" (Best & Marcus, 2009), there is no reason to give a symbolic interpretation of every natural phenomenon; and yet, it is typical to see a flowing river as a metaphor of life's energy or a frozen pond as a poetic figure of death's stillness. Indeed, as Veronica Strang argues, of all the elements in the natural environment, water is "the most suited to convey meaning in every aspect of human life" (Strang, 2004, p. 61). Or, as T.S. McMillin suggests, "since water is essential for life, for being, it might have something to do with meaning as well" (McMillin, 2011, p. xiii). McMillin goes on to suggest that while water indeed is an essential element for life, it might also be a considerable source of literary imagination.

12.3 Winter Minds and Cold Water

In Harding's fiction, water is not only a symbolic resource but also a material entity of its own. Human strivings are constantly situated in the coldness of the environment in *Tinkers*, with "dark and ice pressing down from the north" (T, p. 24) and "cold light from the north" (T, p. 44). The environment is that of "blue snow" and "frozen wood" (T, p. 23), "the dark and the ice" and "the heartbreak of a cold sun" (T, p. 24). These lyrical, often disconnected images, which construct the narrative text, are sometimes enlarged into naturalistic descriptions of the environment as well. Obviously, winter is not the only season there is, but its governing presence in the characters' minds and bodies makes spring and summer something to be anticipated: in the spring "the ice went out of the ponds" (T, p. 38), and "[s]ummer would anneal the chilled earth, but for now the water was so mineral and hard that it seemed to ring" (T, p. 64).

Being in the midst of winter and cold water enables the characters to empty their minds of all-too-human ways of perceiving and interpreting the world, as if they became a kind of pure consciousness. Indeed, Barry Lopez, in his seminal book on the Arctic imagination, argues that in the midst of the Arctic landscape the human mind can be emptied of its "categories and designs" (Lopez, 1999, p. 414). While not situated as north as the Arctic, the harsh winter of Maine provides the characters of *Tinkers* with a physical environment which also constructs their mental landscapes. In these icy surroundings, they develop into "winter minds" with "a wintery wisdom – cold scarlet and opaline minds, brief and burnished, flaring in the metallic blue of dusk" (T, p. 99). This is not only a masculine experience in the narrative, shared by George and his father Howard, because occasionally the viewpoint shifts into a woman's perspective. In the past, George's mother Kathleen experiences the cold and snowy winter in her own way, and the landscape is presented to the reader through her eyes and mind (even though also embedded in George's memory and vision). Thus, the focal point of the narrative is here on Kathleen's vision and understanding: "It is winter because she lies awake with a bare heart, trying to remember a fuller season" (T, pp. 87–88). Winter is the season of bareness, emptiness, and death, whereas summer is the season of fertility, fullness, and life. In the midst of "one freezing January morning," Kathleen is "only aware of the cold" (T, pp. 89–90). There is also an imaginative speculation about "a man who had never seen summer, a winter man" (T, p. 117).

Here, a major American literary reference is Wallace Stevens' famous poem "The Snow Man" (1921), a modernist experiment of seeing a natural landscape in winter as "nothingness" and devoid of any human meaning:

> One must have a mind of winter.
> To regard the frost and the boughs.
> Of the pine trees crusted with snow;
> And have been cold a long time.
> To behold the junipers slagged with ice,
> The spruces rough in the distant glitter.
> Of the January sun; and not to think.
> Of any misery in the sound of the wind,
> In the sound of a few leaves,
> Which is the sound of the land.
> Full of the same wind.
> That is blowing in the same bare place.
> For the listener, who listens in the snow,
> And, nothing himself, beholds.
> Nothing that is not there and the nothing that is (Stevens, 2015, p. 10).

As Robert Pinsky suggests, "the poem mediates the theoretical possibility of seeing nature purely," but in order to achieve this possibility, "one would need a mind enormously purified of human responses, in order to feel that way – figuratively speaking, a cold or wintry mind" (Pinsky, 1976, p. 72). This, of course, is an experiment with imaginative or speculative literature, also suggested by *Tinkers*: how is one able to see water the way it is when one is not present? And, according to the rhetorical trope of apostrophe, how can one speak to one's father when he is not present anymore?

David Gilcrest refers to William Bevis's study of Wallace Stevens, *Mind of Winter*, in which the writer makes the case for "unmediated perception" and argues that Stevens experimented with "a meditative state of consciousness" (Gilcrest, 2002, p. 127). Meditative consciousness would be seen in a certain "restraint" and a melancholy mood, which acknowledges that "the non-human world exists independently of human desire, including the desire for representation" (ibid., p. 132). These kinds of winter minds – with their ability to a *non*-human way of perceiving things – are in *Tinkers* associated with a dying consciousness. Indeed, as George is dying, he reflects the following:

> Human consideration was no longer to be his, for that consideration could be expressed now only by providing physical comfort, and physical comfort was as meaningless to him [...] as it would have been to one of his clocks, laid out in his place to be dusted (T, p. 184).

George needs to ponder whether "death was to fall below some human boundary" (T, pp. 47–48); indeed, on the brink of death, he senses that "such peace was not a human one" (T, p. 183).

The question of survival through a harsh winter is one of the governing senses in the narrative; George, for example, needs to reflect on "how a man could survive one winter alone" (T, p. 37) and how one "endured another winter in the woods" (T, p. 39). The surrounding natural environment is obviously not only a projection of the human mind, but reality in its own terms, even if human language tries its best to name things in nature: "[t]he frozen lake at midnight, the dry hack of the blade on ice so tiny under the wheeling and frozen stars" (T, p. 25). There are constant mentions of the cold, wintry situation in the text. More specifically, the sense of cold water is emphasized in the narrative. George's and his family's world is surrounded by "cold river water" (T, p. 42) and "cold water clear as air" (T, p. 77). Cold water is something one can look at, hear as well as smell: "He smelled cold water and cold, intrepid green. Those early flowers smelled like cold water" (T. p. 60); "it was the mineral smell of cold, raw green" (T, p. 60); "He was enchanted by the smells of wax and cold water" (T, p. 173).

There is a constant attempt in *Tinkers* to try to grasp the very essence of non-human nature beyond human mind and language. This requires a certain kind of relinquishment of the strongly human ways of perceiving and experiencing nature and water: "his own head rising half out of the water, alert, sharp, animallike" (T, p. 106). George aims to connect with his father by memorizing him through Howard's own physical feeling of cold water: "How can I not wonder what it would be like to sit in that cold silver water, that cold stone water up to my chin" (T, p. 129). Here, the sense of cold water as if transmits memories and experiences over generations. George is imagining his father, who for his part tried to feel the presence of his own father through a meditative concentration, imagining things in nature in their pure existence, without any human intrusion:

> By the time the shadows begin to lengthen and creep across the water, the stream has healed itself back around him and he imagines that he will now be able to see the animals and the light and the water the way they are when he is not present, and that that might tell him something about his father. I will have to sit still, like a guru, he thinks. I will have to ignore cramps and the cold. I have to breath very slowly and very quietly, so that my breath does not even stir the water flowing past my chin (T, p. 145).

Here, cold water functions as a portal to the past, connecting George to his father Howard, who is trying to connect to his own father by sitting in the midst of a pond.

We hear Howard's words through George's consciousness: "I felt so sore and so cold, I thought I might lose consciousness" (T. p. 148). Cold water, which keeps one's senses "alert, sharp, animallike," therefore functions as a medium bringing together memories of the three generations.

Howard, traumatized by his own father's disappearance (just as George is traumatized by Howard's absence), comes to a pond with cold water and sits in the middle of it, up to his neck, and stays there for several hours, hoping he might see his father again (see also Bouchard, 2014, p. 460). He also reflects that "the whole event seemed as if it hadn't actually happened outside of my imagination" (T, p. 148). In his *Water and Dreams*, Gaston Bachelard suggests how a contemplation of water can lead us to inner experiences, triggering dynamic and powerful conscious and sub-conscious associations that can form the initial stage of inspiration and artistic creation. As Bachelard poetically presents it, the of waves of the sea can be soothing and comforting, the shiny surface of a pond can provide a site for reflection, and the blue and green shades of water can be tranquil and cooling. Obviously, water can also be associated with fear, danger, and death. For Bachelard, the surface of water forms a boundary between life and death, between conscious and subconscious, and between reality and imagination (Bachelard, 1983, p. 47). Water is also "a natural mirror," encouraging self-reflective contemplation because "it holds up to us a second version of ourselves" (ibid., p. 24). In *Tinkers*, George is seeing himself in "mirrors and windows and metal and water" (T, p. 52) and feeling "the surface of the pond was sleek and reflective" (T, p. 80). Water therefore also provides a site of reflection and meditation for George, as it did for his father Howard.

There is still a feeling in George's mind that he has not finally received any final knowledge of either his father or water: "I have never really known them at all" (T, p. 129). Any ultimate knowledge of other people's minds and non-human nature – or, in this context, consciousness and water – cannot be reached. For the characters of *Tinkers*, water is full of "forgotten songs we never really knew, only think we remember knowing" (T, pp. 128–129). Here, the question of knowing water comes to the forth. Water as if sings its mysterious songs, which remain "unnamable" in human language: "the water [...] in its choral voice stirring all of the old unnamable sorrows in our throats" (T, p. 128). In a larger sense, human knowledge and its connection to non-human nature presents a crucial question in the narrative. It is suggested that nature – including water – cannot be approached and understood through artistic means or scientific concepts but only through being-in-there in nature and water, through experiencing its harshness and coldness. One needs to experience "cold mornings," when "you split frost-laced wood with numb hands," because then "you are still alive, still human, and still open to the beauty of the world" (T, p. 72). Therefore, it is phrased in the narrative that "the true essence, the secret recipe" of nature is something "too fine and subtle" to be observed with one's "*blunt eye*" and "*dumb intention*" (T, p. 54; emphasis original). What is needed is "a heart open to nature" (T, p. 56), because there is something "that was not accessible through words," only "through grass and flowers and light and shadow" (T, p. 62). Here, nature is not so much Emerson's "open book" for the humans to read and interpret, as phrased in his essay "Nature," but reciprocally human beings need to have "a heart open to nature."

12.4 Water, Immersion, and Memory

Narratives are often regarded a fundamental way of organizing human experience, even a means of therapeutic healing of traumatic experiences (see, e.g., Easterlin, 2012, p. 43). In the context of this chapter, the question is whether nature can be in any sense caring, healing, or palliative when it comes to human suffering. The idea of a "cleansing" and "purifying" power of water is of an ancient origin (see, e.g., Bachelard, 1983, p. 143), and in *Tinkers* as well as in *Enon* it is especially cold water in lakes, rivers, and ponds that may have this kind of influence of healing people both in the mental and physical sense. At the same time, water is seen as an agency of its own,[10] indifferent to human needs: "Watch after you leave the water, now cold and regretful"; "Watch the water heal itself of your presence" (T, p. 130); "The water was undisturbed, too; there wasn't a trace of any body entering or leaving it" (T, p. 148). Here, water is not so much healing or palliative for human needs, sickness, and suffering, but water rather heals itself of human presence. After the human body leaves the water, it is again undisturbed – and continues its existence independently of the human mind.

Harding's fiction, both *Tinkers* and its sequel, *Enon*, tell emotionally devastating stories of loss, death, and the fragility of the human life in the northern parts of New England (Maine and Massachusetts). While *Tinkers* tells its story through the consciousness of an old man in the process of dying, *Enon* tells its story through the consciousness of a middle-aged man whose daughter has just died at the beginning of the narrative. In these novels, the natural environment and the geographical place is not only a background, but human thinking and action are strongly immersed in the place with its woods and waters. While the Lake Enon, whose presence governs the narrative of *Enon*, is a fictional one, the actual waters of New England provide the narrative with an experiential and physical landscape in which human action takes place. As it is reflected upon in the novel by the first-person narrator: "The name Enon, spelled Aenon for the first four years of the village's existence, is from the Greek *ainon*, which is from the Hebrew *enayim*, which means double spring, or more generally, a place of abundant water" (E, p. 94–95). As in the Puritan tradition, which is the strongly resonating historical and cultural background in New England, nature is here seen in Biblical terms.

In *Tinkers* and *Enon*, both narrators and characters try to write nature into existence, even as they are constantly being constructed by nature itself. In Bonnie Costello's words, recent studies of the environment, both aesthetic and ecocritical, "find us embedded in a paradox: that we are part of nature, and that nature is our construction" (Costello, 2003, p. 1). Thus, our entanglement with nature is both material and imaginative, as Costello also suggests (ibid., p. 2). Speaking of water, Veronica Strang argues that modern sciences of abstraction have de-materialized water and limited it to a metaphor. She also suggests that "de-materialized" and "deterritorialized" water "denies the reality of local, specific human-environmental

[10] Compare Bruno Latour, who speaks of a river as an agency in "geostory" (Latour, 2014, p. 13).

relationships" (Strang, 2004, p. 246). In Harding's fiction, the physical presence and even the material quality of local waters – such as the Lake Enon with its water "like sheer blue grass, transparent, filled with light, the lake floor lined with clean sand and smooth pebbles" (E, p. 94) – is of utmost importance to the inhabitants of the place. It is not a de-materialized and de-territorialized abstraction to them, and even though water has its metaphorical and symbolic meanings to the people of Enon, it is also an essential part of their living environment.

In *Enon*, the narrator-protagonist is Charlie Crosby, George Washington Crosby's grandson, a bookish and cultivated house painter, whom we glimpsed in *Tinkers* at the dying George's bedside. Harding's second novel is therefore a continuation of the saga of the Crosby family, and the memories and experiences of the past generations also resonate in this "sequel."[11] *Enon* is a more straightforward and more directly emotional narrative as compared to the complex and fragmentary structure of *Tinkers*, however. Still, the motif of cold water and its power of connecting people is also visible in *Enon*. The main story of the narrative focuses on Charlie's painful attempts to live through the trauma of his daughter Kate's sudden death in a car accident at the tender age of thirteen, an event that occurs at the very beginning of the narrative. The masculine narrative line already viable in *Tinkers* – which, down and deep, is about sons and fathers – is especially predominant in *Enon*. The narrative asks the reader to sympathize with the weeping, sweating, and drinking Charlie, who mourns after her dead daughter. It feels like Kate becomes more meaningful to Charlie because of her death – even so that the narrative's focus on Charlie's individual drama requires that Kate dies. The tragic accident results in Charlie's divorce from his wife Susan, who is represented in the text as a somewhat cold and unforgiving person and who has moved back to the West with her parents and sisters ("gigantic Finns from Minnesota" [E, p. 12]). Charlie is somewhere in deep waters of grief after the tragic accident, consuming considerable amounts or alcohol and painkillers, having a cloth soaked in cold water on his forehead, and trying to hold on to life with his memories of Kate:

> I surfaced into consciousness four hours later, sweating and parched. I rose and lurched to the bathroom and ran the cold water tap in the sink until the tepid water in the pipes cleared and the chilled water from underground poured out. I filled the red plastic cup Kate had used for rinsing her mouth when she brushed her teeth and gulped the water down and filled the cup again. I stood for a moment in the dark (E, p. 40).

Here, "surfacing" and "sweating" are connected to the watery experience in Charlie's everyday life. Tap water used for daily activities is also something that keeps Kate alive in his consciousness, as in Charlie's evocative image of Kate's red plastic cup. Later in the narrative, Charlie is weeping in the midst of pools of rainwater:

---

[11] The reader of *Enon* is expected to have read *Tinkers* as well. There are some self-reflexive allusions to the earlier novel in the narrative: "I could tell her and bring up the story about my grandfather. (Kate said once, 'I never met Gramps, but you talk about him so much I feel like he's somebody I know')" (E, p. 5).

> The stark blue sky and the churning, retreating clouds and the cascading sun and the bright green grass and livid blond pith wood gleaming from the broken ends of fallen limbs and the wounds in the sides of the maple trees and the silvery-gray clear rainwater collected into a wide pool in the middle of the backyard corrugating in the wind were all overwhelmingly beautiful and I smiled it at all and sat down in the soaking muddy grass and wept (E, p. 208).

Kate was keen to biking alongside the banks of the Enon River and the shores of the Lake Enon, and it is as if water connected Charlie to his daughter in almost mysterious ways. Still, Charlie now regrets his telling Kate that the lake's "clean" and "pure" water (E, p. 95) is something healthy and life giving, as it has been in the local Puritan tradition, in which water had a spiritual meaning of purification (baptism being one of the two accepted sacraments in Puritanism). Here, as in *Tinkers*, the real, material water does not lend itself to human hopes and beliefs. Instead of these kinds of idealistic conceptions, it remains a natural force and phenomenon indifferent to human strivings. Something, however, draws Charlie to the shores of the Lake Enon, where he senses his connection to water as well as to his daughter. He feels that "[t]he water was cold and clean-tasting, mineral" (E, p. 46).

It may be argued that in the metaphorical conception of "steam of consciousness" water is not necessarily cold and mineral but rather flowing and floating, without strongly haptic and material qualities. In Harding's fiction, it is especially the material presence and sense of cold water, however, that enables the characters' self-perception:

> I half-sat against the rock, gulping breaths, and looked out across Enon Lake. The water near the shore was like sheer blue grass, transparent, filled with light, the lake floor lined with clean sand and smooth pebbles. Breezes etched themselves across the surface farther out, toward the center. I saw my reflection in the water and it angered and embarrassed me. I looked just the way I imagined I would: closer to middle-aged than I wanted to admit, a little heavy in the chops, sweaty, winded, my hair wet around the edges, the rest stood up by the breeze and salt in my sweat (E, p. 94).

Here, the cold lake water serves as a mirror in which the character can see he is sweating (or, in another context, weeping), being a watery creature himself. For the narrator of *Enon*, nature – including water with its reflecting surfaces – is a place to feel connection to his daughter, just as in *Tinkers* water is a medium bringing together three generations of men, George, Howard, and Howard's father.

In *Enon*, the narrative of which is located in Massachusetts, the narrator Charlie Crosby feels "autumn frosts" and "the heave of winter winds and piling snow" (E, p. 66). Once again, as in *Tinkers*, the harsh autumn and winter weather produce a kind of bleakness in the characters' minds. Charlie has the intention of drowning himself in the cold water of the Lake Enon, because he senses his life to be empty and meaningless after Kate's death. The coldness of water – and Charlie's raw experience of cold water through his immersion in it – is, however, also purifying, if not necessarily in the religious sense: "The water was cold and pure and clean. It washed my filthy hands and my filthy face and my filthy hair. [...] I could practically hear the water hiss as I immersed myself in it. [...] I ducked under the surface and sank into the cold quiet water" (E, p. 218).[12] It is through this extreme experience that Kate's presence – and even her reassuring voice ("'It's okay, Dad'") – becomes vivid in Charlie's memory once again. Charlie realizes something profound about the meaning of Kate's life "under the cold water" (E, p. 219). Again, as in *Tinkers*, cold water is not necessarily "healing," but, instead, "water's mercies were brief" (E, p. 219). The human immersion in the non-human nature is not a straightforward business, which would make the real, cold water a warm, human element: "The foreign, submarine world suddenly alarmed me" (E, p. 219). There is no room for human consciousness in the alien realm of water. The harsh experience of cold water at least makes Charlie's thinking sharp and alert to the possibilities of life, including his sense that Kate is still living in his memories and always will be. Charlie grasps life again with all its difficulties until – as the last words of his narrative have it – he simply ceases together with his daughter and "there isn't a soul left in Enon or anywhere else on this awful miracle of a planet to remember either of us" (E, p. 238).

Whatever its real, material presence, water also remains a mysteriously non-human realm in Harding's fiction. This mystical side has something to do with the history of New England and its Puritan origins in Harding's literary imagination. In *Enon*, Charlie Crosby senses the presence of dead people who have been "exiled in an obscure dead water of time, the sort of which Enon is full" (E, p. 45). During his hallucinatory walks in the woods, after the death of his daughter, he even has an imaginary meeting with Sarah Good, a young woman who was hanged as a witch in the infamous Salem trials in 1692 (see Schiff, 2015, pp. 233–238). The colonial and Puritan history of New England – alongside the region's woods and waters – still constitutes human minds and experiences in Harding's fiction. Charlie explores, at least in his mind, the "very old unmarked graves" that the woods of Enon are full of; these are the graves of "anonymous cats and owls, Puritans and Indians, and unnamed infants, getting their bones mixed in the currents of soil and groundwater" (E, p. 83).[13] Indeed, Charlie is interested in the history of Enon, a small town built on the shores of the Lake Enon and the banks of Enon River, reading "the topology of Enon with my fingertips like Braille, tracing brown hills, aqua swamps, curved blue lattices of streams and rivers, the bright moss green of its meadows" (E, p. 75). It is as if he were keeping the human and natural history of Enon alive in his memory – a memory, which always goes back to his daughter Kate, constantly present in

[12] As Bachelard argues in his phenomenologically and psychoanalytically oriented approach to water and imagination, an "immersion" into water and the experience of breaking through the surface of water is reminiscent of our birth. Breaking the surface of the water is also an experience of calm or an experience of panic depending on the individual; in any case, "a leap into water is a leap into the unknown" (Bachelard, 1983, p. 165).

[13] In *Tinkers*, George also imagines his father Howard's encounters in the Maine woods "with the Indians who still linger at the edge of this vanishing wilderness" (Parini, 2004). Native Americans are here seen, perhaps romantically, as having mystical connections to nature (see, e.g., Schweninger, 2008, pp. 2–4).

her absence. In this way, Kate becomes an integral part of history and nature, cold water functioning as a medium through which Charlie may connect to her.

12.5 Conclusion

My analysis of Paul Harding's novels has suggested that human minds function in a close connection to natural environments. More specifically, I have argued that coming to terms with portrayals of fictional minds in natural environments in these narratives requires integrating ecocritical or new materialist perspectives on nonhuman nature with aesthetic or narratological research on consciousness presentation. Therefore, my reading has suggested that we should also take into account the significance of the "stream" (or, more generally, water) in the literary theoretical concept of stream of consciousness. Indeed, the idea of water – its complex material-symbolic entanglements – is what connects Harding's fiction to the Puritan and Transcendentalist tradition of New England literature, while his fiction's naturalist and materialist emphasis simultaneously goes beyond that tradition. Therefore, my reading of Harding's two novels, *Tinkers* and *Enon*, has foregrounded the physical and material presence of water in addition to its symbolic and spiritual meanings.

Even though *Tinkers* and *Enon* are relatively short novels, they are open for several and differing readings; here, in the context of this volume, I have mainly focused on the significance of cold water in the narratives. Natural phenomena are, as if unavoidably, refracted through human consciousness in the novels, and while nonhuman nature is shown to be cruel and indifferent to human strivings, the characters in the narrative try also to find some meaning (and healing) to their vulnerable lives from the harsh beauty of nature. Cold water, especially, functions as a portal to the past and to the dead, connecting the main character of *Tinkers*, George, to his father Howard, as well as helping the narrator of *Enon*, Charlie, to reminiscent his daughter Kate. "Water," "father," and "daughter" are therefore linked together through cold water in experiential and mysterious ways in Paul Harding's fiction.

References

Alaimo, S. (2012). States of suspension: Trans-corporeality at sea. *ISLE, 19*(3), 476–493. https://doi.org/10.1093/isle/iss068

Asselineau, R. (1980). *The transcendentalist constant in American literature*. New York University Press.

Bachelard, G. (1983). *Water and dreams: An essay on the imagination of matter* (trans. E. R. Farrell). The Dallas Institute of Humanities and Culture. (Original work published 1942).

Best, S., & Marcus, S. (2009). Surface reading: An introduction. *Representations, 108*(1), 1–21. https://doi.org/10.1525/rep.2009.108.1.1

Bouchard, L. D. (2014). Belief, revelation, and trust: Faith and the mind's margins in Ian McEwan's *Saturday* and Paul Harding's *tinkers*. *Christianity and Literature, 63*(4), 449–465. https://doi.org/10.1177/2F014833311406300406

Capeloa Gin, I. (2008). Introduction. In I. Capeloa Gin (Ed.), *Fleeting, floating, flowing: Water writing and modernity* (pp. 9–15). Königshausen & Neumann.
Caracciolo, M. (2020). We-narratives and the challenges of nonhuman collectives. *Style, 54*(1), 86–97. https://doi.org/10.5325/style.54.1.0086
Çirakli, M. Z. (2014). Parallel narrative structure in Paul Harding's *tinkers*. *ALLS: Advances in Language and Literary Studies, 5*(5), 227–232. https://doi.org/10.7575/aiac.alls.v.5n.5p.227
Cohn, D. (1978). *Transparent minds: Narrative modes for presenting consciousness in fiction*. Princeton University Press.
Costello, B. (2003). *Shifting ground: Reinventing landscape in modern American poetry*. Harvard University Press.
Costlow, J., Haila, Y., & Rosenholm, A. (Eds.). (2017). *Water in social imagination: From technological optimism to contemporary environmentalism*. Brill/Rodopi.
Easterlin, N. (2012). *A biocultural approach to literary theory and interpretation*. The Johns Hopkins University Press.
Emerson, R. W. (2003). In L. Ziff (Ed.), *Nature and selected essays*. Penguin books.
Gale, R. M. (2010). The naturalism of John Dewey. In M. Cochran (Ed.), *The Cambridge companion to Dewey* (pp. 55–79). Cambridge University Press.
Gilcrest, D. W. (2002). *Greening the lyre: Environmental poetics and ethics*. University of Nevada Press.
Harding, P. (2010). *Tinkers*. Windmill Books.
Harding, P. (2014). *Enon*. Windmill Books.
Henry, M. (2020). Consciousness, information, electromagnetism and water. *Substantia, 4*(1), 23–36. https://doi.org/10.13128/Substantia-645
Hogan, P. C. (2011). *Affective narratology: The emotional structure of stories*. University of Nebraska Press.
Humphrey, R. (1954). *Stream of consciousness in the modern novel*. University of California Press.
James, W. (1950). *The principles of psychology* (Vol. 1). Dover Publications, Inc..
Jarvis, B. (1998). *Postmodern cartographies: The geographical imagination in contemporary American culture*. St. Martin's Press.
Kvernbekk, T., & Frimansson, G. H. (2013). Narrative: A brief introduction. *Scandinavian Journal of Educational Research, 57*(6), 571–573. https://doi.org/10.1080/00313831.2013.798836
Latour, B. (2014). Agency at the time of the Anthropocene. *New Literary History, 45*, 1–18.
Lehtimäki, M., Rosenholm, A., & Strukov, V. (Eds.). (2021). *Visual representations of the Arctic: Imagining shimmering worlds in culture, literature and politics*. Routledge.
Lopez, B. (1999). *Arctic dreams: Imagination and desire in a northern landscape*. The Harvill Press.
Lundin, R. (2005). *From nature to experience: The American search for cultural authority*. Rowman & Littlefield Publishers.
McMillin, T. S. (2011). *The meaning of rivers: Flow and reflection in American literature*. University of Iowa Press.
Parini, J. (2004). *Tinkers* by Paul Harding. *The Guardian*.
Person, L. S. (2008). Transparent-eyeballs: Water and ecological subjectivity in nineteenth- century American fiction. In M. Devine & C. Grewe-Volpp (Eds.), *Words on water: Literary and cultural representations* (pp. 147–158). WVT.
Pinsky, R. (1976). *The situation of poetry: Contemporary poetry and its traditions*. Princeton University Press.
Polger, T. W. (2004). *Natural minds*. The MIT Press.
Porter, C. (1981). *Seeing & being: The plight of the participant observer in Emerson, James, Adams, and Faulkner*. Wesleyan University Press.
Ruland, R., & Bradbury, M. (1991). *From puritanism to postmodernism: A history of American literature*. Penguin Books.
Schiff, S. (2015). *The witches: Salem, 1692*. Little, Brown.
Schweninger, L. (2008). *Listening to the land: Native American literary responses to the landscape*. The University of Georgia Press.

Stevens, W. (2015). In J. N. Serio & C. Beyers (Eds.), *The collected poems of Wallace Stevens*. Vintage Books.
Strang, V. (2004). The meaning of water. .
Sykes, R. (2018). *The quiet contemporary American novel*. Manchester University Press.
Westbrook, P. D. (1988). *A literary history of New England*. Lehigh University Press.
Westling, L. (1996). *The green breast of the New World: Landscape, gender, and American fiction*. University of Georgia Press.

Chapter 13
Speculative Water: Atopic Space and Oceanic Agency in Julie Bertagna's Raging Earth Trilogy

Heidi Hansson and Maria Lindgren Leavenworth

Abstract In this chapter, we take atopias as a conceptual starting point in examinations of Julie Bertagna's speculative Young Adult novels *Exodus* (2002), *Zenith* (2003), and *Aurora* (2011), demonstrating how imaginary and unstable spaces through various processes are rendered un-atopic. In speculative or non-mimetic fiction, the reader is presented with alternatives to ordinary reality, yet texts are grounded in already existing situations, relations and circumstances. In the examined novels, contemporary climate change and attendant anxieties are extrapolated in the depiction of a world covered by water: an extended atopic space. The Arctic fills a specific function in the novels *because* of climate change and Greenland in particular emerges as a possible future home for the surviving humans. The atopic space that the inhospitable and inaccessible Arctic traditionally constitutes, is transformed into space available for new forms of colonization. In earlier examples of speculative fiction set in the Arctic, open water replaces ice in fantasies about conquest or discovery, and there is a large body of fiction connected to flooding caused by climate change. By situating the trilogy in these literary contexts, we demonstrate how both water and the Arctic have been perceived of as unstable, atopic spaces, particularly vulnerable to anthropogenic change.

Keywords Atopia · Arctic · Julie Bertagna · Speculative fiction

"*Imagine the vast, drowned ruin of a world washed clean.*"[1] This appeal to the reader prefaces Julie Bertagna's *Exodus* (Bertagna, 2002), the first installment in her Raging Earth-trilogy. The reading instruction conjures up an aquascape in which effects of climate change have profoundly altered the world. Boundaries between the oceans of the world have been obliterated, merging cold waters with warm,

[1] All formatting in quotations corresponds to the original.

H. Hansson (✉) · M. Lindgren Leavenworth
Umeå University, Umeå, Sweden
e-mail: heidi.hansson@umu.se; maria.lindgren@umu.se

M. Lehtimäki et al. (eds.), *Cold Waters*, Springer Polar Sciences,
https://doi.org/10.1007/978-3-031-10149-6_13

human-made structures and objects are unreachable or unusable, and social, cultural, and economic models are destabilized. Bertagna's trilogy is an effective example of how speculative fiction, which depicts "modes of being that contrast with [...] audiences' understanding of ordinary reality" (Gill, 2013, pp. 72–73), works to depict radical consequences of anthropogenic climate change, that is, the result of human impact on the ecosystems and geology of the earth. In contemporary debates, the northern regions, in particular the melting polar ice cap and rising sea levels, often function as barometers gauging this impact. Speculative climate fiction, or cli-fi, that extrapolate these and other effects of a warming climate thus enter into a dialogue with scientific, ideological and educational discourses, and since it originates in real-world anxieties and aspirations yet remains unrestricted by everyday limitations, the imagined futures can inspire both caution and hope.

However, because of its temporal and global scope, climate change offers significant "representational and imaginative challenges" and is sometimes "deemed unnarratable" (Bracke, 2019, p. 280). As a geological age, the Anthropocene is simply too vast, its implications too far-reaching to be fully conceptualized. Yet, it may be precisely through fiction that readers can explore the reality of living in times of accelerated change. Thus, Astrid Bracke suggests that fiction constitutes "a particularly productive space in which to imagine and think through climate crisis" (Bracke, 2019, p. 280), and Kathryn Yusoff and Jennifer Gabrys argue that speculative (science) fiction should be understood as a "cultural meditation on risk" (Yusoff & Gabrys, 2011, p. 521). By encountering worlds devastated by ice, draughts or floods, readers can imaginatively engage with contemporary risk scenarios in the safe space of fictional speculation.[2] In this way literature becomes "the necessary non-place, or place without the limitations of place, for thinking pasts and futures that are literally uninhabitable" (Hollywood, 2020, p. 33). Place, dwelling and the conditions for human habitation are fundamental questions in speculative climate fiction, not least, or perhaps particularly when the environmental apocalypse features flooding.

Targeting a Young Adult audience, Bertagna's novels provide a particularly rich source for imagining transformations of home into what Siobhan Carroll terms atopias: "'real' natural regions [...] which, because of their intangibility, inhospitality, or inaccessibility, cannot be converted into the locations of affective habitation known as 'place'" (Carroll, 2015, p. 6). In contemporary cli-fi the threat of rising water carries special significance in this process of transforming liveable space into atopia, because of its long association with ends and beginnings in Western thought, and its status as the before and after of the ordered world. As Christopher Connery demonstrates in his historical overview of oceanic tropes, Christian creation myths figure the sea as a site of "disorder and chaos" that always threatens to return to its "omnipresent and undifferentiated" condition (2006, p. 499). The creation moment that turns the world into an inhabitable place is the separation of water from land, but maintaining this state requires a constant "divine and human" vigilance

[2] For an extended discussion about how Bertagna's trilogy exemplifies possibilities of imaginatively exploring issues connected to real-life climate change and (un)sustainability, see Lindgren Leavenworth and Manni (2021).

(Connery, 2006, p. 503). In speculative cli-fi, vigilance in the form of collective environmental responsibility has lapsed, annulling the separation. The dissolution of boundaries and the offer of alternative epistemologies are attractive properties for authors seeking to depict the establishment of new societies and envision alternative systems of thought and action.

The Biblical flood in Genesis 6–9, which returns Earth to a pre-creation state to erase human sin and iniquity, finds a fictional and secular counterpart in *Exodus* and the following novels *Zenith* (2007) and *Aurora* (2011). As an effect of the environmental arrogance of previous generations, the twenty-second-century world is covered by water, which both opens for and necessitates new ways of life. In this water-world, permanent human settlement is precluded or at least incredibly fraught. What used to be mountaintops are shrinking islands, and humans survive aboard ships or in tower cities above the surface of the "world's ocean" (Bertagna, 2007, p. 5). In the far north, Greenland both literally and figuratively emerges as a hope for future life: having been weighed down by glaciers, the melting polar ice makes the enormous island "*bob up like a cork*" and the uncovered land becomes available for new forms of colonization (Bertagna, 2002, p. 178).

In the broad genre of Young Adult (YA) cli-fi, Bertagna's trilogy can be related to two subsets in particular: fictions about flooding and novels in which the Arctic features as a barometer for climate change. Among the former are Marcus Sedgwick's *Floodland* (2000) which, like Bertagna's novels, follows a young girl's search for habitable land in a waterlogged world of isolated islands, Jordana Frankel's *The Ward* (2013) depicting a future Manhattan ravaged by disease in the wake of global flooding and Rebecca Roanhorse's fantasy novel *Trail of Lightning* (2018), in which post-deluge earth is repopulated with gods and monsters from Navajo legend. Examples of YA cli-fi set in the Arctic include Ann Halam's *Siberia* (2005), delineating 13-year-old Sloe's attempts to save animals from complete extinction, Rebecca Stead's *First Light* (2007), in which a community below Greenland's surface are threatened by polar melt, and Cherie Dimaline's *The Marrow Thieves* (2017), in which climate change is a contributing factor to why people can no longer dream. Presenting the opposite trajectory to the movement in Bertagna's novels, the journey south is depicted in in Robert Mann's *Where the Ice Never Melts* (2011) starting in Nunavut and in Paul Greci's *The Wild Lands* (2019) starting in Alaska. Most of these novels relate to the quest tradition, with the significant difference that the conventional, circular form of the quest narrative is arrested since a return home is not possible (Hammer, 2010, p. 48).

The imaginary futures in speculative climate fiction invites readings where the relationship between humans and the environment is reversed and features of the natural world gain new agency. While the focus remains on the experiencing human or at least an experiencing consciousness, the designation climate (change) fiction means that the environment cannot only be the backdrop for characters and plot. The Earth itself takes on the properties of an actor and natural phenomena become what Bruno Latour terms agents "of our common *geostory*" (Latour, 2014, p. 3). This reversal of agency is emphasized through the title of the first section of book one in Bertagna's trilogy: "Savage Earth," underscoring activity and attack as opposed to the victimisation implied by the phrase "savaged Earth." Although the rising sea levels that set off the action are an effect of human-caused climate crisis and in this sense passive and mindless, they actively change the surface of the earth, if not deliberately, at least according to the nature of water. The power of the floods undercuts environmental hierarchies where land as the dwelling-place for humans is more important than water bodies, which negates or at least questions humanity's hegemony over the environment.

With a focus on the protagonist Mara, who has grown up on the diminishing island of Wing in the North Atlantic, and the "gypsea" Tuck, who has never lived with solid land beneath his feet, we will discuss the opposing themes of flooding and dwelling in Bertagna's novels. Central concepts in our analyses are the ideas of oceanic agency, on the one hand, and on the other, the notion of atopia. Carroll's concern is with how eighteenth- and nineteenth-century "[a]topias become visible" as expanded knowledge changes them from unknown or imaginary into being "vulnerable to the logic of state possession and control" (Carroll, 2015, p. 8). Thus, she charts how atopic regions are understood in new ways when they are identified as physically reachable, but still inhospitable to permanent human dwelling. This understanding of atopias undergird our discussions about the fictional future's representations of the ocean and the Arctic, atopias that are closely correlated also in Carroll's discussions. The former is perceived of as "immune to enclosure," the latter as "immune to permanent settlement": both are figured as blank spaces that repel human attempts at inscription and control (Carroll, 2015, pp. 72, 20). Based on imperialist and colonialist thought, such figurations are both anthropocentric and continent-oriented, however, and illustrate that "while the atopic nature of a space may be informed by its material conditions, atopias are cultural constructs" (Carroll, 2015, p. 7). In Bertagna's novels, consequently, the water-covered world for Mara represents an extended atopia where the power balance between ocean and continent is overturned, whereas for Tuck, ice-free Greenland constitutes atopia. As in environmental disaster novels broadly, these "protagonists must cultivate a relation to the destroyed and changed landscape in order to survive" (Yusoff & Gabrys, 2011, p. 521), but in diametrically different ways. In the trilogy, adjustment to the postdiluvian world is imagined as the parallel processes of adapting to atopic spaces created by the rising ocean and finding liveable space in what used to be understood as an ice-bound atopia.

13.1 Mara in Water

Given the current melting of the polar ice caps, cold water flooding has become an all too realistic consequence of anthropogenic climate change. While Bertagna's trilogy depicts these consequences in a dystopian way, it can also be understood as a rejection of the geographical and geopolitical models privileging land masses. The novels are not only an allegory of global warming and a call to action, but a reminder that Earth is a watery planet and that water is an uncontrollable element. Read from an oceanic point of view, Bertagna's story "repositions continental land as circumscribed, minimized, and mere island amid the waters that dominate the globe" (Blum, 2015, p. 30). In the trilogy, this act of repositioning is literal, with islands constituting the only remaining but imminently threatened solid ground.

As the title indicates, *Exodus* begins at the place of leaving, the island of Wing in the North Atlantic Sea made even colder by the arctic melt. The island is thus in the process of transforming from home-place to atopia after the global flooding. The protagonist Mara Bell is from birth used to an unstable environment with hot summers followed by rising water levels caused by "another great meltdown of the ice at the poles" (Bertagna, 2002, p. 13). Possibilities to survive on Wing continuously decrease in a process that emphasises the fragility of human existence and when the "ocean takes a hungry gulp" (Bertagna, 2002, p. 4), the active, encroaching water literally shrinks the habitable world to the formerly isolated but now crowded mountaintops. The century of climate change has also resulted in a lack of communication between groups of survivors. Corresponding to the shrinking physical space, points of contact are focused on the immediate context and after finding that the ocean has engulfed all the islands within reach, the people on Wing have become unable "to think of anything beyond ourselves and our little patch of the world" (Bertagna, 2002, p. 13). To Mara, an unusable phone box on an elevation becomes a symbol of physical and communicative isolation as well as a marker of the rising water levels.

The opening of *Exodus* thus highlights the contrast between habitation and atopia. What distinguishes Mara from her fellow islanders is her need for outside contacts and her ability to inhabit the uninhabitable. The stationary phone box technology has become useless in the flooded world, but Mara has access to a solar-powered "cyberwizz" that allows her to temporarily dwell in the Weave, a mostly abandoned version of the internet (Bertagna, 2002, p. 26). She finds out about New Mungo, a tower city rising above submerged Glasgow, through a chance cyber-meeting with David, a fellow Weave surfer who turns out to be the grandson of the city's founder. The realisation that there might be somewhere to go means that Mara is able to relate to the sea not only as a divider and a threat, as it has become for the island dwellers, but to what Connery refers to as a connector and a possibility (Connery, 2006, p. 496). When a storm hits Wing and the waters reach the phone box this functions as the definitive incentive for Mara to leave the island. One of her final acts is to reach into the phone box and dial the emergency number (Bertagna, 2002, p. 54) in a gesture that draws attention to the futility of old technology in a world where natural forces rule. The permanence of land is superseded by the atopias of water and cyberspace.

Mara actively encourages her fellow islanders to leave Wing in search of New Mungo, but during the arduous journey she is separated from her family and many lives are lost. The ocean is a "ferocious, swallowing beast" (Bertagna, 2002, p. 64), and Bertagna's representation aligns with new developments in hydro-criticism where "an animated ocean has come into being as 'wet matter' rather than inert backdrop" (DeLoughrey, 2019, p. 22). Matching how the land is submerged under the water, the sea takes on the properties of land, with the waves figured as walls, mountains, valleys, and cliff-faces (Bertagna, 2002, p. 83). During the passage, the sea is in charge, causing seasickness, exposing the boat travellers to storms and heaving waves and depriving them of rest. Finally arriving, the Wing evacuees are kept out of New Mungo by the city's sea police and forced to join the cluster of boat refugees outside its walls. Around the artificial construction even the products of the sea are contaminated, with toxic algae and fish causing further suffering and death (Bertagna, 2002, p. 73) and "diseased drinking water" (Bertagna, 2002, p. 333) eventually killing most of the refugees from Wing. As an inversion of the cleansing function of water, the toxicity is particularly sinister and Bertagna attaches to a long tradition in literature where swamps, sulphurous pools or acid rain and its equivalents are prominent in the geography of hell. Thus, both the physical conditions and the temporary and involuntary social organization of the refugee camp epitomize maritime atopia in its most menacing form. As in the nautical narratives analysed by Carroll, Bertagna imagines the sea surrounding the elevated city as "an exceptional form of space opposing the stability and social order of land" (Carroll, 2015, p. 78). The disordered nature of the ocean, or what Carroll terms its "atopic fluidity" (Carroll, 2015, p. 76) means that it cannot be effectively patrolled however, and Mara eventually manages to enter the city by diving in and allowing herself to be transported by the sea, "caught in the churning foam" of the wake of a ship (Bertagna, 2002, p. 94).

New Mungo has been created in response to the flooding and is built on top of Glasgow's medieval cathedral church.[3] This draws attention to the break in ideological and/or spiritual continuity caused by the present flood. The heart of New Mungo is the Cybercath, an inversion of the ancient stone cathedral "created out of light and air" (Bertagna, 2002, p. 222), but in contrast to the symbolic transcendence suggested by the old church, its tower city counterpart epitomises consumerism, as an enormous financial centre where cybertraders are the most regarded citizens (Bertagna, 2002, p. 224). The reliance on cyber-technology turns the city into atopia, the main difference from the island of Wing being that the tower city atopia is not only human-caused but actually human-built.

Originally constructed according to principles of "natural engineering" New Mungo and its sister-cities are initially environmentally sustainable, powered "by sun, wind, and waves" (Bertagna, 2011, p. 48). The linked cities however transform into a sinister empire where the hierarchical tower construction enforces a separation of those deemed worthy to survive from refugees and outcasts. For the tower citizens, the water-dwellers as well as the water itself constitute nature's threat to ordered and controlled existence, and they both figuratively and literally look down at the masses of boats denied entrance. Thus, New Mungo can be seen as a "mimetic" representation of contemporary "distancing strategies" between affluent cultures

[3] New Mungo is engineered by Caledon Stone and stewarded by his son Mungo Stone. The symbolic names place the trilogy's foundations in real-life Scotland – Caledonia – and together with scattered references to Christian and pre-Christian traditions, they emphasise its spiritual dimensions. Caledon is referred to as the "*creator of the New World. The Grand Father of All*" (Bertagna, 2002, p. 263), and the name of both his son and the city recalls Glasgow's patron saint.

and those that "will be the first to feel the effects of global warming" (Hammer, 2010, p. 43). Like Wing, the city is physically circumscribed, but unlike the island, it is entirely artificial: entertained by endless virtual displays and surrounded by likewise virtual nature, the tower residents are kept uninformed of the catastrophe beneath them and unexposed to the natural world. The citizens' non-attachment to nature leads to a lack of connection with their immediate context, that, in turn, results in a "lack of civic engagement and personal growth" (Ostry, 2013, p. 102). Their engineered, surveilled and detached existence functions as a marked contrast to the choices made by Mara and the people who follow her across the waters.

Beneath the towers is the archipelago of the Netherworld where the roofs of Glasgow's tallest buildings form islets that are temporary homes for communities that in different ways have adapted or been adapted to the new circumstances. One group of children, in a double entendre referred to as urchins, live partly on the roofs of Glasgow's highest buildings, partly in the water. They have gills and webbed hands and feet, but no language and no names. In the last novel it becomes clear that the urchins are the result of a scientific "Amphibian Experiment" designed to transition humankind into a maritime existence (Bertagna, 2011, p. 50), but they also resemble Charles Kingsley's water babies (1863), drowned children who live underwater while they undergo physical and spiritual cleansing. Whether the human adjustment should be seen as a form of displaced punishment for the environmental disaster, following the Kingsley tradition, or as an attempt to counteract its effects through bio-engineering, in line with the principles of speculative fiction, water is imagined as the element of rebirth. The urchins' counterpart are the Treenesters who consciously adapt by instead avoiding the water and making their home in the sparse canopies in trees growing on top of the submerged buildings, where they tell stories about the destruction of the world, or "the age of tree crime" (Bertagna, 2002, p. 173). Whereas the lives of the New Mungo citizens represent the efforts to control the environment through geo-engineering and a replacement of the natural world, the lives of the urchins and the Treenesters thus exemplify possible adjustments through bio-engineering and biological adaptation.

Places in the water cannot be named other than through their landed boundaries, which is one reason why historically, "the ocean not only resisted conversion into place, but also retained a degree of epistemological uncertainty" (Carroll, 2015, p. 78). In a rapidly flooding world, processes of naming in Bertagna's novels illustrate a wish to remember and hold on to drowned places. The Treenesters take the names of areas in Glasgow like Candleriggs, Partick and Gorbals, thus turning themselves into "a bunch of walking placenames, the living limbs of the lost city" (Bertagna, 2002, p. 132). Imitating the Treenesters' manner of honouring the past, Mara names urchins after North Atlantic islands such as Wing, Jura and Skye and after "lost villages and hamlets when she runs out of islands" (Bertagna, 2002, p. 139). The re-used place names emphasise the importance to forge continuous links to what now lies beneath the sea and what is history, that thematically runs through the trilogy. As matter that cannot be permanently inscribed, water in all its forms is outside time and in a sense without a history, which makes it alien to the acts of remembering that Mara and the Treenesters adopt.

13.2 Atopic Greenland

The naming episodes draw attention to the lost connection to land in the flooded world, but they also reverse and critique the place-claiming methods of imperialist explorers and map-makers, and, by extension, the sense of land ownership that contributed to the environmental disaster in the novels. With the Arctic as an example, Chauncey Loomis (1971, p. 326) notes that "famous Victorian names were scattered [...] like confetti" across the polar region in the mid-nineteenth century as part of the process to transform the area from atopia to colonial possession. What the explorers perceived of as nameless, often threatening space was controlled by naming and ordered into the worldview of the exploring nation. In Bertagna's trilogy, the rising water is the advancing force which in a contrasting process obliterates signs of humanity and the fiction that geographical naming constitutes.

There is a paradox at the centre of historical Arctic colonisation since the principle of *terra nullius* makes the territory the "idealized object of an idealized imperialism," at the same time as the harsh environment imaginatively places it permanently outside colonial space (Carroll, 2015, p. 20). In her novels, Bertagna revalues the Arctic and the North, not because of their attributes of northernness and climate stabilisers, but as the new temperate zone after the climate collapse. When Mara tries to convince surviving refugees, urchins and Treenesters in New Mungo to accompany her, her argument is that "there *is* land in the north, in a place called Greenland—a huge land that was covered in ice until the meltdown which flooded the oceans. It has lots of high land and must be free from the ice now. It's the kind of place we could build a new life" (Bertagna, 2002, p. 206). Ice is figured as confining and thaw as liberating in a reversal of the value system that informs the environmental message of the trilogy and undercuts the representation of flooding as a threat. Ironically, the global melt that turns Scotland and Europe into atopias is what renders the previously atopic Arctic habitable.

Fiona McCullough argues that for Bertagna, the Arctic is anything but a coincidental choice as a destination. Despite being the "possibly unlikeliest of imaginative terrains" it offers the author a release from constraints only possible at the periphery (Bertagna, 2007, p. 82). There is also a logical movement from the atopia of the flooded world to the atopia of the Arctic in the trilogy, since just as the ocean has for centuries been regarded as a "space that no nation can truly possess" (Carroll, 2015, p. 73), "[t]he north appears as politically and culturally unencumbered, supranational space [and it leaves] room for new, malleable identities to emerge" (McCulloch, 2007, p. 82). This is naturally a fiction in itself, ignoring the heterogeneity of the Arctic, its role in national politics, and its history of displacement of indigenous peoples. What McCullough gestures to are discourses permeating both fiction and non-fiction, written by authors not indigenous to or residing permanently in the Arctic, in which the area functions as an empty stage where speculative potentials can be realized, in a fictional counterpart to the *terra nullius* doctrine of the age of exploration.

In imagining an ice-free Arctic Bertagna attaches to a long tradition of both concretely searching for and fictionally depicting a temperate or even tropical polar zone with access to both the North Pole and the Northwest Passage. The late nineteenth and early twentieth centuries saw an abundance of speculative fiction in which authors not only discovered the famed Open Polar Sea and warmer climate around the Pole, but forgotten civilizations above or below ground. As Francis Spufford remarks, there was a longstanding desire to imagine "*something* at the poles" since the reality was rather that of "an expanse of ice significant only by geographical convention" (Spufford, 1996, p. 76). If shipwreck is understood as the ultimate oceanic atopia (Carroll, 2015, p. 81), it seems appropriate that finding societies beyond the atopic Poles is frequently the result of maritime distress, often caused by malfunctioning technology (Hansson, 2018, p. 52). Thus, respectively hapless or intrepid explorers in Mary J. Bradley Lane's *Mizora: A Prophecy* (1880–1881) and William R. Bradshaw's *The Goddess of Atvatabar* (1892) encounter advanced subterranean societies after accidents at sea, which they make their home for decades and for forever, respectively. In reality, visitors' figurative or literal marks on the actual Arctic landscape were slight and impermanent: "[s]igns were torn down by the elements; discoveries were challenged by other explorers; and the histories of expeditions were lost along with the men who perished" (Carroll, 2015, p. 20). Bradshaw's and Lane's novels conform to this pattern: the Arctic is depicted as bleak, hostile and unwelcoming, and "not to be inhabited for long even in text" (Lindgren Leavenworth, 2020, p. 7). With journeys swiftly narrated, the atopic Arctic "functions in both novels as a space of transition between real and imagined" (Lindgren Leavenworth, 2020, p. 3), whereas the imagined civilizations offer permanent habitation.

Fin de siècle speculative fiction also illustrates the idea of changing arctic atopic space to habitation through human agency. In Jules Verne's *The Purchase of the North Pole* (*Sans dessous dessous*, 1899), an enormous cannon is to be fired from the top of Mount Kilimanjaro to tilt the axis of the world and increase the temperature in the far north. The ensuing melt will allow access to vast coal deposits beneath the North Pole, anticipating the actual and fictional struggles for arctic oil by a century. In Mark Twain's *The American Claimant*, enforced climate change will instead transform "the entire Arctic Circle [into] a summer resort" (Twain, 1892, p. 272). In novels like these, Steve Asselin remarks, "anthropogenic climate change is not the inadvertent effect of capitalistic ventures but the very goal of such enterprises" (Asselin, 2018, p. 443). When the earth is transformed to correspond to human desires, the Arctic becomes a commodity.

In what almost reads as a comment on such commodification of climate and place, Bertagna briefly refers to "Weatherengineering – satellite-controlled weather patterning to make the Arctic winters more temperate for the invaders" (Bertagna, 2011, p. 92). This project, however, is revealed as a hoax, designed by David to create financial havoc, and part of a series of "fairy tales" designed to this end (Bertagna, 2011, p. 92). As in many other twenty-first-century speculative novels, firmly anchored in "today's omnipresent, fundamental, multiple risk space" (Buell, 2014,

p. 277), it is anthropogenic climate change, the slow, steady human impact on the Earth's systems, rather than quick and instantly transformative ventures that is the real threat.

Somewhat problematically, the Greenland Mara and her group reach is still conceived of as primarily open to outside settlement, in a continuation of the idea of "the empty Arctic." There are initially no references to a surviving indigenous population, and the waterfront is occupied by rivalling groups whose cultural identity remains unknown. "Kalaallit Nunaat. The land of the people" with its connotations of democracy and inclusion, is replaced by a frontier world in which one group of people attempts to enslave another (Bertagna, 2002, p. 178), mirroring the more regulated enslavement in the tower cities. Leaving the dangerous beaches, Mara and her followers have to encounter the atopia of the underworld, first by escaping into empty caves, and then by crawling through an ice tunnel to the island's interior.

On the journey through the atopic space of the mountain, the group is guided by a "green wind" and led to a copse of trees by a vast body of sweet water where they begin to build their new home (Bertagna, 2007, p. 230). The worldwide extinction of trees is described as a major contribution to the floods (Bertagna, 2002, p. 295), and the saplings work as a sign that Mara and her followers can re-establish some kind of balance by rooting "themselves to the Earth: in time the land can be cultivated and transformed to a permanent home" (Bertagna, 2007, p. 323). Cultivation entails enclosure and introduces a new cycle of human domination of nature, however. The Treenesters have brought tools to work the land, and seeds which will introduce foreign species that will change the eco-system. Even for purposes of survival, and with the aim to create an ecological balance, Mara's group continues a long real-world history of appropriation of land. Furthermore, landlocked Candlewood is not a pastoral haven for all its inhabitants, in particular the urchins who have followed Mara and who no longer have use of their engineered bodies. Fifteen years later, the urchin Wing, "always … on the edges of the human world [still] tries to satisfy [his craving for] the sea. His first home" (Bertagna, 2011, pp. 283, 30). Rather than representing Candlewood and the Arctic as ideal for all settlers, Bertagna emphasises that collective adaptation is a troubled process.

The bio-engineered urchins represent a first step towards enforced adaptation, and in *Aurora*, the final instalment of the trilogy, there are references to new experiments by which the empire "wants to engineer humans designed for life in the Far North, for colonizing space" (Bertagna, 2011, p. 52). For leaders of the empire, residing in the sky cities, the Arctic has been perceived of as uninhabitable on a par with outer space, but with resources that are crucial to survival and expansion. The process of transforming atopia to home threatens to place the Arctic outside imperial reach, however, and in the quest for fuel, minerals and metals, war breaks out when air ships from the tower cities descend in the newly established settlements. The people of the coastal community Ilira are told that empire forces "have driven out the people of the Far North – Inuits [*sic*] who have lived there from time out of mind" and that the prospect of new land "risen from the White Age of ice" will cause additional dispossessions (Bertagna, 2011, pp. 181, 183). To justify the expansion, the Arctic is once again categorised as "empty [with] just a handful of savage tribes to reorganize, resettle, or where necessary, reduce – the empire's code words for exile, enslavement, extermination" (Bertagna, 2011, p. 261). The discouraging message seems to be that the temporary return of agency and power to the natural world and the ensuing human adaptation will again be replaced by imperial aspirations for domination and control, possibly ushering in a new ecological disaster that may even reach beyond the limits of the Earth.

13.3 Information Flows

The perception of Greenland as possible recourse is the result of Mara reconstituting and reclaiming knowledge, which can be seen in relation to how information broadly flows in the trilogy. Discussing representations of anthropogenic climate change in contemporary flood fictions, Bracke notes that "narratives, books, and knowledge are continually at risk of being lost" in a process "intimately tied in with the wider collapse of (Western) civilization," a disintegration which is often echoed in the fragmentation of narrative style (2019, p. 284). In Bertagna's post-apocalyptic water world, fragmentation of knowledge is apparent in the concealment of the truth about the world's destruction. It is only Mara's access to snippets of history digitally stored in the Weave that enables her to relate footage of past floods and storms to what is "happening on [Wing's] doorstep" (Bertagna, 2002, p. 47), and to recognize the lapsed responsibility and betrayal of the preceding generations via a hundred-year-old message in vain calling on "*all countries*" to immediately "*stabilize emissions of carbon dioxide*" to avoid a global ecological collapse (Bertagna, 2007, p. 206). The mainly deserted Weave, figured as a cityscape with boulevards and towering structures, is one example of what Alice Curry terms a "knowledgescape" in the trilogy, and Mara's ability to navigate it and make sense of its scattered data constitutes "the epistemological equivalent of survival of the fittest" (Curry, 2013, p. 19). Seemingly more coherent strands of information are transmitted in Noospace, a new version of the internet used to connect New Mungo with other sky cities in the world, but the system de-values and suppresses traditional sources of information. When Mara searches for "story," the Noos' return is that the meaning of the term is "[f]*alsehood. Lie*" and when she enters the search term "books," the reply is that the term is "[d]*efunct*" (Bertagna, 2002, p. 255). As part of the Empire's system of control, Noospace thus isolates the inhabitants both from their present and from history.

With both cyberspaces figured as incomplete and unreliable, Mara turns towards older sources of information, and in the Netherworld, she finds that books are not as obsolete as the Noos insist. As Curry notes, Bertagna presents the book as "a medium less susceptible to epistemic atrophy" (Curry, 2013, p. 19), but in the remains of the Glasgow University library, books are materially threatened by the ever-rising water levels. Books as artefacts are also unfamiliar to Mara and to understand and traverse this knowledgescape is a lengthy process, drawing attention to how much has been lost in the collapse and reconstitution of society. While Mara recognizes that the printed volumes have their electronic counterparts in the Weave, these have "shrunk to virtually nothing in cyberspace" (Bertagna, 2002, p. 153) and have paled in comparison with her adventurous zipping through the deserted virtual avenues and the fast-paced, albeit fragmented, information retrieval. When Mara opens a book in the library, "the words are settled and calm... Yet out of their stillness emerges a story" (Bertagna, 2002, p. 153). The story revolves around the northern "*Athapascans* [who] *have co-existed in fine balance with the land and its animals*" and introduces to Mara the idea of a habitable place "at the very top of the word" (Bertagna, 2002, pp. 154, 155). Water has already destroyed part of the page, however, and leaves her with only the vague notion that the vast Arctic is the place she needs to find. In her continued search, even the material space of the library underlines lost knowledge, as the categorization of sections carry little meaning: "Philosophy? What could that be?" (Bertagna, 2002, p. 176). It is only by chance that Gorbals, drawn to a name suggesting "a land of trees," opens a volume and finds that the Arctic is a continent and that Greenland, "strangely [...] not green at all" is located there (Bertagna, 2002, pp. 177, 178). In this way, scraps and fragments come together to form a more coherent whole, which increasingly tugs Mara towards knowledge and stories that are almost, but not entirely, lost under water.

This turn to past knowledge and to the concrete practice of reading makes natural phenomena and movements increasingly legible on the journey north. Observations of "the September geese, flocking through hidden corridors in the sky" as well as of "the narwhals, navigating the icebergs, horns thrust at the North Star" reveal alternatives to land-based roads and thoroughfares and suggest intuitive ways of steering a course (Bertagna, 2007, p. 95). Becoming more attuned to natural patterns not only forges links to the Athapascans whose way of life is to be emulated, but connects Mara more strongly to the Treenesters who actively remember the destruction of the world. The stories they relay, like the drawings and writings on the cave walls they find while attempting to reach Greenland's interior, provide a non-interrupted knowledge flow. As Carroll notes, subterranean atopias' "association with human histories lost or hidden to the sovereign gaze" commonly function as imaginative settings for temporary acts of resistance (Carroll, 2015, p. 148). The preserved narratives, a form of protest against the suppression truth, solidify Mara's belief that a continued transmission of stories is crucial to prevent history from repeating itself.

The stories about the global collapse have consequently been preserved only in places that just allow for temporary visits—the atopias of submerged Glasgow, the underground caves, or the abandoned cyberspace of the Weave. While Mara primarily reconnects with what Connery terms "the terrestrial character of knowing" or a grounded kind of knowledge based on physical traces, David exemplifies the "cyber-body, de-materialized, networked, and connected" and "the latest version of earth knowing that supersedes the sea" (Connery, 2006, p. 508). Both transform their knowledge into stories to galvanise action. Mara encourages a continued practice of storytelling to avoid a future lapse of environmental responsibility, whereas David marshals the story fragments in the Weave to spearhead a resistance movement in New Mungo. The uprising starts in Noospace and becomes a "global Surge" when stories of the past are transmitted through "the global soundwave" of old radio

(Bertagna, 2011, pp. 88, 139). Bracke notes that in the fictions she examines, "floods seep into the very works themselves, breaking apart language and eroding narrative coherence" (2019, p. 283). In Bertagna's novels, and *Aurora* in particular, water instead provides narrative coherence. Aquatic imagery based both on the idea of water as agent and on the experience of living in a water-world emphasise how humanity's survival is contingent on an undammed access to information. In the guise of the "Midnight Storyteller" (Bertagna, 2011, p. 143) David encourages people to remember the past and come together, and the stories become "lifebelts in a storm" at sea, and "an exotic harbour where frenzied Nooworlders come to anchor" (Bertagna, 2011, p. 143). Finally, the surge of discontent and protest is as unstoppable as flooding water, "breaking through [the] walls" of the empire's plans (Bertagna, 2011, p. 245).

13.4 Tuck on Land

Mara's voyage from New Mungo to Greenland on the ship *Arkiel* is a frightening phase, with the ship a small, mobile atopia that offers only precarious safety. At the same time, the preoccupation with finding land circumvents the material presence of the ocean by causing natural phenomena to be interpreted as human-made structures: the aurora borealis is perceived of as a city in the sky with "towers, the great arch of a bridge, pillars" and icebergs appear as "strange, wrecked castles of an icy realm [whose] collapsed towers, spires and arches are an uncanny echo of the drowned city's ruined cathedral and the university tower" (Bertagna, 2007, pp. 82, 89). The illusions of buildings recall the attempts by early Arctic explorers to claim space by creating replicas of London landmarks in the ice (Craciun, 2010, pp. 698–699), with the important difference that in Bertagna's fiction the builders are the air and ice and not the humans. The obliteration of differences between earth, sea and ice is further emphasised when Greenland is apprehended as a great, ominous "tidal wave" on the horizon (Bertagna, 2007, p. 98).

The description of Mara's temporary oceanic existence in *Zenith* alternates with chapters focalising Tuck who was born on a ship and grew up in Pomperoy, a floating city of interlinked vessels anchored to a deep-sea rig. Constantly shifting in relation to its oceanic surroundings, Pomperoy is the epitome of atopia, in this regard coinciding with how Michel Foucault defines heterotopia. According to Foucault, the ship is the "heterotopia *par excellence*" since a "boat is a floating piece of space, a place without a place, that exists by itself, that is closed in on itself and at the same time is given over to the infinity of the sea" (Foucault, 1986, p. 27). Importantly, the idea of the ship as heterotopia or atopia does not equal placelessness, but rather foregrounds a sense of impermanence that extends to its occupants. As Carroll remarks, people who are attracted to and associated with the temporary "[o]ccupation of atopic space" are traditionally mobile: "explorers, exiles, refugees, bandits, and mutineers" (Carroll, 2015, pp. 6–7). Tuck belongs to several of these categories as "a motley mix of gypsea, pirate and bridger" (Bertagna, 2007, p. 144).

The terms draw attention to transitory and transgressive ways of life, and to skills needed aboard the fundamentally unstable construction of Pomperoy. Although the floating city is the dwelling for the generations after the ecological collapse, it is constantly at the mercy of the winds and the waves that threaten to transform it into the ultimate atopia of a shipwreck (Carroll, 2015, p. 87).

The threat of shipwreck is realised when *Arkiel* eventually rams Pomperoy and sinks the barge that has been Tuck's primary home. The constituent small boats are de-attached from each other and transformed into an armada that follows Mara's group to Greenland. Thus, Tuck's trajectory is the opposite of Mara's: where she must negotiate unstable water to reach the goal of a *terra firma*, Tuck struggles with a never before experienced "stillness" when he sets foot on solid ground for the first time in his life (Bertagna, 2007, p. 141). His thoughts and reactions illustrate how he and his fellow citizens of Pomperoy have undergone a process of epistemological adaptation to a watery existence. They read the world through undulating water, the sounds of waves lapping against the hulls, and of the wind in the sails, and regard their "patch of ocean" as home where they may be safe from the terrors of land (Bertagna, 2007, p. 34). Unlike Mara who sees reaching solid ground as the promise of survival, Tuck does not initially plan on landing on the frightening beaches of Greenland and uses the distorted form "Urth" as a curse (Bertagna, 2007, p. 26). Where Mara and her followers imagine the waves during their passage from Wing as high mountains, Tuck experiences the equivalent of seasickness when "around him Land rises up in mountainous waves that seem to surge and swell" (Bertagna, 2007, p. 157). Once this movement of the ground ceases, however, the solidity of land becomes an anomaly that suggests to Tuck that "[m]aybe it's dead. What if that's why the Land sunk down into the sea—because it died?" (Bertagna, 2007, p. 157). Water is the constant in Tuck's existence, whereas land is unpredictable and unstable. He consequently continues to see Greenland as an atopic space and distrusts its ability to sustain life.

The depiction of Tuck allows Bertagna to approach flooding from an oceanic perspective: for him and the other inhabitants of Pomperoy, it is the discovery of and embarkation on land that function as points of crisis. A life lived on the ocean challenges "the terrestrial character of knowing" that Mara rediscovers (Connery, 2006, p. 508), and has shaped Tuck's physical traits: his body movements are compared to a "sail in the wind," his voice has a "salt-scraped ocean roll" and his eyes are full of "sea-glitter" (Bertagna, 2007, pp. 172, 185). He accompanies Mara and her group to the caves, but distanced from the open sky and the ocean, he experiences a "great rift" in his identity (Bertagna, 2007, p. 199). Although "his Landcestors […] must have left strong, Earthy footprints on his soul" that make it possible for him to leave the ocean, to "turn Lander" means betrayal of the community of gypseas and the adaptation to life on the sea that they have undergone (Bertagna, 2007, pp. 144, 176). Consequently, Tuck's existence on land is uneasy, and instead of the relative safety that Mara finds in the caves, he conceives of the darkness inside them as "an unnatural, monstrous thing" (Bertagna, 2007, p. 168). In Ilira, Tuck reverts to his heritage as bridger and horizontally interlinks ships in the fjord. Whereas the vertical structure of New Mungo fosters hierarchy and detachment, literally rising above the water, the layout of Ilira instead connects the land and the sea. Although Tuck is not a particularly sympathetic character, his essentially transient nature becomes an asset that makes him able to negotiate the demands of the new world and adapt to a state in-between earth and water. The juxtaposition of Mara and Tuck contrasts the need for solidity and shelter with fluidity and openness, illustrating different responses to the global catastrophe and to what constitutes salvation. The adaptation otherwise at focus in the novels is accompanied by the more conservative suggestion that natural environments forcefully shape humans and impede or at least temporarily arrest adjustment processes.

13.5 Conclusion

The oceanic atopia in Carroll's analyses is "an exemplary space of atypicality, the iconic space of anti-land that threatens to undo the social structures of the shore" (Carroll, 2015, pp. 186–187). Fictional deluges in post-apocalyptic speculative fiction often employ this watery agency to envision more productive relations between humans and nature. The rising sea certainly represents "disorder and chaos" in Bertagna's novels but the coming together of people across the flooded world suggests the more benevolent agency of "[t]he ocean as connector" (Connery, 2006, pp. 499, 496). One of the functions of this recurring trope in Western thought, Connery argues, is that it productively questions "older ideological or civilizational divides" (Connery, 2006, p. 508). Uniting against the empire's plans for future colonisation of land in the last instalment in the trilogy, people gather around "radios in boat camps and in the ruins of drowned cities, in gypsea pirate fleets and precarious mountain hamlets, in the floating caravans of nomads who fish the sea-logged plains of the Earth" (Bertagna, 2007, p. 142). Terrestrial national or cultural boundaries have been replaced by an aquascape in which new loyalties are forged.

This movement, intended to empower the generations that are to inhabit the future, represents an ideological adaptation to the permanently changed conditions. It complements the three distinctive models of coping with flooding that Bertagna outlines: attempts to deny and control the climate collapse by building and policing tower cities that rise above the water, biological adaptation to life in and on the ocean, and the re-establishment of a pre-technological lifestyle characterized by a harmonic and dynamic relationship between people and the natural environment. All three models illustrate ways of transforming imaginary and unstable spaces into places of dwelling: the ocean as well as roofs of high buildings become home, and the previously peripheral and inhospitable Arctic emerges as one of few remaining *terra firma*. In *Aurora* it is clear that a return to an antediluvian world is impossible and that new forms of environmental and social responsibility are required.

In order for the future, speculative world to be understandable, and for anthropogenic climate change to remain narratable, Bertagna relies on well-known tropes: the ocean with its ties to ends and beginnings, the empty Arctic, and repeated colonial practices. Even though the world she depicts requires new relations between humans and nature and produces new epistemologies, speculative cli-fi is always uncomfortably close to real world conditions. In the preface to *Exodus*, Bertagna asks her readers to envision a world permanently altered by climate change and to imagine themselves "*at the fragile moment before the devastation begins. Is this where we stand right now, right here on the brink?*" (Bertagna, 2002, n. p.). The metafictional question draws attention to the fact that the speculative fiction that follows depicts a worst-case scenario that can still be prevented. *Aurora*, published almost a decade later, is dedicated to "*the Kiribati Islanders whose plight inspired*" the trilogy and includes an epigraph from the Pacific culture: "To be part of a nation that might be under the sea gives me a feeling that I am from nowhere" (Bertagna, 2011, n. p.). Despite the sustained depiction of a water world, this paratextual comment foregrounds the understanding that a nowhere, an anti-land, is an anomaly or an atopia. In the trilogy Bertagna speculates that this future of non-belonging will extend also to Scotland, continental Europe and eventually the entire world. The novels' emphasis on human responsibility to protect the world's home-places from turning into atopias and conversely, to prepare for a future where atopias need to be reclaimed as home may thus be read as a call to action in the present.

References

Asselin, S. (2018). A climate of competition: Climate change as political economy in speculative fiction, 1889–1915. *Science Fiction Studies, 45*(3), 440–453. https://doi.org/10.5621/sciefictstud.45.3.0440

Bertagna, J. (2002). *Exodus*. Young Picador.

Bertagna, J. (2007). *Zenith*. Young Picador.

Bertagna, J. (2011). Aurora. McMillan.

Blum, H. (2015). Terraqueous planet: The case for oceanic studies. In A. J. Elias & C. Moraru (Eds.), *The planetary turn: Relationality and geoaesthetics in the twenty-first century* (pp. 25–36). Northwestern University Press.

Bracke, A. (2019). Flooded futures: The representation of the Anthropocene in twenty-first-century British flood fictions. *Critique: Studies in Contemporary Fiction, 60*(3), 278–288. https://doi.org/10.1080/00111619.2019.1570911

Buell, F. (2014). Global warming as literary narrative. *Philological Quarterly, 93*(3), 261–294.

Carroll, S. (2015). *An empire of air and water: Uncolonizable space in the British imagination* (pp. 1750–1850). University of Pennsylvania Press.

Connery, C. (2006). There was no more sea: The supersession of the ocean, from the bible to cyberspace. *Journal of Historical Geography, 32*, 494–511. https://doi.org/10.1016/j.jhg.2005.10.005

Craciun, A. (2010). The frozen ocean. *PMLA, 125*(3), 693–702.

Curry, A. (2013). Knowledge: Navigating the visual ecology – Information literacy and the 'knowledgescape' in young adult fiction. In Y. Wu, K. Mallan, & R. McGillis (Eds.), *(Re)imagining the world: Children's literature's response to changing times* (pp. 15–26). Springer.

DeLoughrey, E. (2019). Toward a critical ocean studies for the Anthropocene. *English Language Notes, 57*(1), 21–36.

Foucault, M. (1986). Of other spaces (trans: J. Miskowiec). *Diacritics, 16*, 22–27. https://doi.org/10.2307/464648

Gill, R. B. (2013). The uses of genre and the classification of speculative fiction. *Mosaic, 46*(2), 71–85. https://doi.org/10.1353/mos.2013.0021

Hammer, Y. (2010). Confronting ecological futures: Global environmental crises in contemporary survival quests for young adults. *Barnboken – tidskrift för barnlitteraturforskning, 2*, 34–50.

Hansson, H. (2018). The Arctic in literature and the popular imagination. In M. Nuttal, T. R. Christensen, & M. J. Siegert (Eds.), *The Routledge handbook of the polar regions* (pp. 45–56). Routledge.

Hollywood, A. (2020). Dystopia, utopia, atopia. *Modern Theology, 36*(1), 31–52. https://doi.org/10.1111/moth.12574

Latour, B. (2014). Agency at the time of the Anthropocene. *New Literary History, 45*, 1–18. https://doi.org/10.1353/nlh.2014.0003

Lindgren Leavenworth, M. (2020). The atopic Arctic in lost world novels. *Translocal: Culturas Contemporâneas Locais e Urbanas, 5*, 1–10. https://translocal.cmfunchal.pt/2019/05/02/revista05/. Accessed May 28, 2021.

Lindgren Leavenworth, M., & Manni, A. (2021). Climate fiction and young learners' thoughts – A dialogue between literature and education. *Environmental Education Research, 27*(5), 727–742. https://doi.org/10.1080/13504622.2020.1856345

Loomis, C. (1971). *Weird and tragic shores: The story of Charles Francis hall, explorer*. The Modern Library.

McCulloch, F. (2007). A new home in the world: Scottish devolution, nomadic writing, and supranational citizenship in Julie Bertagna's *exodus* and *zenith*. *Ariel, 38*(4), 69–96.

Ostry, E. (2013). On the brink: The role of young adult culture in environmental degradation. In B. Basu, K. R. Broad, & C. Hintz (Eds.), *Contemporary dystopian fiction for young adults: Brave new teenagers* (pp. 101–114). Routledge.

Spufford, F. (1996). *I may be some time: Ice and the English imagination*. Faber and Faber.

Twain, M. (1892). *The American claimant*. Charles L. https://babel.hathitrust.org/cgi/pt?id=mdp.39015066650238&view=1up&seq=7. Accessed May 4, 2020

Yusoff, K., & Gabrys, J. (2011). Climate change and the imagination. *WIREs Climate Change, 2*, 516–534. https://doi.org/10.1002/wcc.117

Index

M. Lehtimäki et al. (eds.), *Cold Waters*, Springer Polar Sciences, https://doi.org/10.1007/978-3-031-10149-6